Engineering Applications

Engineering Applications

Analytical and Numerical Calculation with MATLAB

Mihai Dupac
Bournemouth University
UK

Dan B. Marghitu
Auburn University
USA

The right of Mihai Dupac and Dan B. Marghitu to be identified as the authors of this work has been asserted in accordance with law.

Registered Office
John Wiley & Sons, Inc., 111 River Street, Hoboken, NJ 07030, USA

Editorial Office
111 River Street, Hoboken, NJ 07030, USA

For details of our global editorial offices, customer services, and more information about Wiley products visit us at www.wiley.com.

Wiley also publishes its books in a variety of electronic formats and by print-on-demand. Some content that appears in standard print versions of this book may not be available in other formats.

Limit of Liability/Disclaimer of Warranty
MATLAB® is a trademark of The MathWorks, Inc. and is used with permission. The MathWorks does not warrant the accuracy of the text or exercises in this book. This work's use or discussion of MATLAB® software or related products does not constitute endorsement or sponsorship by The MathWorks of a particular pedagogical approach or particular use of the MATLAB® software.

While the publisher and authors have used their best efforts in preparing this work, they make no representations or warranties with respect to the accuracy or completeness of the contents of this work and specifically disclaim all warranties, including without limitation any implied warranties of merchantability or fitness for a particular purpose. No warranty may be created or extended by sales representatives, written sales materials or promotional statements for this work. The fact that an organization, website, or product is referred to in this work as a citation and/or potential source of further information does not mean that the publisher and authors endorse the information or services the organization, website, or product may provide or recommendations it may make. This work is sold with the understanding that the publisher is not engaged in rendering professional services. The advice and strategies contained herein may not be suitable for your situation. You should consult with a specialist where appropriate. Further, readers should be aware that websites listed in this work may have changed or disappeared between when this work was written and when it is read. Neither the publisher nor authors shall be liable for any loss of profit or any other commercial damages, including but not limited to special, incidental, consequential, or other damages.

Library of Congress Cataloging-in-Publication Data

Names: Dupac, Mihai, author. | Marghitu, Dan B., author.
Title: Engineering applications : analytical and numerical calculation with
    MATLAB / Mihai Dupac, Bournemouth University. Dan B. Marghitu, Auburn University.
Description: First edition. | Hoboken, NJ, USA : John Wiley & Sons, Inc.,
    2021. | Includes bibliographical references and index.
Identifiers: LCCN 2020032949 (print) | LCCN 2020032950 (ebook) | ISBN
    9781119093626 (hardback) | ISBN 9781119093633 (Adobe pdf) | ISBN
    9781119093640 (epub)
Subjects: LCSH: Engineering–Data processing. | MATLAB.
Classification: LCC TA345.5.M42 D87 2021 (print) | LCC TA345.5.M42
    (ebook) | DDC 620.00285/53–dc23
LC record available at https://lccn.loc.gov/2020032949
LC ebook record available at https://lccn.loc.gov/2020032950

Cover Design: Wiley
Cover Image: cherezoff/Getty images

Set in 9.5/12.5pt STIXTwoText by SPi Global, Chennai, India
Printed and bound by CPI Group (UK) Ltd, Croydon, CR0 4YY

10 9 8 7 6 5 4 3 2 1

Contents

1

Forces

1.1 Terminology and Notation

A force exerted on a body tends to change the state of the body, that is, if the body is rigid the force tends to move the body, but when the body is elasto-plastic the force tends to deform the body.

A force can be defined as a vector quantity that is defined by magnitude and direction. The *direction* of a force is specified by its *orientation* (also known as the line of action) and *sense*. The *magnitude* of a force is a positive *scalar*. A *scalar* is a number expressed in specific units of measure.

Vectors (forces) are usually denoted by boldface letters. If the starting point O and the end point S of a vector (force) are given, the vector (force) could be denoted by $\mathbf{F}_{OS}$ or more simply $\mathbf{F}$. It is also usual to denote the magnitude of the vector (force) by F_{OS} or by $|\mathbf{F}_{OS}|$. Some other notations for vectorial quantities could be $\vec{F}_{OS}$, $\overrightarrow{OS}$, or $\overline{F}_{OS}$.

Graphically a force $\mathbf{F}_{OS}$ is represented by a straight arrow as shown in Figure 1.1. The point O is named the *application point* or the *origin* of the force $\mathbf{F}_{OS}$ and the line passing through O and S is named the action line of $\mathbf{F}_{OS}$.

There are some possible operations regarding vectors.

Equality of forces
Two forces $\mathbf{F}_1$ and $\mathbf{F}_2$ are equal to each other when they have the same magnitude and direction, that is

$$\mathbf{F}_1 = \mathbf{F}_2. \tag{1.1}$$

If the forces $\mathbf{F}_1$ and $\mathbf{F}_2$ are equal but are acting at different locations on the same body it will not cause identical motion.

Multiplication of a Force by a Scalar
The product between a force $\mathbf{F}$ and a scalar d written as $d\mathbf{F}$, is a force having the same orientation as $\mathbf{F}$, the same sense as if $\mathbf{F}$ if $d > 0$ and opposite sense if $d < 0$, and the magnitude $|d||\mathbf{F}|$.

Engineering Applications: Analytical and Numerical Calculation with MATLAB, First Edition.
Mihai Dupac and Dan B. Marghitu.
© 2021 John Wiley & Sons Ltd. Published 2021 by John Wiley & Sons Ltd.

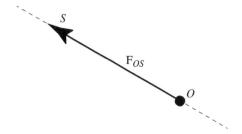

Figure 1.1 Vector representation.

Zero Force
A *zero force*, usually denoted by **0**, has a zero magnitude and an undefined direction.

Unit Vector (Force)
A *unit vector* **u** has its magnitude equal to unity, that is, $|\mathbf{u}| = 1$. Any force **F** can be written as a product of a unit vector **u** having the same orientation and sense as the force **F** and its magnitude $|\mathbf{F}|$ or equivalent

$$\mathbf{u} = \mathbf{F} \, \frac{1}{|\mathbf{F}|} = \frac{\mathbf{F}}{|\mathbf{F}|} \,. \tag{1.2}$$

Addition of Forces
The sum of a two forces $\mathbf{F}_1$ and $\mathbf{F}_2$ is a new force $\mathbf{F} = \mathbf{F}_1 + \mathbf{F}_2$ named resultant. The sum of the forces $\mathbf{F}_1$ and $\mathbf{F}_2$ is the force **F** represented graphically by the diagonal of the parallelogram shown in Figure 1.2 with its tail connecting the tail of the force $\mathbf{F}_1$ and head connecting the head of the force $\mathbf{F}_2$.

The sum $\mathbf{F}_1 + (-\mathbf{F}_2)$ is named the *difference* of the two forces as shown in Figure 1.3.

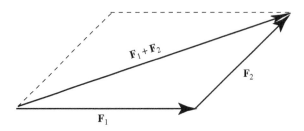

Figure 1.2 Parallelogram law of vector addition.

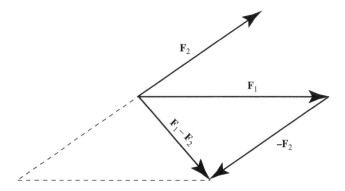

Figure 1.3 Parallelogram law of vector subtraction.

1.2 Resolution of Forces

If the unit vectors $\mathbf{i}$, $\mathbf{j}$, $\mathbf{k}$ have the same application point (origin) and are perpendicular to each other, as shown in Figure 1.4, they form a *Cartesian reference frame*.

Any force $\mathbf{F}$ can be expressed with respect to the unit vectors $\mathbf{j}$, $\mathbf{i}$, $\mathbf{k}$ by $\mathbf{F} = F_x\mathbf{i} + F_y\mathbf{j} + F_z\mathbf{k}$ where F_x, F_y, and F_z are the $\mathbf{i}$, $\mathbf{j}$, $\mathbf{k}$ components of the force.

The magnitude of $\mathbf{F}$ can be written as

$$|\mathbf{F}| = \sqrt{F_x^2 + F_y^2 + F_z^2}.$$

Addition and subtraction of forces could be easily manipulated using the resolution of forces into components. Considering the forces $\mathbf{F} = F_x\,\mathbf{i} + F_y\,\mathbf{j} + F_z\,\mathbf{k}$ and $\mathbf{P} = P_x\,\mathbf{i} + P_y\,\mathbf{j} + P_z\,\mathbf{k}$, one can calculate

$$\mathbf{F} + \mathbf{P} = (F_x + P_x)\,\mathbf{i} + (F_y + P_y)\,\mathbf{j} + (F_z + P_z)\,\mathbf{k}$$

and

$$\mathbf{F} - \mathbf{P} = (F_x - P_x)\,\mathbf{i} + (F_y - P_y)\,\mathbf{j} + (F_z - P_z)\,\mathbf{k}.$$

1.3 Angle Between Two Forces

The angles between the forces $\mathbf{F}$ and $\mathbf{P}$, and respectively $\mathbf{F}$ and $\mathbf{R}$ – in the range between $0°$ and $360°$ – are usually denoted by Greek letters such as α and θ, as shown in Figure 1.5.

The direction of a force $\mathbf{F} = F_x\mathbf{i} + F_y\mathbf{j} + F_z\mathbf{k}$ in a Cartesian frame is given by the direction cosines (Figure 1.6) of the angles between by the force and the associated unit vectors $\mathbf{i}$, $\mathbf{j}$, $\mathbf{k}$, written as $\cos\alpha = \dfrac{F_x}{|\mathbf{F}|}$; $\cos\beta = \dfrac{F_y}{|\mathbf{F}|}$; $\cos\gamma = \dfrac{F_z}{|\mathbf{F}|}$.

A unit force $\mathbf{F_u}$ (of magnitude 1) having the same direction as $\mathbf{F}$ can be written as

$$\mathbf{F_u} = \frac{F_x}{|\mathbf{F}|}\mathbf{i} + \frac{F_y}{|\mathbf{F}|}\mathbf{j} + \frac{F_z}{|\mathbf{F}|}\mathbf{k} = \cos\alpha\,\mathbf{i} + \cos\beta\,\mathbf{j} + \cos\gamma\,\mathbf{k}.$$

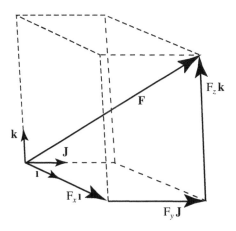

Figure 1.4 Resolution of a force.

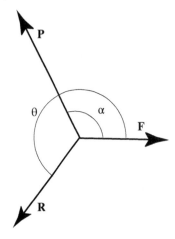

Figure 1.5 The angles α and θ between the forces **F** and **P**, and respectively **F** and **R**.

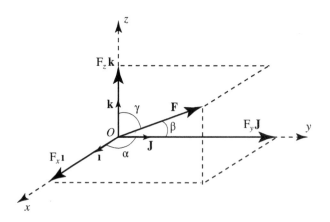

Figure 1.6 Direction cosines.

1.4 Force Vector

The position force (vector) $\mathbf{F}_{OS}$ shown in Figure 1.7 of a point $S(x_S,\ y_S,\ z_S)$ relative to a point $O(x_O,\ y_O,\ z_O) = O(0,\ 0,\ 0)$ can be written as

$$\mathbf{F}_{OS} = F_{x_S}\,\mathbf{1} + F_{y_S}\,\mathbf{J} + F_{z_S}\,\mathbf{k}. \tag{1.3}$$

The position force (vector) $\mathbf{F}_{MS}$ shown in Figure 1.7 of the point $S(x_S,\ y_S,\ z_S)$ relative to a point $(x_M,\ y_M,\ z_M)$ is calculated with

$$\mathbf{F}_{MS} = (F_{x_S} - F_{x_M})\,\mathbf{1} + (F_{y_S} - F_{y_M})\,\mathbf{J} + (F_{z_S} - F_{z_M})\,\mathbf{k}. \tag{1.4}$$

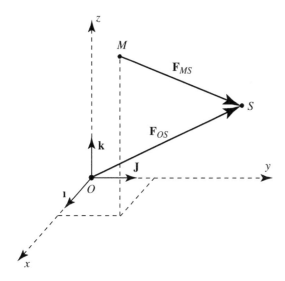

Figure 1.7 Position forces (vectors).

1.5 Scalar (Dot) Product of Two Forces

Definition. The dot product of two forces $\mathbf{F} = F_x\mathbf{i} + F_y\mathbf{j} + F_z\mathbf{k}$ and $\mathbf{P} = P_x\mathbf{i} + P_y\mathbf{j} + P_z\mathbf{k}$ is

$$\mathbf{F} \cdot \mathbf{P} = |\mathbf{F}|\ |\mathbf{P}|\cos(\theta) = F_xP_x + F_yP_y + F_zP_z \tag{1.5}$$

where θ is the angle between the forces $\mathbf{F}$ and $\mathbf{P}$.

1.6 Cross Product of Two Forces

The cross product of two forces $\mathbf{F}$ and $\mathbf{P}$ is another force defined by (Figure 1.8)

$$\mathbf{F} \times \mathbf{P} = |\mathbf{F}|\ |\mathbf{P}|\sin(\mathbf{F}, \mathbf{P})\mathbf{u} \tag{1.6}$$

where $\mathbf{u}$ is a unit force normal to $\mathbf{F}$ and $\mathbf{P}$ having its direction given by the right-hand rule.
The magnitude of the cross product is given by

$$|\mathbf{F} \times \mathbf{P}| = |\mathbf{F}|\ |\mathbf{P}|\sin(\mathbf{F}, \mathbf{P}).$$

When $\mathbf{F} = F_x\,\mathbf{i} + F_y\,\mathbf{j} + F_z\,\mathbf{k}$, and $\mathbf{P} = P_x\,\mathbf{i} + P_y\,\mathbf{j} + P_z\,\mathbf{k}$, the cross product $\mathbf{F} \times \mathbf{P}$ can be calculated using

$$\mathbf{F} \times \mathbf{P} = \begin{vmatrix} \mathbf{i} & \mathbf{j} & \mathbf{k} \\ F_x & F_y & F_z \\ P_x & P_y & P_z \end{vmatrix} = (F_y\,P_z - F_z\,P_y)\mathbf{i} + (F_z\,P_x - F_x\,P_z)\mathbf{j} + (F_x\,P_y - F_y\,P_x)\mathbf{k}. \tag{1.7}$$

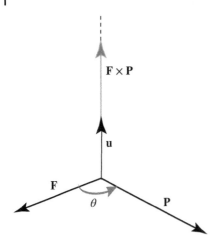

Figure 1.8 Cross product of two forces **F** and **P**.

1.7 Examples

Example 1.1

Figure 1.9 shows three forces $\mathbf{F}_1$, $\mathbf{F}_2$, and $\mathbf{F}_3$, and the angles of the forces with the horizontal $\theta_1 = \pi/6$, $\theta_2 = \pi/3$, and $\theta_3 = \pi$. The forces have the magnitudes $F_1 = 1$ kN, $F_2 = 3$ kN, and $F_3 = 2$ kN. Find the resultant of the planar forces and the angle of the resultant with the horizontal.

Solution

The input data are introduced in MATLAB with:

```
clear; clc; close all
F1 = 1; % kN
F2 = 3; % kN
F3 = 2; % kN
% angle of force F1_ with x-axis
theta1 = pi/6;
% angle of force F2_ with x-axis
```

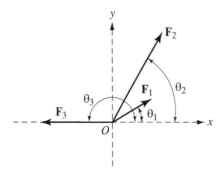

Figure 1.9 Graphical representation the forces $\mathbf{F}_1$, $\mathbf{F}_2$ and $\mathbf{F}_3$.

```
theta2 = pi/3;
% angle of force F3_ with x-axis
theta3 = pi;
```

The components of the forces on x and y axes are

$$\mathbf{F}_i = F_i \cos\theta_i \mathbf{I} + F_i \sin\theta_i \mathbf{J}, \ i = 1, \ 2, \ 3,$$ (1.8)

or in MATLAB:

```
% components of forces F1_, F2_, and F3_
F1x = F1*cos(theta1);
F1y = F1*sin(theta1);
F1_ = [F1x F1y];

F2x = F2*cos(theta2);
F2y = F2*sin(theta2);
F2_ = [F2x F2y];

F3x = F3*cos(theta3);
F3y = F3*sin(theta3);
F3_ = [F3x F3y];
```

The numerical values are:

```
F1_ = [ 0.866   0.500] (kN)
F2_ = [ 1.500   2.598] (kN)
F3_ = [-2.000   0.000] (kN)
```

The resultant is calculated with

$$\mathbf{R} = \sum \mathbf{F}_i = \mathbf{F}_1 + \mathbf{F}_2 + \mathbf{F}_3,$$ (1.9)

and the angle of the horizontal with the horizontal axis is

$$\phi = \tan^{-1}\frac{\mathbf{R}\cdot\mathbf{J}}{\mathbf{R}\cdot\mathbf{I}}.$$ (1.10)

With MATLAB the resultant and the angle are calculated with:

```
R_  = F1_+F2_+F3_;
phi = atand(R_(2)/R_(1));
```

and the results are

```
% R_  = F1_+F2_+F3_ = [ 0.366   3.098] (kN)
% phi = atan(Ry, Rx) = 83.262 (deg)
```

The MATLAB representation of the forces is shown in Figure 1.10 and it is obtained with:

```
sa = 4;
hold on
axis([-sa sa -sa sa])
axis square
```

```
quiver(0,0,F1_(1),F1_(2),0,'Color','k','LineWidth',1.2)
quiver(0,0,F2_(1),F2_(2),0,'Color','k','LineWidth',1.2)
quiver(0,0,F3_(1),F3_(2),0,'Color','k','LineWidth',1.2)
quiver(0,0,R_(1),R_(2),0,'Color','r','LineWidth',2)

text(F1_(1),F1_(2),' F_1',...
    'fontsize',14,'fontweight','b')
text(F2_(1),F2_(2),' F_2',...
    'fontsize',14,'fontweight','b')
text(F3_(1),F3_(2),' F_3',...
    'fontsize',14,'fontweight','b')
text(R_(1),R_(2),' R',...
    'fontsize',14,'fontweight','b')
grid on
xlabel('x'), ylabel('y'),
```

Example 1.2

Figure 1.11 shows a system of spatial forces with the magnitudes $F_1 = 15$ N, $F_2 = 30$ N, $F_3 = 10$ N, and $F_4 = 15$ N. The parallelepiped has the dimensions $a = 2$ m, $b = 3$ m, and $c = 5$ m. Find:

(a) the resultant of the system of forces;
(b) the angle between the forces $\mathbf{F}_1$ and $\mathbf{F}_2$;
(c) the projection of the force $\mathbf{F}_1$ on the force $\mathbf{F}_4$; and
(d) calculate $\mathbf{F}_1 \cdot (\mathbf{F}_2 \times \mathbf{F}_3)$.

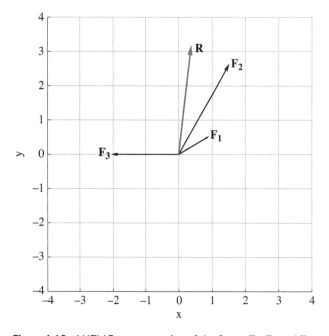

Figure 1.10 MATLAB representation of the forces $\mathbf{F}_1$, $\mathbf{F}_2$ and $\mathbf{F}_3$.

Solution

1)

(a) The input data in MATLAB are:

```
clear; clc; close all
a = 2; % m
b = 3; % m
c = 5; % m

F1 = 15; % N
F2 = 30; % N
F3 = 10; % N
F4 = 15; % N
```

A Cartesian reference frame xyz is selected as shown in Figure 1.11. The position force (vector) of the application point of the force $\mathbf{F}_1$ is

$$\mathbf{r}_E = a\mathbf{\imath} + b\mathbf{\jmath} + c\mathbf{k}. \tag{1.11}$$

The position vector of the application point of the force $\mathbf{F}_2$ is

$$\mathbf{r}_D = b\mathbf{\jmath} + c\mathbf{k}. \tag{1.12}$$

The position (vector) of the application point of the force $\mathbf{F}_3$ is

$$\mathbf{r}_F = a\mathbf{\imath} + c\mathbf{k}. \tag{1.13}$$

The position vector of the application point of the force $\mathbf{F}_4$ is

$$\mathbf{r}_B = a\mathbf{\imath} + b\mathbf{\jmath}. \tag{1.14}$$

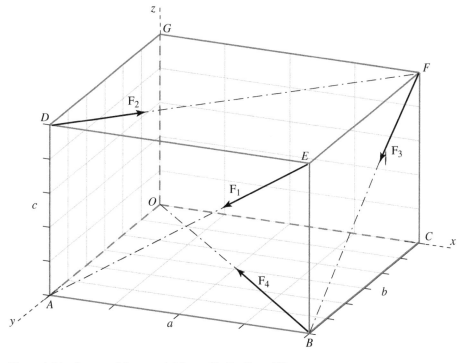

Figure 1.11 System of four spatial forces $\mathbf{F}_1$, $\mathbf{F}_2$, $\mathbf{F}_3$ and $\mathbf{F}_4$.

The position vectors of the application points of the spatial forces are defined in MATLAB as:

```
% position force of F1_
rE_ = [a b c];
% position force of F2_
rD_ = [0 b c];
% position force of F3_
rF_ = [a 0 c];
% position force of F4_
rB_ = [a b 0];
```

Next the position vectors of points A, G, and C, shown in Figure 1.11, are defined in MATLAB as:

```
% position force of A
rA_ = [0 b 0];
% position force of G
rG_ = [0 0 c];
% position force of C
rC_ = [a 0 0];
```

The unit vector of the force $\mathbf{F}_1$ is

$$\mathbf{u}_1 = \frac{\mathbf{r}_A - \mathbf{r}_E}{|\mathbf{r}_A - \mathbf{r}_E|}, \tag{1.15}$$

and the force $\mathbf{F}_1$ is $\mathbf{F}_1 = F_1\,\mathbf{u}_1$. The unit vector of the force $\mathbf{F}_2$ is

$$\mathbf{u}_2 = \frac{\mathbf{r}_F - \mathbf{r}_D}{|\mathbf{r}_F - \mathbf{r}_D|}, \tag{1.16}$$

and the force $\mathbf{F}_2$ is $\mathbf{F}_2 = F_2\,\mathbf{u}_2$. The unit vector of the force $\mathbf{F}_3$ is

$$\mathbf{u}_3 = \frac{\mathbf{r}_B - \mathbf{r}_F}{|\mathbf{r}_B - \mathbf{r}_F|}, \tag{1.17}$$

and the force $\mathbf{F}_3$ is $\mathbf{F}_3 = F_3\,\mathbf{u}_3$. The unit vector of the force $\mathbf{F}_4$ is

$$\mathbf{u}_4 = \frac{-\mathbf{r}_B}{|\mathbf{r}_B|}, \tag{1.18}$$

and the force $\mathbf{F}_4$ is $\mathbf{F}_4 = F_4\,\mathbf{u}_4$. The unit vectors and the forces are calculated in MATLAB with

```
u1_ = (rA_-rE_)/norm(rA_-rE_);
F1_ = F1*u1_;

u2_ = (rF_-rD_)/norm(rF_-rD_);
F2_ = F2*u2_;

u3_ = (rB_-rF_)/norm(rB_-rF_);
F3_ = F3*u3_;

u4_ = (-rB_)/norm(-rB_);
F4_ = F4*u4_;
```

The numerical results are:

```
% unit forces
% u1 = [-0.371,0,-0.928]
% u2 = [ 0.555,-0.832,0]
% u3 = [0,  0.514,-0.857]
% u4 = [-0.555,-0.832,0]
%
% forces
% F1_ = [ -5.57,0,  -13.9]  (N)
% F2_ = [  16.6,   -25,0]  (N)
% F3_ = [0,   5.14,  -8.57]  (N)
% F4_ = [ -8.32,  -12.5,0]  (N)
```

The plot of the unit vectors is shown in Figure 1.12, and it is obtained using the commands:

```
axis([0 a 0 b  0 c])
grid on, hold on
axis ij
xlabel('x'), ylabel('y'), zlabel('z')

text(0,0,0,'O','fontsize',12,'fontweight','b')

quiver3(rE_(1),rE_(2),rE_(3),u1_(1),u1_(2),u1_(3),0,...
    'Color','k','LineWidth',1.5)
text(rE_(1)+u1_(1),rE_(2)+u1_(2),rE_(3)+u1_(3),'u_1',...
    'fontsize',12,'fontweight','b')
```

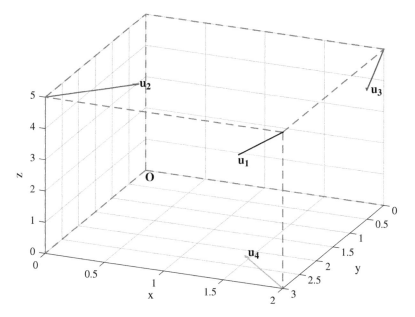

Figure 1.12 Plot of unit vectors $\mathbf{u}_1$, $\mathbf{u}_2$, $\mathbf{u}_3$ and $\mathbf{u}_4$

```
quiver3(rD_(1),rD_(2),rD_(3),u2_(1),u2_(2),u2_(3),0,...
    'Color','r','LineWidth',1.5)
text(rD_(1)+u2_(1),rD_(2)+u2_(2),rD_(3)+u2_(3),'u_2',...
    'fontsize',12,'fontweight','b')

quiver3(rF_(1),rF_(2),rF_(3),u3_(1),u3_(2),u3_(3),0,...
    'Color','b','LineWidth',1.5)
text(rF_(1)+u3_(1),rF_(2)+u3_(2),rF_(3)+u3_(3),'u_3',...
    'fontsize',12,'fontweight','b')

quiver3(rB_(1),rB_(2),rB_(3),u4_(1),u4_(2),u4_(3),0,...
    'Color','g','LineWidth',1.5)
text(rB_(1)+u4_(1),rB_(2)+u4_(2),rB_(3)+u4_(3),' u_4',...
    'fontsize',12,'fontweight','b')

line([rE_(1) rD_(1)],[rE_(2) rD_(2)],[rE_(3) rD_(3)],...
    'LineStyle','--','LineWidth',1.5)
line([rE_(1) rF_(1)],[rE_(2) rF_(2)],[rE_(3) rF_(3)],...
    'LineStyle','--','LineWidth',1.5)
line([rE_(1) rB_(1)],[rE_(2) rB_(2)],[rE_(3) rB_(3)],...
    'LineStyle','--','LineWidth',1.5)

line([rF_(1) rG_(1)],[rF_(2) rG_(2)],[rF_(3) rG_(3)],...
    'LineStyle','--','LineWidth',1.5)
line([rD_(1) rG_(1)],[rD_(2) rG_(2)],[rD_(3) rG_(3)],...
    'LineStyle','--','LineWidth',1.5)
line([rD_(1) rA_(1)],[rD_(2) rA_(2)],[rD_(3) rA_(3)],...
    'LineStyle','--','LineWidth',1.5)

line([0 rC_(1)],[0 rC_(2)],[0 rC_(3)],...
    'LineStyle','--','LineWidth',1.5)
line([0 rA_(1)],[0 rA_(2)],[0 rA_(3)],...
    'LineStyle','--','LineWidth',1.5)
line([0 rG_(1)],[0 rG_(2)],[0 rG_(3)],...
    'LineStyle','--','LineWidth',1.5)

view(23, 30)
```

The resultant of the spatial system, the magnitude of the resultant, and the direction cosines of the the resultant are calculated with:

```
R_  = F1_+F2_+F3_+F4_;
modR = norm(R_);
uR_  = R_/modR;
% R_=F1+F2+F3+F4=[ 2.750,-32.297,-22.502] (N)
% |R|=  39.5 (N)
% direction cosines=uR_=R_/|R|=[ 0.070,-0.819,-0.570]
```

2)

(b) The angle between the forces F_1 and F_2 is calculated in MATLAB with:

```
c12 = dot(F1_,F2_)/(F1*F2);
phi12= acosd(c12);

% F1.F2 = |F1||F2| cos(phi12)
% phi12 =    102 (deg)
```

3)

(c) The projection of the force F_1 on the force F_4 is obtained with

```
prF1F4 = dot(F1_, u4_);
% projection of F1 on F4 = F1_.u4_ =  3.09
```

4)

(d) The term $F_1 \cdot (F_2 \times F_3)$ is calculated in MATLAB:

```
T = dot(F1_,cross(F2_,F3_));
% F1_ . (F2_ x F3_) = -2384.809
```

Example 1.3

Show that $(p + q) \times (p - q) = 2\,q \times p$.

Solution

The two symbolical forces p and q are defined in MATLAB with:

```
syms px py pz qx qy qz

p_ = [px py pz];
q_ = [qx qy qz];
```

The left-hand side (LHS) is $(p + q) \times (p - q)$ which is

```
LHS = cross(p_+q_, p_-q_);
```

The right-hand side (RHS) is $2\,q \times p$ which is

```
RHS = 2*cross(q_, p_);
```

The difference LHS−RHS is calculated with

```
simplify(LHS-RHS)
% [ 0, 0, 0]
```

and the equality is true. If p and q are the adjacent sides of a parallelogram then $p + q = d_1$ and $p - q = d_2$ are the two diagonals of a parallelogram.

The area of the parallelogram is the magnitude of the cross product $q \times p$. Notice that the area of a parallelogram is given by half of the magnitude of the cross product of its diagonals.

2

Moments and Couples

2.1 Types of Moments

Moments can be classified as internal moments, external moments, and reaction moments.

External Moments
An *external moment* can be defined as a moment acting on the surface of the body and transmitted to the body by contact, i.e. by a force acting external to the body. An external moment applied at a point is called a *concentrated moment*, and applied over a surface it is called a *distributed moment*.

Internal Moments
Internal forces acting on a body are independent of whether or not external forces are applied on the body. When an external force is applied to a body, internal forces occur inside the object causing stresses. When an external moment is applied to a body, the result will be either torsion or bending. The process introduces internal moments in the body. Internal moments about an axis are either *torsional moments*, i.e. *torque*, or respectively *bending moments*, depending on the axis direction with respect to the imaginary cut surface, either normal or respectively tangent.

Reaction Moments
Moments can be developed at the supports in order to resist movement (rotation) due to the applied external forces or moments.

2.2 Moment of a Force About a Point

Consider a force $\mathbf{F}$ and a given point P located on the plane T (Figure 2.1) defined by the axes x and y of the Cartesian reference frame xyz. The moment $\mathbf{M}_P^{\mathbf{F}}$ of the force $\mathbf{F}$ about the point P is defined by

$$\mathbf{M}_P^{\mathbf{F}} = \mathbf{r}_{PS} \times \mathbf{F}, \tag{2.1}$$

where $\mathbf{r}_{PS}$ is the vector position of the point S, Ω is the line of action of the position vector $\mathbf{r}_{PS}$, and Δ is the action line of the force $\mathbf{F}$ (Figure 2.1).

Engineering Applications: Analytical and Numerical Calculation with MATLAB, First Edition.
Mihai Dupac and Dan B. Marghitu.
© 2021 John Wiley & Sons Ltd. Published 2021 by John Wiley & Sons Ltd.

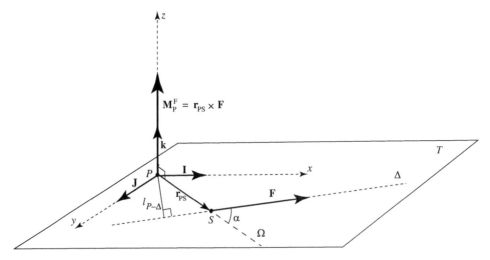

Figure 2.1 Moment of a force **F** about a point P.

The position vector $\mathbf{r}_{PS}$ and the force **F** can be formulated in terms of the reference frame xyz by

$$\mathbf{r}_{PS} = r_{PSx}\mathbf{\iota} + r_{PSy}\mathbf{J} + r_{PSz}\mathbf{k},$$
$$\mathbf{F} = F_{Px}\mathbf{\iota} + F_{Py}\mathbf{J} + F_{Pz}\mathbf{k}, \tag{2.2}$$

where $\mathbf{r}_{PS}$ has its direction defined from the point P to any point on the line Δ (S in this case), $\mathbf{\iota}$, $\mathbf{J}$, and $\mathbf{k}$ are unit vectors along the x, y and z axes, r_{PS_x}, r_{PS_y}, and r_{PS_z} are the magnitudes of the cartesian components of $\mathbf{r}_{PS}$, and F_{Px}, F_{Py}, and F_{Pz} are the magnitudes of the cartesian components of **F**.

From Equations (2.1) and (2.2) the moment $\mathbf{M}_P^F$ is calculated as

$$\mathbf{M}_P^F = \mathbf{r}_{PS} \times \mathbf{F}$$

$$= \begin{vmatrix} \mathbf{\iota} & \mathbf{J} & \mathbf{k} \\ r_{PS_x} & r_{PS_y} & r_{PS_z} \\ F_{Px} & F_{Py} & F_{Pz} \end{vmatrix}$$

$$= (r_{PS_y} F_{Pz} - r_{PS_z} F_{Py})\mathbf{\iota} + (r_{PS_z} F_{Px} - r_{PS_x} F_{Pz})\mathbf{J} + (r_{PS_x} F_{Py} - r_{PS_y} F_{Px})\mathbf{k}. \tag{2.3}$$

If the coordinates of S and P are x_S, y_S, z_S and x_P, y_P, z_P, the position vector $\mathbf{r}_{PS}$ can be expressed by $\mathbf{r}_{PS} = (x_S - x_P)\mathbf{\iota} + (y_S - y_P)\mathbf{J} + (z_S - z_P)\mathbf{k}$ and the the moment $\mathbf{M}_P^F$ can be calculated with

$$\mathbf{M}_P^F = \mathbf{r}_{PS} \times \mathbf{F} = \begin{vmatrix} \mathbf{\iota} & \mathbf{J} & \mathbf{k} \\ x_S - x_P & y_S - y_P & z_S - z_P \\ F_{Px} & F_{Py} & F_{Pz} \end{vmatrix}.$$

The moment $\mathbf{M}_P^F$ action line (z axis in this case) is perpendicular to the plane containing the position vector $\mathbf{r}_{PS}$ and the force **F** (plane Pxy in this case), and the sense of the moment $\mathbf{M}_P^F$ is identified by the right-hand rule.

The moment $\mathbf{M}_P^F$ is equal to zero if either the applied force **F** is zero or the line of action of the force **F** passes through the point P. The moment magnitude of $\mathbf{M}_P^F$ denoted by M_P^F or by $|\mathbf{M}_P^F|$ could be calculated with

$$M_P^F = |\mathbf{M}_P^F| = |\mathbf{r}_{PS}| \, |\mathbf{F}| \, \sin \alpha = r_{PS} \, F \, \sin \alpha, \tag{2.4}$$

where α is the angle between the action line Δ of the force $\mathbf{F}$ and the action line Ω of the vector position $\mathbf{r}_{PS}$, and r_{PS} is the magnitude of the of the position vector $\mathbf{r}_{PS}$. The distance $l_{P-\Delta}$ from the point P to the action line Δ is calculated with

$$l_{P-\Delta} = |\mathbf{r}_{PS}| \, \sin \alpha = r_{PS} \, \sin \alpha. \tag{2.5}$$

From Equations (2.4) and (2.5) the magnitude of the moment $\mathbf{M}_P^F$ is calculated as

$$|\mathbf{M}_P^F| = F \, r_{PS} \, \sin \alpha = F \, l_{P-\Delta} \tag{2.6}$$

Consider next another point N located on the line Δ where the force $\mathbf{F}$ is acting (Figure 2.2). The moment about the point P can now be calculated as

$$\mathbf{M}_P^F = \mathbf{r}_{PN} \times \mathbf{F}, \tag{2.7}$$

where $\mathbf{r}_{PN}$ is the position vector of the point N, and Γ is the action line of the position vector $\mathbf{r}_{PN}$ (Figure 2.2).

The magnitude of the moment $\mathbf{M}_P^F$ can be calculated with

$$M_P^F = |\mathbf{r}_{PN}| \, |\mathbf{F}| \, \sin \beta = r_{PN} \, F \, \sin \beta, \tag{2.8}$$

where β is the angle between the action line Δ and the line of action Γ of the position vector $\mathbf{r}_{PN}$, and r_{PN} is the magnitude of the position vector $\mathbf{r}_{PN}$. The distance $l_{P-\Delta}$ from P to the action line Δ can now be calculated with

$$l_{P-\Delta} = |\mathbf{r}_{PN}| \, \sin \beta = r_{PN} \, \sin \beta. \tag{2.9}$$

From Equations (2.8) and (2.9) the magnitude of the moment $\mathbf{M}_P^F$ can calculated as

$$|\mathbf{M}_P^F| = F \, r_{PN} \, \sin \beta = F \, l_{P-\Delta}$$

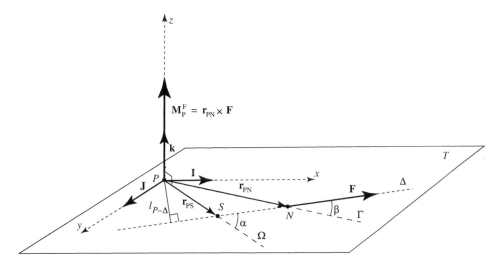

Figure 2.2 Moment of a force $\mathbf{F}$ about a point P.

which is the same with the magnitude obtained in Equation (2.4), that is, the moment calculated in Equation (2.4) is independent of the point where $\mathbf{F}$ is applied.

Since the position vector $\mathbf{r}_{PS}$ could be written as $\mathbf{r}_{PS} = \mathbf{r}_{PN} + \mathbf{r}_{NS}$ one can calculate

$$\mathbf{M}_P^{\mathbf{F}} = \mathbf{r}_{PS} \times \mathbf{F} = (\mathbf{r}_{PN} + \mathbf{r}_{NS}) \times \mathbf{F} = \mathbf{r}_{PN} \times \mathbf{F} + \mathbf{r}_{NS} \times \mathbf{F} = \mathbf{r}_{PN} \times \mathbf{F}, \tag{2.10}$$

where $\mathbf{r}_{NS} \times \mathbf{F} = \mathbf{0}$, thus

$$\mathbf{M}_P^{\mathbf{F}} = \mathbf{r}_{PS} \times \mathbf{F} = (\mathbf{r}_{PN} + \mathbf{r}_{NS}) \times \mathbf{F} = \mathbf{r}_{PN} \times \mathbf{F} + \mathbf{r}_{NS} \times \mathbf{F} = \mathbf{r}_{PN} \times \mathbf{F}, \tag{2.11}$$

that is, another way to show that the moment calculated in Equation (2.4) is independent of the point where the force $\mathbf{F}$ is applied.

The moment of a system of k forces $\mathbf{F}_i$, $i = \overline{1, k}$, about a point P can be calculated as

$$\mathbf{M}_P^{\mathbf{F}_i} = \sum_{i=1}^{k} \mathbf{M}_P^{\mathbf{F}_i}.$$

2.3 Moment of a Force About a Line

The moment $\mathbf{M}_\Omega^{\mathbf{F}}$ of a force $\mathbf{F}$ about a line Ω (Figure 2.3) can be calculated in two steps, i.e. using two equations. On the first step, the moment about an arbitrary point N lying on the line Ω is calculated (first equation) using

$$\mathbf{M}_N^{\mathbf{F}} = \mathbf{r}_{NS} \times \mathbf{F}. \tag{2.12}$$

The moment $\mathbf{M}_\Omega^{\mathbf{F}}$ of the force $\mathbf{F}$ in Equation (2.12) could be written using the cross product definition by

$$\mathbf{M}_N^{\mathbf{F}} = \mathbf{r}_{NS} \times \mathbf{F} = \begin{vmatrix} \mathbf{\imath} & \mathbf{\jmath} & \mathbf{k} \\ r_{NS_x} & r_{NS_y} & r_{NS_z} \\ F_{Px} & F_{Py} & F_{Pz} \end{vmatrix} = \begin{vmatrix} \mathbf{\imath} & \mathbf{\jmath} & \mathbf{k} \\ x_S - x_N & y_S - y_N & z_S - z_N \\ F_{Px} & F_{Py} & F_{Pz} \end{vmatrix}, \tag{2.13}$$

where x_N, y_N, z_N and x_S, y_S, z_S are the coordinates of the points N and S in the xyz cartesian reference frame, $\mathbf{r}_{NS} = (x_S - x_N)\mathbf{\imath} + (y_S - y_N)\mathbf{\jmath} + (z_S - z_N)\mathbf{k}$ is the position vector of the point S on the action line Δ of the force $\mathbf{F}$ relative to a point N located on the line Ω, and $\mathbf{\imath}$, $\mathbf{\jmath}$, and $\mathbf{k}$ are the unit vectors of the xyz reference frame.

In the second step, the projection of the moment $\mathbf{M}_N^{\mathbf{F}}$ on the line Ω is calculated using the dot product (second equation) by

$$\mathbf{M}_\Omega^{\mathbf{F}} = (\mathbf{M}_N^{\mathbf{F}} \cdot \mathbf{u}_\Omega)\mathbf{u}_\Omega \tag{2.14}$$

where $\mathbf{u}_\Omega$ is a unit vector collinear (or parallel) to Ω, as shown in Figure 2.3.

The two equations can be combined into a triple scalar product, that is the moment $\mathbf{M}_\Omega^{\mathbf{F}}$ about the line Ω (Figure 2.3) is

$$\mathbf{M}_\Omega^{\mathbf{F}} = (\mathbf{M}_N^{\mathbf{F}} \cdot \mathbf{u}_\Omega)\mathbf{u}_\Omega = ((\mathbf{r}_{NS} \times \mathbf{F}) \cdot \mathbf{u}_\Omega)\mathbf{u}_\Omega = [\mathbf{u}_\Omega, \mathbf{r}_{NS}, \mathbf{F}]\mathbf{u}_\Omega. \tag{2.15}$$

The magnitude of $\mathbf{M}_\Omega^{\mathbf{F}}$ can be calculated by

$$|\mathbf{M}_\Omega^{\mathbf{F}}| = |[\mathbf{u}_\Omega, \mathbf{r}_{NS}, \mathbf{F}]|$$

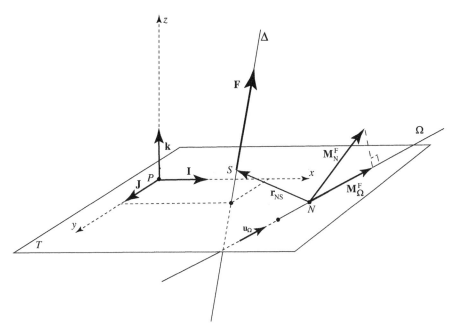

Figure 2.3 Moment of a force about a line.

$$= \begin{vmatrix} u_{\Omega_x} & u_{\Omega_y} & u_{\Omega_z} \\ r_{NS_x} & r_{NS_y} & r_{NS_z} \\ F_{Px} & F_{Py} & F_{Pz} \end{vmatrix}$$

$$= \begin{vmatrix} u_{\Omega_x} & u_{\Omega_y} & u_{\Omega_z} \\ x_S - x_N & y_S - y_N & z_S - z_N \\ F_{Px} & F_{Py} & F_{Pz} \end{vmatrix},$$

where u_{Ω_x}, u_{Ω_y} and u_{Ω_z} are the x, y and z components of the unit vector $\mathbf{u}_\Omega$ in the xyz reference frame. If the resulting moment is negative its direction is opposite to the direction defined by the unit vector $\mathbf{u}_\Omega$.

If Ω and Δ are perpendicular to each other, and $d_{\Delta-\Omega}$ is the distance (shortest) between the lines (Figure 2.4), one can calculate the moment $\mathbf{M}_\Omega^F$ using

$$|\mathbf{M}_\Omega^F| = |[\mathbf{u}_\Omega, \mathbf{r}_{NS}, \mathbf{F}]| = |\mathbf{u}_\Omega \cdot (\mathbf{r}_{NS} \times \mathbf{F})| = |\mathbf{u}_\Omega \cdot (|\mathbf{r}_{NS}||\mathbf{F}| \sin(\mathbf{r}_{NS}, \mathbf{F})\mathbf{u}_\Omega)|$$
$$= |\mathbf{r}_{NS}||\mathbf{F}| = d_{\Delta-\Omega}|\mathbf{F}|.$$

This case is shown in Figure 2.4 where the line of action Δ of the force $\mathbf{F}$ is parallel to the z axis (and perpendicular to the plane xy) and the line Ω lies in the xy plane, i.e. the lines Δ and Ω are perpendicular to.

If Ω intersects the action line Δ of the force $\mathbf{F}$, the position vector $\mathbf{r}$ can be selected such as $\mathbf{r} = \mathbf{0}$ resulting $\mathbf{M}_\Omega^F = \mathbf{0}$.

If the line Ω is parallel to the action line Δ of the force $\mathbf{F}$, then the triple scalar product $[\mathbf{u}_\Omega \, \mathbf{r}, \mathbf{F}]\mathbf{u}_\Omega = 0$ and therefore $\mathbf{M}_\Omega^F = \mathbf{0}$.

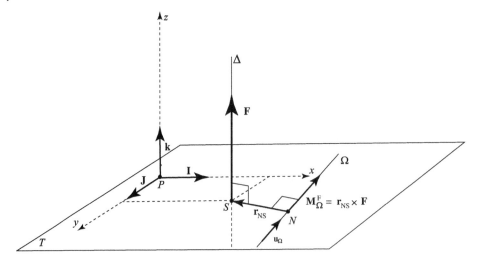

Figure 2.4 Moment of a force about a line perpendicular to the action line of the force.

The moment of a system of k forces $\mathbf{F}_i$, $i = \overline{1, k}$ about an action line Ω can be calculated using

$$\mathbf{M}_\Omega^{\mathbf{F}_i} = \sum_{i=1}^{n} \mathbf{M}_\Omega^{\mathbf{F}_i}.$$

2.4 Couples

A *simple couple* is a system of two parallel forces having the same magnitude and opposite sense. The moment of a couple about a point is defined by

$$\mathbf{M} = \mathbf{r} \times \mathbf{F}, \tag{2.16}$$

where $\mathbf{r}$ is a vector from any point on the action line of one force of the couple to any point of the action line of the other force defining the couple.

The magnitude of a couple is calculated using

$$|\mathbf{M}| = l_{\Delta-\Omega} \, |\mathbf{F}| = l_{\Delta-\Omega} \, F,$$

where $l_{\Delta-\Omega}$ is the distance between the action lines Δ and Ω of the two forces of the couple.

The moment $\mathbf{M}$ (Figure 2.5) can be calculated as the sum of the moments of the parallel forces $\mathbf{F}$ and $-\mathbf{F}$ about N by

$$\mathbf{M} = \mathbf{M}_N^{-F} + \mathbf{M}_N^{F}$$
$$= \mathbf{0} + \mathbf{r}_{NP} \times \mathbf{F}$$
$$= \mathbf{r}_{NP} \times \mathbf{F}.$$

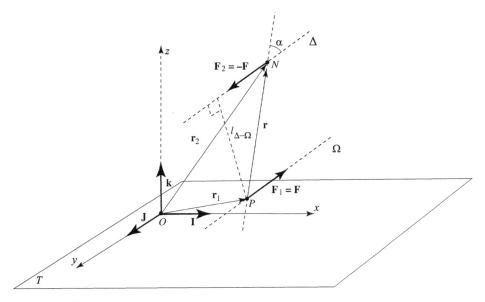

Figure 2.5 Simple couple of two parallel forces.

The moment **M** (Figure 2.5) can be also calculated with respect to the point O by

$$\mathbf{M} = \mathbf{M}_O^{-F} + \mathbf{M}_O^{F}$$
$$= \mathbf{r}_{ON} \times (-\mathbf{F}) + \mathbf{r}_{OP} \times \mathbf{F}$$
$$= (\mathbf{r}_{OP} - \mathbf{r}_{ON}) \times \mathbf{F}$$
$$= \mathbf{r}_{NP} \times \mathbf{F}$$

where $\mathbf{r}_{NP} = \mathbf{r}_{OP} - \mathbf{r}_{ON}$, as shown in Figure 2.5.

The magnitude of the moment can be calculated using the definition of cross product by

$$|\mathbf{M}| = |\mathbf{r} \times \mathbf{F}| = |\mathbf{r}||\mathbf{F}| \sin(\mathbf{F}, \mathbf{v}) = l_{\Delta-\Omega}|\mathbf{F}|$$

where $l_{\Delta-\Omega} = |\mathbf{r}| \sin(\mathbf{F}, \mathbf{r}) = |\mathbf{r}| \ |\sin \alpha|$ is the distance between the action lines of Δ and Ω of the two forces of the couple (Figure 2.5).

2.5 Examples

Example 2.1
The horizontal cantilever beam OC shown in shown in Figure 2.6 has the length l and mass m. A force $\mathbf{F}_C$ making an angle α with the horizontal direction is exerted at the free end $C(x_C, 0)$ of beam, and a counter-clockwise moment $\mathbf{M}_B$ is acting at the point $B(x_B, 0)$ of the beam, where x_B is the distance from the support O to the point B and x_C is the distance from the support O to the point C. Considering the magnitudes F_C and M_B of the applied force and moment, calculate the reaction force and the reaction moment at the support O.

Numerical application: $x_O = 0$ m, $y_O = 0$ m, $x_B = 3$ m, $y_B = 0$ m, $x_C = 8$, $y_C = 0$ m, $F_C = 3500$ N, $\alpha = 35°$, $m = 250$ kg, $g = 9.81$ m s^{-2}, $M_B = 10 \times 10^3$ N m.

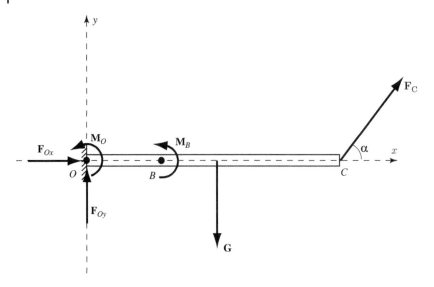

Figure 2.6 Horizontal cantilever beam under the action of moments and forces.

Solution

To calculate the reaction force at O, e.g. the vertical and horizontal components at O, the next equilibrium equations are written

$$\sum \mathbf{F} = \mathbf{F}_O + \mathbf{F}_C + \mathbf{G} = \mathbf{0} \tag{2.17}$$

or equivalent

$$\sum \mathbf{F}_x = \mathbf{F}_{Ox} + \mathbf{F}_{Cx} = \mathbf{0}$$
$$= F_{Ox}\mathbf{l} + F_C \cos \alpha \mathbf{l}$$
$$\sum \mathbf{F}_y = \mathbf{F}_{Oy} + \mathbf{F}_{Cy} + \mathbf{G}$$
$$= F_{Oy}\mathbf{J} + F_C \sin \alpha \mathbf{J} - mg\mathbf{J} = \mathbf{0} \tag{2.18}$$

where the component of $\mathbf{F}_C$ on the Ox direction is $F_{Cx} = F_C \sin \alpha$ and the component on the Oy direction is $F_{Cy} = F_C \cos \alpha$.

In MATLAB, the forces acting on the beam and the equilibrium equation are introduced using:

```
clear all; clc; close all

syms xO xB xC yO yB yC
rO_=[xO,yO,0];
rB_=[xB-xO,yB-yO,0];
rC_=[xC-xO,yC-yO,0];

syms FC alpha
FCx=FC*cos(alpha);
FCy=FC*sin(alpha);
```

```
FC_=[FCx,FCy,0];
fprintf('FCx=%s \n',char(FCx))
fprintf('FCy=%s \n\n',char(FCy))

syms m g F_Ox F_Oy
G_=[0,-m*g,0];
FO_=[F_Ox,F_Oy,0];
sumF_ = FO_+ FC_ + G_;
SFy = sumF_(2); SFx = sumF_(1);

fprintf('sum of forces on x and on y:\n')
fprintf(' sumFx = %s\n sumFy = %s\n\n',...
    char(SFx),char(SFy))
```

From Equation (2.18) one can calculate

$$F_{Ox} = -F_C \cos \alpha$$
$$F_{Oy} = -F_C \sin \alpha + mg. \tag{2.19}$$

The Ox and Oy reaction forces at the support O are calculated and printed in MATLAB with:

```
F_Ox=solve(SFx, F_Ox);
F_Oy=solve(sumF_(2), F_Oy);
fprintf('reaction F_Oy and F_Ox are: \n')
fprintf('F_Ox = %s \n F_Oy = %s \n\n',...
    char(F_Ox), char(F_Oy))
```

The reaction force at the support O can be written as

$$\mathbf{F}_O = -F_C \cos \alpha \mathbf{\imath} + (-F_C \sin \alpha + mg)\mathbf{\jmath}.$$

The sum of moments of all forces and moments about the support O is zero, that is

$$\sum \mathbf{M} = \mathbf{M}_O + \mathbf{M}_B + \mathbf{r}_G \times \mathbf{G} + \mathbf{r}_C \times \mathbf{F}_C = 0 \tag{2.20}$$

or equivalent

$$\sum M = M_O + M_B + x_G G_y + x_C F_{Cy}$$
$$= M_O + M_B - \frac{x_C}{2} mg + x_C F_C \sin \alpha = 0 \tag{2.21}$$

where $x_G = \dfrac{x_C}{2} = \dfrac{l}{2}$.

The equilibrium of all forces and moments about the support O are written in MATLAB with:

```
syms M_Oz MB
MO_=[0,0,M_Oz];
MB_=[0,0,MB];
sumMO_=MO_+MB_...
+cross(0.5*rC_,G_)+cross(rC_,FC_);
SMO=sumMO_(3);
```

```
fprintf('equilibrium about O:\n')
fprintf('SMO = %s \n\n', char(SMO));
```

From Equation (2.21) one can calculate

$$M_O = -M_B + \frac{l}{2}mg - lF_C \sin \alpha.$$

The reaction moment at the support O is calculated and printed in MATLAB with:

```
M_Oz=solve(SMO, M_Oz);
fprintf('The moment MO is \n')
fprintf('MO = %s \n\n', char(M_Oz));
```

The numerical data are set in MATLAB with:

```
items_s = {xO,yO,xB,yB,xC,yC,...
            FC,alpha,m,g,MB};
items_t = {0,0,3,0,8,0,...
3500,35*pi/180,250,9.81,10*10^3};
```

In MATLAB the numerical results are computed using:

```
F_Ox=subs(F_Ox,items_s,items_t);
F_Oy=subs(F_Oy,items_s,items_t);
M_Oz=subs(M_Oz,items_s,items_t);
fprintf('The reaction forces and the ');
fprintf('moment at O are \n')
fprintf(' F_Ox = %g (kN)\n F_Oy = %g (kN)\n M_Oz = %g (kN m)
   \n\n',...
     F_Ox/1000, F_Oy/1000, M_Oz/1000);
```

The numerical results (reaction forces and moment at O) are obtained in MATLAB as:

```
F_Ox = -2.86703  (kN)
F_Oy = 0.444982  (kN)
M_Oz = -16.2501  (kN m)
```

The MATLAB graphical representation of the input forces acting on the of the beam is shown in Figure 2.7.

Example 2.2
The vertical beam OA shown in Figure 2.8 is fixed at the base O. The beam is supporting a uniform equilateral triangular plate ABC with the mass m and the sides AB, BC and CA having the lengths $d_{AB} = d_{BC} = d_{CA} = l$. The mass of the beam is negligible. The distance between the base O and the end A is $d_{OA} = 3l$. A force F making an angle α with the horizontal direction is applied by a worker at the point C of the plate by the use of an inextensible cable. Calculate the moment and the reaction at support O when the force F_C is applied as shown (Figure 2.9).

Numerical application: $l = 1$ m, $F_C = 350$ N, $\alpha = 35°$, $m = 25$ Kg, $g = 9.81$ m s^{-2}.

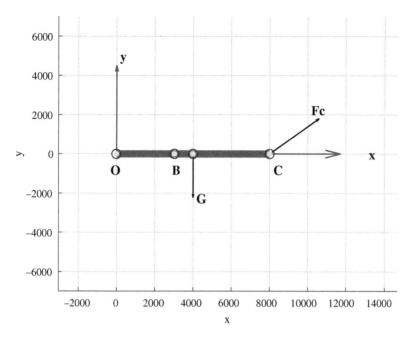

Figure 2.7 The MATLAB graphical representation of the forces acting on the beam.

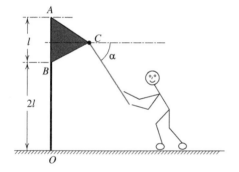

Figure 2.8 Vertical beam supporting a uniform equilateral triangular plate.

Solution

Since this is a uniform triangular plate with the weight $G = mg$ and $g = 9.81$ m s^{-2}, its mass centre $H(x_G, y_G) = H(^l\sqrt{3}/6, 2l + 1/2)$ is the same as its centroid. A Cartesian reference frame Oxy with the y axis directed along the beam is considered. To calculate the reaction force at O, e.g. the vertical and horizontal components at O, the next equilibrium equations are written

$$\sum \mathbf{F} = \mathbf{F}_O + \mathbf{F}_C + \mathbf{G} = 0 \tag{2.22}$$

or equivalent

$$\sum \mathbf{F}_x = \mathbf{F}_{Ox} + \mathbf{F}_{Cx} = 0$$
$$= F_{Ox}\mathbf{I} + F_C \cos \alpha \mathbf{I}$$
$$\sum \mathbf{F}_y = \mathbf{F}_{Oy} + \mathbf{F}_{Cy} + \mathbf{G}$$
$$= F_{Oy}\mathbf{J} - F_C \sin \alpha \mathbf{J} - mg\mathbf{J} = 0 \tag{2.23}$$

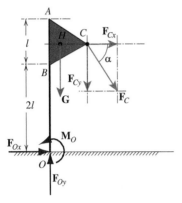

Figure 2.9 Free body diagram of the vertical beam supporting a uniform equilateral triangular plate ABC.

where the component of $\mathbf{F}_C$ on the Ox direction is $F_{Cx} = F_C \sin \alpha$ and the component on the Oy direction is $F_{Cy} = F_C \cos \alpha$.

The equilibrium equations are introduced in MATLAB with:

```
clear all; clc; close all

syms x_A x_B x_C x_G yA yB yC yG l
rA_=[x_A,y_A,0];
rB_=[x_B,y_B,0];
rC_=[x_C,y_C,0];
rG_=[x_G,y_G,0];

syms FC alpha
FCx=FC*cos(alpha);
FCy=-FC*sin(alpha);
FC_=[FCx,FCy,0];
fprintf('FCx=%s \n',char(FCx))
fprintf('FCy=%s \n\n',char(FCy))

syms m g F_Ox F_Oy
G_=[0,-m*g,0];
FO_=[F_Ox,F_Oy,0];
sumF_ = FO_+ FC_ + G_;
SFy = sumF_(2); SFx = sumF_(1);
fprintf('sum of forces on x and on y:\n')
fprintf('sumFx = %s \n', char(SFx))
fprintf('sumFy = %s \n\n', char(SFy))
```

From Equation (2.23) one can calculate

$$F_{Ox} = -F_C \cos \alpha$$
$$F_{Oy} = mg + F_C \sin \alpha. \tag{2.24}$$

The reactions F_{Oy} and F_{Ox} are computed in MATLAB with:

```
F_Ox=solve(SFx, F_Ox);
F_Oy=solve(sumF_(2), F_Oy);
fprintf('reaction F_Ox and F_Oy are: \n')
fprintf(' F_Ox = %s\n F_Oy = %s\n\n',...
    char(F_Ox),char(F_Oy))
```

Using the F_{Ox} and F_{Oy} components in Equation (2.24) the reaction force at the support O can be written as

$$\mathbf{F}_O = -F_C \cos\alpha\mathbf{1} + (mg + F_C \sin\alpha)\mathbf{J}.$$

The equilibrium of all forces and moments about the support O can be written as

$$\sum \mathbf{M} = \mathbf{M}_O + \mathbf{r}_G \times \mathbf{G} + \mathbf{r}_C \times \mathbf{F}_C = 0 \tag{2.25}$$

where

$$\mathbf{r}_C \times \mathbf{F}_C = \begin{vmatrix} \mathbf{1} & \mathbf{J} & \mathbf{k} \\ x_C & y_C & 0 \\ F_{Cx} & F_{Cy} & 0 \end{vmatrix}$$

$$= \begin{vmatrix} \mathbf{1} & \mathbf{J} & \mathbf{k} \\ x_C & y_C & 0 \\ F_C \cos\alpha & -F_C \sin\alpha & 0 \end{vmatrix}$$

$$= \begin{vmatrix} x_C & y_C \\ F_C \cos\alpha & -F_C \sin\alpha \end{vmatrix} \mathbf{k}$$

$$= -F_C(x_C \sin\alpha + y_C \cos\alpha)\mathbf{k} \tag{2.26}$$

and

$$\mathbf{r}_G \times \mathbf{G} = \begin{vmatrix} \mathbf{1} & \mathbf{J} & \mathbf{k} \\ x_G & y_G & 0 \\ 0 & -G_y & 0 \end{vmatrix}$$

$$= \begin{vmatrix} x_G & y_G \\ 0 & -G_y \end{vmatrix} \mathbf{k}$$

$$= -x_G G_y \mathbf{k}. \tag{2.27}$$

The equilibrium about the support O is written in MATLAB with:

```
syms M_Oz MB
MO_=[0,0,M_Oz];
sumMO_=MO_+...
+cross(rG_,G_)+cross(rC_,FC_);
SMO=sumMO_(3);
fprintf('sum of moments about O:\n')
fprintf('SMO = %s \n\n', char(SMO));
```

From Equations (2.25, 2.26) and (2.27) one can write

$$\sum \mathbf{M} = \mathbf{M}_O - F_C(x_C \sin\alpha + y_C \cos\alpha) - x_G G_y \mathbf{k} = 0$$

that is, the moment about the support O can be calculated as

$$\mathbf{M}_O = (F_C(x_C \sin \alpha + y_C \cos \alpha)\mathbf{k} + x_G G_y)\mathbf{k}. \tag{2.28}$$

In MATLAB the magnitude of the moment $\mathbf{M}_O$ is calculated with:

```
M_Oz=solve(SMO, M_Oz);
fprintf('The moment MO is \n')
fprintf('MO = %s \n\n', char(M_Oz));
```

In the Oxy reference frame one can calculate

$$x_C = \sqrt{l^2 - \frac{l^2}{4}} = l\frac{\sqrt{3}}{2}, \ y_C = 2l + \frac{l}{2} = \frac{5l}{2}, \ x_G = \frac{x_C}{3} = l\frac{\sqrt{3}}{6}. \tag{2.29}$$

The x_C, y_C and x_G coordinates are calculated in MATLAB with:

```
%x_A=0; x_B=0; x_C=l*sqrt(3)/2; x_G=l*sqrt(3)/6;
%y_A=3*l; y_B=2*l; y_C=2*l+l/2; y_G=2*l+l/2;
items_sxy = {x_A,x_B,x_C,x_G,y_A,y_B,y_C,y_G};
items_txy = {0,0,l*sqrt(3)/2,l*sqrt(3)/6,...
    3*l,2*l,2*l+l/2,2*l+l/2};
```

From Equations (2.28) and (2.29 one can calculate the moment about the base O as

$$M_O = \left(\frac{5}{2}F_C l \cos \alpha + \frac{\sqrt{3}}{2}F_C l \cos \alpha + \frac{\sqrt{3}}{6}glm \right) \mathbf{k}$$

and in MATLAB:

```
M_Oz=subs(M_Oz,items_sxy,items_txy);
fprintf('or equivalent with respect to l the moment MO is \n')
fprintf('MO = %s \n\n', char(M_Oz));
```

The input numerical data are introduced in MATLAB with:

```
items_s = {l,FC,alpha,m,g};
items_t = {1,350,35*pi/180,25,9.81};
```

The numerical results are calculated and printed in MATLAB using:

```
F_Ox=subs(F_Ox,items_s,items_t);
F_Oy=subs(F_Oy,items_s,items_t);
M_Oz=subs(M_Oz,items_s,items_t);
fprintf('The reaction forces and the ');
fprintf('moment at O are \n')
fprintf(' F_Ox = %g (N)\n F_Oy = %g (N)\n M_Oz = %f (N m)
    \n\n',...
    F_Ox,F_Oy,M_Oz);
```

The numerical results are obtained in MATLAB as:

```
The reaction forces and the reaction moment at O are
 F_Ox = -286.703 (N)
```

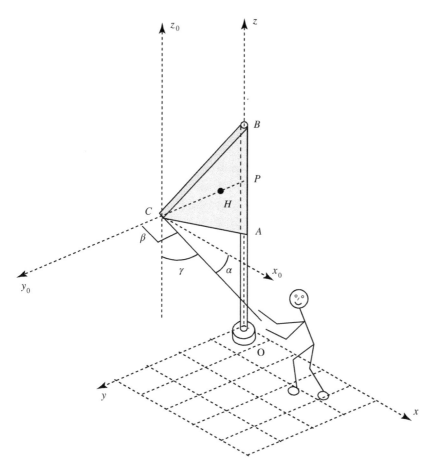

Figure 2.10 Vertical rod supporting a uniform equilateral triangular plate.

```
F_Oy = 446.002 (N)
M_Oz = 961.411733 (N m)
```

Example 2.3

The vertical rod (*OB*) of length l shown in Figure 2.10 is fixed at the base *O*. The mass of the rod is negligible. The rod is supporting at its upper end a uniform equilateral triangular plate *ABC* with mass m and sides *AB*, *BC* and *CA* of lengths $d_{AB} = d_{BC} = d_{CA} = d$. A force F is applied by a worker at the vertex *C* of the plate by the use of an inextensible cable. The dimensions of the rod, triangular plate, and the location, magnitude and direction of the force applied by the worker are depicted by the free body diagram in Figure 2.11. Calculate the moment about base *O* of the cable force acting on point *C* and the reaction force at the base *O*.

Numerical application: $l = 1$ m, $d = 1$ m, $F_C = 350$ N, $\alpha = 35°$, $\beta = 90°$, $\gamma = 45°$, $m = 25$ kg, $g = 9.81$ m s^{-2}.

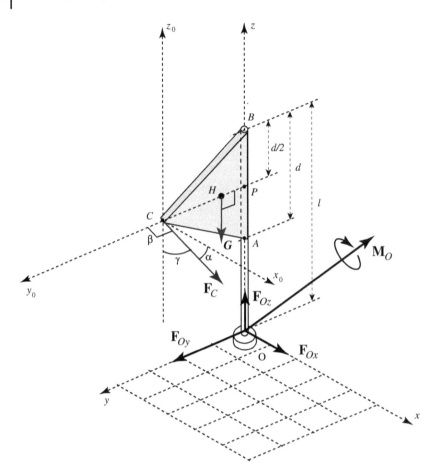

Figure 2.11 Free body diagram of the vertical rod supporting a uniform equilateral triangular plate.

Solution

Since this is an uniform plate with the weight $G = mg$ and $g = 9.81$ m s^{-2}, its mass centre $H(x_H, y_H, z_H)$ is the same as its centroid. The triangular plate is located in the Oyz plane as shown in Figure 2.10.

Using equilibrium principle, the following equations are written

$$\sum \mathbf{F} = \mathbf{F}_O + \mathbf{F}_C + \mathbf{F}_G = \mathbf{0} \tag{2.30}$$

or equivalent

$$\sum F_x = F_{Ox} + F_{Cx} = 0$$
$$\sum F_y = F_{Oy} + F_{Cy} = 0$$
$$\sum F_z = F_{Oz} - F_{Cz} - mg = 0 \tag{2.31}$$

where the component of $\mathbf{G}$ on the Oz direction is $\mathbf{G}_z = \mathbf{G} = -mg\mathbf{k}$.

The equilibrium is implemented in MATLAB with:

```
clear all; clc; close all

syms xH yH zH xC yC zC xB yB zB xA yA zA l d
rG_=[xH,yH,zH];
rC_=[xC,yC,zC];
rB_=[xB,yB,zB];
rA_=[xA,yA,zA];

syms FC alpha beta gamma
FCx=FC*cos(alpha);
FCy=FC*cos(beta);
FCz=-FC*cos(gamma);
FC_=[FCx,FCy,FCz];

syms m g F_Ox F_Oy F_Oz
G_=[0,0,-m*g];
FO_=[F_Ox,F_Oy,F_Oz];
sumF_ = FO_+ FC_ + G_;
SFx = sumF_(1);
SFy = sumF_(2);
SFz = sumF_(3);
fprintf('sumFx = %s\n sumFy = %s\n sumFz = %s\n\n',...
      char(SFx), char(SFy),char(SFz))
```

From Equation (2.31)one can calculate

$$F_{Ox} = -F_{Cx} = -F_C \cos \alpha$$
$$F_{Oy} = -F_{Cy} = -F_C \cos \beta$$
$$F_{Oz} = F_{Cz} + mg = F_C \cos \gamma + mg \tag{2.32}$$

where $\cos \gamma$, $\cos \beta$ and $\cos \alpha$ are the direction cosines of the force $\mathbf{F}_C$ applied by the worker.

The components F_{Ox}, F_{Oy} and F_{Oz} of the reaction force $\mathbf{F}_O$ at the base O are calculated in MATLAB with:

```
F_Ox=solve(SFx, F_Ox);
F_Oy=solve(sumF_(2), F_Oy);
F_Oz=solve(sumF_(3), F_Oz);
fprintf('reaction F_Ox, F_Oy and F_Oz are: \n')
fprintf('reaction F_Ox and F_Oy are: \n')
fprintf(' F_Ox = %s\n F_Oy = %s\n F_Oz = %s\n\n',...
    char(F_Ox),char(F_Oy),char(F_Oz))
```

Equilibrium of all the moments about O gives

$$\sum \mathbf{M} = -\mathbf{M}_O + \mathbf{M}_{F_C/O} + \mathbf{M}_{G/O} = 0 \tag{2.33}$$

or equivalent

$$\mathbf{M}_O = \mathbf{M}_{F_C/O} + \mathbf{M}_{G/O}$$
$$= \mathbf{r}_{OC} \times \mathbf{F}_C + \mathbf{r}_H \times \mathbf{G} = 0 \tag{2.34}$$

where

$$\mathbf{r}_{OC} \times \mathbf{F}_C = \begin{vmatrix} \imath & \jmath & \mathbf{k} \\ x_C & y_C & z_C \\ F_{Cx} & F_{Cy} & -F_{Cz} \end{vmatrix} = \begin{vmatrix} \imath & \jmath & \mathbf{k} \\ x_C & y_C & z_C \\ F_C \cos\alpha & F_C \cos\beta & -F_C \cos\gamma \end{vmatrix}$$
$$= (-y_C F_{Cz} - z_C F_{Cy})\imath - (-x_C F_{Cz} - z_C F_{Cx})\jmath + (x_C F_{Cy} - y_C F_{Cx})\mathbf{k}$$
$$= (-y_C F_C \cos\gamma - z_C F_C \cos\beta)\imath - (-x_C F_C \cos\gamma - z_C F_C \cos\alpha)\jmath$$
$$+ (x_C F_C \cos\beta - y_C F_C \cos\alpha)\mathbf{k} \tag{2.35}$$

$$\mathbf{r}_H \times \mathbf{G} = \begin{vmatrix} \imath & \jmath & \mathbf{k} \\ x_H & y_H & z_H \\ 0 & 0 & -mg \end{vmatrix} = \begin{vmatrix} \imath & \jmath & \mathbf{k} \\ x_H & y_H & z_H \\ 0 & 0 & -mg \end{vmatrix}$$
$$= \begin{vmatrix} y_H & z_H \\ 0 & -mg \end{vmatrix} \imath - \begin{vmatrix} x_H & z_H \\ 0 & -mg \end{vmatrix} \jmath = -y_H mg\imath + x_H mg\jmath. \tag{2.36}$$

The moment about point O is calculated in MATLAB with:

```
syms M_Ox M_Oy M_Oz
MO_ = [M_Ox,M_Oy,M_Oz];
sumMO_=MO_+...
+cross(rG_,G_)+cross(rC_,FC_);
SM_Ox=sumMO_(1);
SM_Oy=sumMO_(2);
SM_Oz=sumMO_(3);
```

From Equations (2.34), (2.35) and (2.36) one can calculate

$$\mathbf{M}_O = (-y_C F_C \cos\gamma - z_C F_C \cos\beta - y_H mg)\imath - (-x_C F_C \cos\gamma - z_C F_C \cos\alpha + x_H mg)\jmath$$
$$+ (x_C F_C \cos\beta - y_C F_C \cos\alpha)\mathbf{k} \tag{2.37}$$

or equivalent

$$M_{Ox} = y_C F_C \cos\gamma + z_C F_C \cos\beta + y_H mg$$
$$M_{Oy} = x_C F_C \cos\gamma + z_C F_C \cos\alpha - x_H mg$$
$$M_{Oz} = y_C F_C \cos\alpha - x_C F_C \cos\beta. \tag{2.38}$$

In MATLAB the x, y and z components of the moment $\mathbf{M}_O$ are calculated with:

```
M_Ox=solve(SM_Ox, M_Ox);
M_Oy=solve(SM_Oy, M_Oy);
M_Oz=solve(SM_Oz, M_Oz);

fprintf('The moment MO is \n')
fprintf(' M_Ox=%s\n M_Oy=%s\n M_Oz=%s\n\n',...
```

```
        char(M_Ox),char(M_Oy),char(M_Oz));
```

The coordinates are calculated in MATLAB using:

```
items_sxy = {xA,yA,zA,xH,yH,zH,xC,yC,zC};
items_txy = {0,0,1-d,0,d*sqrt(3)/6,1-d/2,0,d*sqrt(3)/2,1-d/2};
```

Using the x_H, x_A, x_C, y_H, y_A, y_C, and z_H, z_A, z_C coordinates the components of the moment M_O are calculated in MATLAB with:

```
M_Ox=subs(M_Ox,items_sxy,items_txy);
M_Oy=subs(M_Oy,items_sxy,items_txy);
M_Oz=subs(M_Oz,items_sxy,items_txy);
fprintf('or equivalent with respect to l and d the moment MO is
    \n')
fprintf(' M_Ox=%s\n M_Oy=%s\n M_Oz=%s\n\n',...
    char(M_Ox),char(M_Oy),char(M_Oz));
```

The input numerical data are introduced in MATLAB with:

```
items_s = {l,d,FC,alpha,beta, gamma,m,g};
items_t = {1,1,350,35*pi/180,90*pi/180,45*pi/180,25,9.81};
```

The numerical results are calculated and printed in MATLAB using:

```
F_Ox=subs(F_Ox,items_s,items_t);
F_Oy=subs(F_Oy,items_s,items_t);
F_Oz=subs(F_Oz,items_s,items_t);
M_Ox=subs(M_Ox,items_s,items_t);
M_Oy=subs(M_Oy,items_s,items_t);
M_Oz=subs(M_Oz,items_s,items_t);
fprintf('The reaction forces and the ');
fprintf('reaction moment at the support O are \n')
fprintf(' F_Ox=%g (N)\n F_Oy=%g (N)\n F_Oz=%g (N)\n',...
    F_Ox,F_Oy,F_Oz);
fprintf(' M_Ox = %f (N m)\n', M_Ox);
fprintf(' M_Oy = %f (N m)\n', M_Oy);
fprintf(' M_Oz = %f (N m)\n\n', M_Oz);
```

The numerical results are obtained in MATLAB as:

```
The reaction forces and the reaction moment at the support O
    are
 F_Ox = -286.703 (N)
 F_Oy = 0 (N)
 F_Oz = 492.737 (N)
 M_Ox = 285.127929 (N m)
 M_Oy = -143.351608 (N m)
 M_Oz = 248.292268 (N m)
```

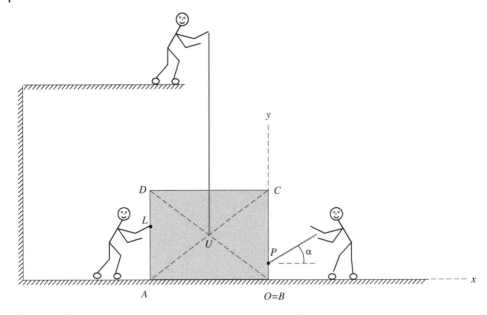

Figure 2.12 A rectangular heavy plate under the action of three workers.

Example 2.4

Three workers are trying to move a rectangular heavy plate $ABCD$ as shown in Figure 2.12. The mass of the of the plate is m and the sides lengths are $d_{AB} = d_{DC} = 3l$ and $d_{AD} = d_{BC} = 2l$ where the length l is given. One worker is pushing the plate from the left side with a force $\mathbf{F}_L$, one is using a cable to pull the plate from its right side with a force $\mathbf{F}_P$ while the last one is using a cable to lift the plate from above with a force $\mathbf{F}_U$. The dimensions of the plate, the location, magnitude and direction of the forces applied by the workers are depicted by the free body diagram in Figure 2.13. Determine an equivalent force-couple system (Figure 2.13), i.e. the resultant $\mathbf{F}_O$ of the forces and the moment $\mathbf{M}_O$, about the point $O = B$.

Numerical application: $l = 1$ m, $F_L = 1500$ N, $F_U = 600$ N, $F_P = 1200$ N, $\alpha = 35°$, $m = 250$ kg, $g = 9.81$ m s^{-2}.

Solution

The rectangular heavy plate $ABCD$ of weight $G = mg$ is uniform, so its mass centre is located at its centroid U. A Cartesian reference frame Oxy with the origin at the point $O = B$ of the plate is considered. To calculate the reaction force at $O = B$, e.g. the vertical and horizontal components at O, the next equilibrium equations are written

$$\sum \mathbf{F} = \mathbf{F}_O + \mathbf{F}_L + \mathbf{F}_P + \mathbf{F}_U + \mathbf{G} = 0 \tag{2.39}$$

or equivalent

$$\sum \mathbf{F}_x = \mathbf{F}_{Ox} + \mathbf{F}_{Lx} + \mathbf{F}_{Px} = 0$$
$$= \mathbf{F}_{Ox} + \mathbf{F}_{Lx}\mathbf{1} + \mathbf{F}_P \cos \alpha_1 = 0$$
$$\sum \mathbf{F}_y = \mathbf{F}_{Oy} + \mathbf{F}_{Uy} + \mathbf{F}_G + \mathbf{F}_{Py}$$

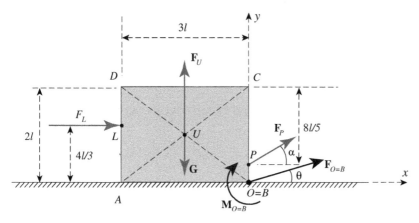

Figure 2.13 Free body diagram of the rectangular heavy plate.

$$= \mathbf{F}_{Oy} + F_{Uy}\mathbf{J} - mg\mathbf{J} + F_P \sin \alpha \mathbf{J} = 0 \tag{2.40}$$

where $\mathbf{F}_{Ux} = 0$, $F_{Ly} = 0$, and the components of $\mathbf{F}_P$ on the Ox and Oy directions are $F_{Px} = F_P \cos \alpha$ and $F_{Py} = F_P \sin \alpha$ respectively.

The equilibrium is implemented in MATLAB with:

```
clear all; clc; close all

syms xP yP xU yU xL yL xG yG l
rP_ = [0,yP,0];
rU_ = [-xU,yU,0];
rL_ = [-xL,yL,0];
rG_ = [-xG,-yG,0];

syms FP FL FU alpha
FPx=FP*cos(alpha); FPy=FP*sin(alpha);
FP_ = [FPx,FPy,0];
FLx=FL; FLy=0;
FL_ = [FLx,FLy,0];
FUx=0; FUy=FU;
FU_ = [FUx,FUy,0];

syms m g F_Ox F_Oy
G_ = [0,-m*g,0];
FO_ = [F_Ox,F_Oy,0];

sumF_ = FO_ + FL_ + FP_ + FU_ + G_;
SFx = sumF_(1);
SFy = sumF_(2);

fprintf('sumFx = %s \n', char(SFx))
```

```
fprintf('sumFy = %s \n\n', char(SFy))
```

From Equation (2.40) one can calculate

$$\mathbf{F}_{Ox} = -F_{Lx}\mathbf{1} - F_P \cos \alpha \mathbf{1}$$

$$\mathbf{F}_{Oy} = -F_{Uy}\mathbf{J} + mg\mathbf{J} - F_P \sin \alpha \mathbf{J}.$$

The magnitude of the Ox and Oy components of $\mathbf{F}_O$ resultant are calculated and printed in MATLAB with:

```
F_Ox=solve(SFx, F_Ox);
F_Oy=solve(sumF_(2), F_Oy);
fprintf('reaction F_Ox and F_Oy are: \n')
fprintf('F_Ox = %s \n', char(F_Ox))
fprintf('F_Oy = %s \n\n', char(F_Oy))
```

The reaction force at the support $O = B$ can be written as

$$\mathbf{F}_O = \mathbf{F}_{Ox}\mathbf{1} + \mathbf{F}_{Oy}\mathbf{J} = \sum(-F_{Lx} - F_P \cos \alpha)\mathbf{1} + (-F_{Uy} + mg - F_P \sin \alpha)\mathbf{J}.$$

The equilibrium about the support O is written as

$$\sum \mathbf{M} = \mathbf{M}_O + \mathbf{r}_G \times \mathbf{G} + \mathbf{r}_P \times \mathbf{F}_P + \mathbf{r}_U \times \mathbf{F}_U + \mathbf{r}_L \times \mathbf{F}_L = 0 \tag{2.41}$$

where

$$\mathbf{r}_P \times \mathbf{F}_P = \begin{vmatrix} \mathbf{1} & \mathbf{J} & \mathbf{k} \\ x_P & y_P & 0 \\ F_{Px} & F_{Py} & 0 \end{vmatrix} = \begin{vmatrix} \mathbf{1} & \mathbf{J} & \mathbf{k} \\ 0 & y_P & 0 \\ F_P \cos \alpha & -F_P \sin \alpha & 0 \end{vmatrix} = \begin{vmatrix} 0 & y_P \\ F_P \cos \alpha & -F_P \sin \alpha \end{vmatrix} \mathbf{k}$$

$$= -F_P y_C \cos \alpha \mathbf{k} \tag{2.42}$$

$$\mathbf{r}_U \times \mathbf{F}_U = \begin{vmatrix} \mathbf{1} & \mathbf{J} & \mathbf{k} \\ -x_U & y_U & 0 \\ F_{Ux} & F_{Uy} & 0 \end{vmatrix} = \begin{vmatrix} \mathbf{1} & \mathbf{J} & \mathbf{k} \\ -x_U & y_U & 0 \\ 0 & F_{Uy} & 0 \end{vmatrix} = \begin{vmatrix} -x_U & y_U \\ 0 & F_U \end{vmatrix} \mathbf{k} = -F_U x_U \mathbf{k} \tag{2.43}$$

$$\mathbf{r}_L \times \mathbf{F}_L = \begin{vmatrix} \mathbf{1} & \mathbf{J} & \mathbf{k} \\ -x_L & y_L & 0 \\ F_{Lx} & F_{Ly} & 0 \end{vmatrix} = \begin{vmatrix} \mathbf{1} & \mathbf{J} & \mathbf{k} \\ -x_L & y_L & 0 \\ F_{Lx} & 0 & 0 \end{vmatrix} = \begin{vmatrix} -x_L & y_L \\ F_L & 0 \end{vmatrix} \mathbf{k} = -F_L y_L \mathbf{k} \tag{2.44}$$

$$\mathbf{r}_G \times \mathbf{G} = \begin{vmatrix} \mathbf{1} & \mathbf{J} & \mathbf{k} \\ -x_G & y_G & 0 \\ 0 & -G_y & 0 \end{vmatrix} = \begin{vmatrix} -x_G & y_G \\ 0 & -G_y \end{vmatrix} \mathbf{k} = x_G G_y \mathbf{k}. \tag{2.45}$$

From Equations (2.41)–(2.45) one can write

$$\sum \mathbf{M} = \mathbf{M}_O - F_P y_P \cos \alpha \mathbf{k} - F_U x_U \mathbf{k} - F_L y_L \mathbf{k} + x_G G_y \mathbf{k} = 0$$

that is, the moment about the support O is

$$\mathbf{M}_O = (F_P y_P \cos \alpha + F_U x_U + F_L y_L - x_G G_y)\mathbf{k}.$$

In MATLAB, the magnitude of the moment $\mathbf{M}_O$ is computed with:

```
syms M_Oz MB
MO_ = [0,0,M_Oz];
```

```
sumMO_=MO_+cross(rG_,G_)+cross(rP_,FP_)+...
    +cross(rU_,FU_)+cross(rL_,FL_);
SMO=sumMO_(3);
fprintf('SMO = %s \n\n', char(SMO));
M_Oz=solve(SMO, M_Oz);
fprintf('Moment MO is \n')
fprintf('MO = %s \n\n', char(M_Oz));
```

Replacing $x_L = 3l$, $y_L = \frac{4}{3}l$, $x_G = \frac{3}{2}l$, $y_G = l$, $x_U = \frac{3}{2}l$, $y_U = l$, $x_P = 0$, $y_P = \frac{2}{5}l$, in the moment M_O above one can calculate

$$M_O = \left(\frac{4}{3}lF_L + \frac{3}{2}lF_U - \frac{3}{2}mgl + \frac{2}{5}lF_P \cos \alpha \right) \mathbf{k}.$$

The $x_L, y_L, x_G, y_G, x_U, y_U, x_P, y_P$ coordinates are introduced in MATLAB using:

```
items_sxy = {xL,yL,xG,yG,xU,yU,yP,yP};
items_txy = {3*1,4*1/3,3*1/2,2*1/2,3*1/2,2*1/2,0,2*1/5};
```

and the moment M_O is calculated in MATLAB with:

```
M_Oz=subs(M_Oz,items_sxy,items_txy);
fprintf('or equivalent with respect to l')
fprintf(' the moment MO is \n')
fprintf('MO = %s \n\n', char(M_Oz));
```

In MATLAB the input data is:

```
items_s = {l,FL,FU,FP,alpha,m,g};
items_t = {1,1500,600,1200,35*pi/180,250,9.81};
```

The numerical results are ccomputed and printed in MATLAB using:

```
F_Ox=subs(F_Ox,items_s,items_t);
F_Oy=subs(F_Oy,items_s,items_t);
M_Oz=subs(M_Oz,items_s,items_t);
fprintf('The reaction forces and the ');
fprintf('moment at O are \n')
fprintf(' F_Ox=%g (N)\n F_Oy=%g (N)\n M_Oz=%g (N m)\n\n',...
    F_Ox,F_Oy, M_Oz);
```

The numerical results, namely the reaction forces and the reaction moment at the support O, are obtained in MATLAB as:

```
F_Ox = -2482.98 (N)
F_Oy = 1164.21 (N)
M_Oz = -385.557 (N m)
```

Example 2.5

Three workers are trying to rotate a rectangular box $OABCDENH$ of mass m, height $h = d_{OE}$, length $l = d_{OC}$ and depth $d = d_{OA}$, as shown in Figure 2.14. One worker located in the Oyz

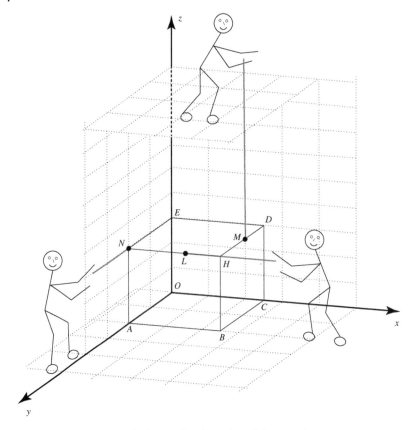

Figure 2.14 A rectangular box under the action of three workers.

plane is using a cable connected to the corner N to pull the box in the Oy direction with a force $\mathbf{F}_N$, one is using a cable connected to the point L (the middle of the side NH) to pull the box in the Ox direction with a force $\mathbf{F}_L$, while the last one is using a cable connected to the point M (where $d_{MD} = \frac{d}{3}$) to lift the box from above with a force $\mathbf{F}_M$. The dimensions of the box, the application point, magnitude and direction of the forces applied by the workers are depicted by the free body diagram in Figure 2.15. Determine the reaction force at the supoport O and the moment of the applied forces (Figure 2.15) about the point O.

Numerical application: $l = 2$ m, $d = 1$ m, $h = 1$ m, $m = 250$ kg, $g = 9.81$ m s^{-2}, $F_N = 400$ N, $F_L = 600$ N, $F_M = 500$ N.

Solution
The rectangular box $OABCDENH$ of weight $G = mg$ shown in Figure 2.14 is uniform, so its centre of mass is located at its centroid given by the intersection of its diagonals. A Cartesian frame $Oxyz$ with the origin at O and the Ox, Oy, and Oz axes along the OC, OA and OE directions is considered. To calculate the reaction force at O, the next equilibrium equations is considered

$$\sum \mathbf{F} = \mathbf{F}_O + \mathbf{F}_N + \mathbf{F}_L + \mathbf{F}_M + \mathbf{G} = 0 \tag{2.46}$$

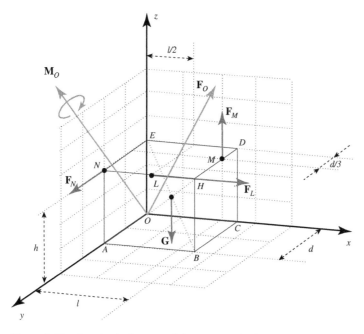

Figure 2.15 Free body diagram of the rectangular box.

or equivalent

$$\sum F_x = F_{Ox} + F_L = 0$$
$$= F_{Ox} + F_L \mathbf{1} = 0$$
$$\sum F_y = F_{Oy} + F_N$$
$$= F_{Oy} + F_N \mathbf{J} = 0$$
$$\sum F_z = F_{Oz} + F_M - G = 0$$
$$= F_{Oz} + F_M \mathbf{k} - mg\mathbf{k} = 0. \tag{2.47}$$

The equilibrium equations are written in MATLAB with:

```
clear all; clc; close all
syms xA xB xC xH yA yB yC yH zA zB zC zH
syms xD xE xN xG yD yN yG yE zD zE zN zG
syms xM xL yM yL zM zL l d h
rA_=[xA,yA,zA]; rB_=[xB,yB,zB];
rC_=[xC,yC,zC]; rD_=[xD,yD,zD];
rE_=[xE,yE,zE]; rN_=[xN,yN,zN];
rH_=[xH,yH,zH]; rG_=[xG,yG,zG];
rM_=[xM,yM,zM]; rL_=[xL,yL,zL];

syms FN FL FM
FNx=0; FNy=FN; FNz=0;
FN_=[FNx,FNy,FNz];
```

```
FLx=FL; FLy=0; FLz=0;
FL_=[FLx,FLy,FLz];
FMx=0; FMy=0; FMz=FM;
FM_=[FMx,FMy,FMz];

syms m g F_Ox F_Oy F_Oz
G_=[0,0,-m*g];
FO_=[F_Ox,F_Oy,F_Oz];
sumF_ = FO_+ FN_ + FL_ + FM_ + G_;
S_Fx = sumF_(1);
S_Fy = sumF_(2);
S_Fz = sumF_(3);
fprintf(' sumFx=%s\n sumFy=%s\n sumFz=%s\n\n',...
    char(S_Fx), char(S_Fy),char(S_Fz));
```

From Equation (2.47) one can calculate

$$\mathbf{F}_{Ox} = -F_L\mathbf{i}$$
$$\mathbf{F}_{Oy} = -F_N\mathbf{j}$$
$$\mathbf{F}_{Oz} = (mg - F_M)\mathbf{k}. \tag{2.48}$$

The magnitude of the *Ox*, *Oy* and *Oz* components of the resultant $\mathbf{F}_O$ are calculated and printed in MATLAB with:

```
F_Ox=solve(sumF_(1), F_Ox);
F_Oy=solve(sumF_(2), F_Oy);
F_Oz=solve(sumF_(3), F_Oz);
fprintf('reaction F_Ox, F_Oy, F_Oz are: \n')
fprintf(' F_Ox=%s\n F_Oy=%s\n F_Oz=%s\n\n',...
    char(F_Ox),char(F_Oy),char(F_Oz));
```

From Equation (2.46), Equation (2.47) and Equation (2.48) the reaction force at the support *O* can be written as

$$\mathbf{F}_O = \mathbf{F}_{Ox}\mathbf{i} + \mathbf{F}_{Oy}\mathbf{j} + \mathbf{F}_{Oz}\mathbf{k}$$
$$= -F_L\mathbf{i} - F_N\mathbf{j}(mg - F_M)\mathbf{k}.$$

The equilibrium about the support *O* is written as

$$\sum \mathbf{M} = \mathbf{M}_O + \mathbf{r}_G \times \mathbf{G} + \mathbf{r}_N \times \mathbf{F}_N + \mathbf{r}_L \times \mathbf{F}_L + \mathbf{r}_M \times \mathbf{F}_M = 0 \tag{2.49}$$

where

$$\mathbf{r}_N \times \mathbf{F}_N = \begin{vmatrix} \mathbf{i} & \mathbf{j} & \mathbf{k} \\ 0 & y_N & z_N \\ 0 & F_N & 0 \end{vmatrix} = \begin{vmatrix} y_N & z_N \\ F_N & 0 \end{vmatrix}\mathbf{i} = -F_N z_N \mathbf{i} \tag{2.50}$$

$$\mathbf{r}_G \times \mathbf{G} = \begin{vmatrix} \mathbf{i} & \mathbf{j} & \mathbf{k} \\ x_G & y_G & z_G \\ 0 & 0 & -G_z \end{vmatrix} = -G_z \begin{vmatrix} \mathbf{i} & \mathbf{j} \\ x_G & y_G \end{vmatrix}$$
$$= -G_z(y_G\mathbf{i} - x_G\mathbf{j}) = -mgy_G\mathbf{i} + mgx_G\mathbf{j} \tag{2.51}$$

$$\mathbf{r}_L \times \mathbf{F}_L = \begin{vmatrix} \mathbf{1} & \mathbf{J} & \mathbf{k} \\ x_L & y_L & z_L \\ F_L & 0 & 0 \end{vmatrix} = F_L \begin{vmatrix} \mathbf{J} & \mathbf{k} \\ y_L & z_L \end{vmatrix} \mathbf{1}$$

$$= F_L(z_L \mathbf{J} - y_L \mathbf{k}) = F_L z_L \mathbf{J} - F_L y_L \mathbf{k} \tag{2.52}$$

$$\mathbf{r}_M \times \mathbf{F}_M = \begin{vmatrix} \mathbf{1} & \mathbf{J} & \mathbf{k} \\ x_M & y_M & z_M \\ 0 & 0 & F_M \end{vmatrix} = F_M \begin{vmatrix} \mathbf{1} & \mathbf{J} \\ x_M & y_M \end{vmatrix} \mathbf{1}$$

$$= F_M(y_M \mathbf{1} - x_M \mathbf{J}) = F_M y_M \mathbf{1} - F_M x_M \mathbf{J}. \tag{2.53}$$

The equilibrium is computed in MATLAB with:

```
syms M_Ox M_Oy M_Oz
MO_ = [M_Ox, M_Oy, M_Oz];
sumMO_ = MO_ + cross(rG_, G_)...
    + cross(rN_, FN_) + cross(rM_, FM_)...
    + cross(rL_, FL_);
SM_Ox = sumMO_(1);
SM_Oy = sumMO_(2);
SM_Oz = sumMO_(3);
```

Replacing Equations (2.50)–(2.53) in Equation (2.49) one can write the equilibrium about the support O as

$$\sum \mathbf{M} = \mathbf{M}_O - mgy_G \mathbf{1} + mgx_G \mathbf{J} - F_N z_N \mathbf{1} + F_L z_L \mathbf{J} - F_L y_L \mathbf{k} + F_M y_M \mathbf{1} - F_M x_M \mathbf{J} = 0 \tag{2.54}$$

that is, the moment about the support O is

$$\mathbf{M}_O = mgy_G \mathbf{1} - mgx_G \mathbf{J} + F_N z_N \mathbf{1} - F_L z_L \mathbf{J} + F_L y_L \mathbf{k} - F_M y_M \mathbf{1} + F_M x_M \mathbf{J}$$

$$= (mgy_G + F_N z_N - F_M y_M) \mathbf{1} + (-mgx_G - F_L z_L + F_M x_M) \mathbf{J} + F_L y_L \mathbf{k}. \tag{2.55}$$

From Equation (2.55) one can write

$$M_{Ox} = mgy_G + F_N z_N - F_M y_M$$

$$M_{Oy} = -mgx_G - F_L z_L + F_M x_M$$

$$M_{Oz} = F_L y_L. \tag{2.56}$$

In MATLAB the components of the moment $\mathbf{M}_O$ are calculated and printed with:

```
M_Ox = solve(SM_Ox, M_Ox);
M_Oy = solve(SM_Oy, M_Oy);
M_Oz = solve(SM_Oz, M_Oz);
fprintf('The moment MO is \n')
fprintf(' M_Ox=%s\n M_Oy=%s\n M_Oz=%s\n\n',...
    char(M_Ox), char(M_Oy), char(M_Oz));
```

Replacing $x_N = 0$, $y_N = d$, $z_N = h$, $x_G = \frac{1}{2}l$, $y_G = \frac{1}{2}d$, $z_G = \frac{1}{2}h$, $x_L = \frac{1}{2}l$, $y_L = d$, $z_L = h$, $x_M = l$, $y_M = \frac{1}{3}d$ and $z_M = h$ in Equation (2.56) one can calculate

$$M_{Ox} = mg\frac{d}{2} + F_N h - F_M \frac{d}{3}$$

$$M_{Oy} = -mg\frac{l}{2} - F_L h + F_M l$$

$$M_{Oz} = F_L d.$$

The Ox, Oy and Oz coordinates of points A, B, C, D, E, N are introduced in MATLAB using:

```
items_sxy = {xA,yA,zA,xB,yB,zB,xC,yC,zC,xD,yD,zD,...
    xE,yE,zE,xN,yN,zN,xH,yH,zH,xL,yL,zL,...
    xG,yG,zG,xM,yM,zM};
items_txy = {0,'d',0,'l','d',0,'l',0,0,'l',0,'h',...
    0,0,'h',0,'d','h','l','d','h','l'/2,'d',...
    'h','l'/2,'d'/2,'h'/2,'l','d'/3,'h'};
```

The magnitude of the Ox, Oy and Oz components of the moment $\mathbf{M}_O$ are now calculated in MATLAB with:

```
M_Ox=subs(M_Ox,items_sxy,items_txy);
M_Oy=subs(M_Oy,items_sxy,items_txy);
M_Oz=subs(M_Oz,items_sxy,items_txy);
fprintf('or equivalent with respect ')
fprintf('to l and d the moment MO is \n')
fprintf(' M_Ox=%s\n M_Oy=%s\n M_Oz=%s\n\n',...
    char(M_Ox),char(M_Oy),char(M_Oz));
```

The input data is given in MATLAB with:

```
items_s = {l,d,h,m,g,FN,FL,FM};
items_t = {2,1,1,250,9.81,400,600,500};
```

The numerical results are computed and printed in MATLAB using:

```
F_Ox=subs(F_Ox,items_s,items_t);
F_Oy=subs(F_Oy,items_s,items_t);
F_Oz=subs(F_Oz,items_s,items_t);
M_Ox=subs(M_Ox,items_s,items_t);
M_Oy=subs(M_Oy,items_s,items_t);
M_Oz=subs(M_Oz,items_s,items_t);
fprintf('The reaction forces and the ');
fprintf('reaction moment at the support O are \n')
fprintf(' F_Ox=%g (N)\n F_Oy=%g (N)\n F_Oz=%g (N)\n',...
    F_Ox,F_Oy,F_Oz);
fprintf(' M_Ox = %f (N m)\n', M_Ox);
fprintf(' M_Oy = %f (N m)\n', M_Oy);
fprintf(' M_Oz = %f (N m)\n\n', M_Oz);
```

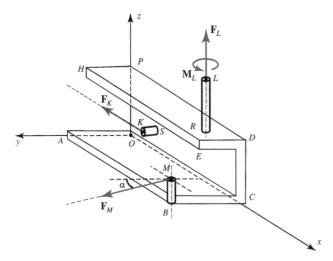

Figure 2.16 Structural member with pins.

The numerical results are obtained in MATLAB as:

```
F_Ox = -600 (N)
F_Oy = -400 (N)
F_Oz = 1952.5 (N)
M_Ox = 1459.583333 (N m)
M_Oy = -2052.500000 (N m)
M_Oz = 600.000000 (N m)
```

Example 2.6
The couple $\mathbf{M}_L$ and the forces $\mathbf{F}_K$, $\mathbf{F}_L$, and $\mathbf{F}_M$ are acting on the pins SK, RL and BM connected to the structural member as shown in Figure 2.16. The pins SK, RL and BM having the heights (lengths) h_{SK}, h_{RL} and h_{BM}, are perpendicular to the faces $OCDP$, $HEDP$ and $OABC$ respectively. The dimensions of the structural member, and the location and direction of the forces are depicted in the free body diagram shown in Figure 2.17. The structural member is assumed to be of negligible weight. Determine an equivalent resultant force $\mathbf{F}_O$ and couple moment $\mathbf{M}_O$ (Figure 2.17) about the point O.

Numerical application: $l = 3$ m, $d_b = 1$ m, $d_{up} = 0.75$ m, $u = 1$ m, $h_{RL} = 0.3$ m, $h_{BM} = 0.1$ m, $h_{SK} = 0.1$ m, $F_K = 400$ N, $F_L = 500$ N, $F_M = 600$ N, $\alpha = 35°$, $d_R = 0.25$ m, $l_R = 0.25$ m, $u_S = 0.25$ m, $M_L = 200$ N m.

Solution
To calculate the reaction force at O, the next equilibrium equation is considered

$$\sum \mathbf{F} = \mathbf{F}_O + \mathbf{F}_K + \mathbf{F}_L + \mathbf{F}_M = 0 \tag{2.57}$$

or equivalent

$$\sum \mathbf{F}_x = \mathbf{F}_{Ox} + \mathbf{F}_{Mx} - \mathbf{F}_K = 0$$
$$= \mathbf{F}_{Ox} + F_M \sin \alpha \mathbf{1} - F_K \mathbf{1} = 0$$

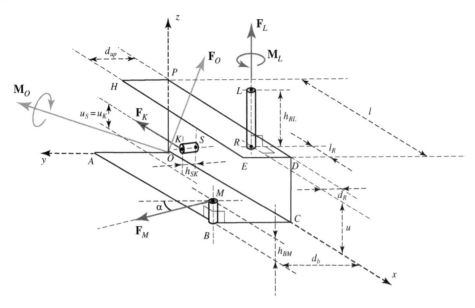

Figure 2.17 Free body diagram of the structural member with pins.

$$\sum \mathbf{F}_y = \mathbf{F}_{Oy} + \mathbf{F}_{My}$$
$$= \mathbf{F}_{Oy} + F_M \cos \alpha \mathbf{j} = 0$$
$$\sum \mathbf{F}_z = \mathbf{F}_{Oz} + \mathbf{F}_L = 0$$
$$= \mathbf{F}_{Oz} + F_L \mathbf{k} = 0. \qquad (2.58)$$

The equilibrium is addressed in MATLAB with:

```
clear all; clc; close all

syms x_A x_B x_C y_A y_B y_C z_A z_B z_C
syms x_D x_E x_H y_D y_E y_H z_D z_E z_H
syms x_P y_P z_P x_R y_R z_R x_L y_L z_L
syms x_M y_M z_M x_S y_S z_S x_K y_K z_K
syms l d_b d_up l_R d_R u
syms h_RL h_BM h_SK l_S u_S

rA_ = [x_A,y_A,z_A]; rB_ = [x_B,y_B,z_B];
rC_ = [x_C,y_C,z_C]; rD_ = [x_D,y_D,z_D];
rE_ = [x_E,y_E,z_E]; rH_ = [x_H,y_H,z_H];
rP_ = [x_P,y_P,z_P]; rS_ = [x_S,y_S,z_S];
rR_ = [x_R,y_R,z_R]; rL_ = [x_L,y_L,z_L];
rK_ = [x_K,y_K,z_K]; rM_ = [x_M,y_M,z_M];

syms FK FL FM alpha
FKx=-FK; FKy=0; FKz=0;
```

```
FK_=[FKx,FKy,FKz];
FLx=0; FLy=0; FLz=FL;
FL_=[FLx,FLy,FLz];
FMx=FM*sin(alpha); FMy=FM*cos(alpha); FMz=0;
FM_=[FMx,FMy,FMz];

syms F_Ox F_Oy F_Oz
FO_=[F_Ox,F_Oy,F_Oz];
sumF_ = FO_+ FK_ + FL_ + FM_;
SFx = sumF_(1);
SFy = sumF_(2);
SFz = sumF_(3);
fprintf('sum of forces on x and on y:\n')
fprintf(' sumFx=%s\n sumFy=%s\n sumFz=%s\n\n',...
    char(SFx),char(SFy),char(SFz))
```

From Equation (2.58) one can write

$$\mathbf{F}_{Ox} = -F_M \sin \alpha \mathbf{\imath} + F_K \mathbf{\imath}$$
$$\mathbf{F}_{Oy} = -F_M \cos \alpha \mathbf{J}$$
$$\mathbf{F}_{Oz} = -F_L \mathbf{k}. \tag{2.59}$$

The magnitude of the $\mathbf{F}_{Ox}$, $\mathbf{F}_{Oy}$ and $\mathbf{F}_{Oz}$ components of the reaction force $\mathbf{F}_O$ at the base O are calculated in MATLAB with:

```
F_Ox=solve(sumF_(1), F_Ox);
F_Oy=solve(sumF_(2), F_Oy);
F_Oz=solve(sumF_(3), F_Oz);
fprintf('reaction F_Ox and F_Oy are: \n')
fprintf(' F_Ox=%s\n F_Oy=%s\n F_Oz=%s\n\n',...
    char(F_Ox),char(F_Oy),char(F_Oz))
```

From Equations (2.57) and (2.59) the reaction force at the support O can be written as

$$\mathbf{F}_O = (-F_M \sin \alpha + F_K)\mathbf{\imath} - F_M \cos \alpha \mathbf{J} - F_L \mathbf{k}.$$

The equilibrium about the support O is written as

$$\sum \mathbf{M} = \mathbf{M}_O + \mathbf{M}_L + \mathbf{r}_L \times \mathbf{F}_L + \mathbf{r}_M \times \mathbf{F}_M + \mathbf{r}_K \times \mathbf{F}_K = 0 \tag{2.60}$$

where

$$\mathbf{r}_K \times \mathbf{F}_K = \begin{vmatrix} \mathbf{\imath} & \mathbf{J} & \mathbf{k} \\ x_K & y_K & z_K \\ -F_K & 0 & 0 \end{vmatrix} = -F_K \begin{vmatrix} \mathbf{J} & \mathbf{k} \\ y_K & z_K \end{vmatrix} = -F_K(z_K \mathbf{J} - y_K \mathbf{k}) \tag{2.61}$$

and where $z_K = s_K$ and $x_K = x_S = l_S$. Also, one can calculate

$$\mathbf{r}_L \times \mathbf{F}_L = \begin{vmatrix} \mathbf{\imath} & \mathbf{J} & \mathbf{k} \\ x_L & y_L & z_L \\ 0 & 0 & F_L \end{vmatrix} = F_L \begin{vmatrix} \mathbf{\imath} & \mathbf{J} \\ x_L & y_L \end{vmatrix} = F_L(y_L \mathbf{\imath} - x_L \mathbf{J}) \tag{2.62}$$

$$\mathbf{r}_M \times \mathbf{F}_M = \begin{vmatrix} \mathbf{1} & \mathbf{J} & \mathbf{k} \\ x_M & y_M & z_M \\ F_{Mx} & F_{My} & 0 \end{vmatrix} = F_{Mx} \begin{vmatrix} \mathbf{J} & \mathbf{k} \\ y_M & z_M \end{vmatrix} - F_{My} \begin{vmatrix} \mathbf{1} & \mathbf{k} \\ x_M & z_M \end{vmatrix}$$

$$= F_{Mx}(z_M \mathbf{J} - y_M \mathbf{k}) - F_{My}(z_M \mathbf{1} - x_M \mathbf{k})$$

$$= F_M \sin \alpha (z_M \mathbf{J} - y_M \mathbf{k}) - F_M \cos \alpha (z_M \mathbf{1} - x_M \mathbf{k}.) \tag{2.63}$$

The equilibrium is computed in MATLAB with:

```
syms M_Ox M_Oy M_Oz MLx MLy MLz ML
MO_ = [M_Ox,M_Oy,M_Oz];
MLx=0; MLy=0; MLz=ML;
ML_ = [MLx,MLy,MLz];
sumMO_=MO_ + ML_...
        + cross(rK_,FK_) + cross(rM_,FM_)...
        + cross(rL_,FL_);
SM_Ox=sumMO_(1);
SM_Oy=sumMO_(2);
SM_Oz=sumMO_(3);
```

Replacing Equations (2.61)– (2.63) in Equation (2.60) one can calculate

$$\mathbf{M}_O = -\mathbf{M}_L - \mathbf{r}_L \times \mathbf{F}_L - \mathbf{r}_M \times \mathbf{F}_M - \mathbf{r}_K \times \mathbf{F}_K$$

$$= -\mathbf{M}_L - F_L(y_L \mathbf{1} - x_L \mathbf{J}) - F_M \sin \alpha (z_M \mathbf{J} - y_M \mathbf{k}) + F_M \cos \alpha (z_M \mathbf{1} - x_M \mathbf{k})$$

$$+ F_K(z_K \mathbf{J} - y_K \mathbf{k})$$

$$= -\mathbf{M}_L + (F_M z_M \cos \alpha - F_L y_L)\mathbf{1} + (F_L x_L - F_M z_M \sin \alpha + F_K z_K)\mathbf{J}$$

$$+ (F_M y_M \sin \alpha - F_M x_M \cos \alpha - F_K y_K)\mathbf{k} \tag{2.64}$$

where

$$\mathbf{M}_{Ox} = (F_M z_M \cos \alpha - F_L y_L)\mathbf{1}$$

$$\mathbf{M}_{Oy} = (F_L x_L + F_K z_K - F_M \sin \alpha z_M)\mathbf{J}$$

$$\mathbf{M}_{Oz} = (F_M y_M \sin \alpha - F_K y_K - F_M x_M \cos \alpha)\mathbf{k} - \mathbf{M}_L. \tag{2.65}$$

In MATLAB the maginutude of the components of the moment $\mathbf{M}_O$ are calculated with:

```
M_Ox=solve(SM_Ox, M_Ox);
M_Oy=solve(SM_Oy, M_Oy);
M_Oz=solve(SM_Oz, M_Oz);
fprintf('The moment MO is \n')
fprintf('M_Ox=%s\n M_Oy=%s\n M_Oz=%s\n\n',...
    char(M_Ox), char(M_Oy),char(M_Oz));
```

Replacing

$$x_A = 0, \ y_A = d_b, \ z_A = 0, \ x_B = l, \ y_B = d_b, \ z_B = 0,$$

$$x_C = l, \ y_C = 0, \ z_C = 0, \ x_D = l, \ y_D = 0, \ z_D = u,$$

$$x_E = l, \ y_E = d_u p, \ z_E = u, \ x_H = 0, \ y_H = d_u p, \ z_H = u,$$

$$x_P = 0, \ y_P = 0, \ z_P = u, \ x_M = l, \ y_M = d_b, \ z_M = h_{BM},$$
$$x_K = l_S, \ y_K = h_{SK}, \ z_K = u_S, \ x_S = l_S, \ y_S = 0, \ z_S = u_S,$$
$$x_R = l - l_R, \ y_R = d_R, \ z_R = u, \ x_L = l - l_R, \ y_L = d_R, \ z_L = u + h_{RL}$$

in Equation (2.65) one can calculate

$$M_{Ox} = F_M h_{BM} \cos \alpha - F_L d_R$$
$$M_{Oy} = F_K u_S + F_L(l - l_R) - F_M h_{BM} \sin \alpha$$
$$M_{Oz} = F_M d_b \sin \alpha - F_K h_{SK} - F_M l \cos \alpha - M_L.$$

The Ox, Oy and Oz coordinates of the points A, B, C, D, E, H, P, M, K, S, R and L are introduced in MATLAB using:

```
items_sxy = {x_A,y_A,z_A,x_B,y_B,z_B,x_C,y_C,z_C,x_D,y_D,...
    z_D,x_E,y_E,z_E,x_H,y_H,z_H,x_P,y_P,z_P,x_M,y_M,z_M,...
    x_K,y_K,z_K,x_S,y_S,z_S,x_R,y_R,z_R,x_L,y_L,z_L};
items_txy = {0,d_b,0,l,d_b,0,l,0,0,1,0,u,...
    1,d_up,u,0,d_up,u,0,0,u,...
    1,d_b,h_BM,l_S,h_SK,u_S,...
    l_S,0,u_S,1-l_R,d_R,u,...
    1-l_R,d_R,u+h_RL};
```

The magnitude of the components of the moment $\mathbf{M}_O$ are now calculated in MATLAB with:

```
M_Ox=subs(M_Ox,items_sxy,items_txy);
M_Oy=subs(M_Oy,items_sxy,items_txy);
M_Oz=subs(M_Oz,items_sxy,items_txy);
fprintf('or after computation the moment MO is \n')
fprintf('M_Ox=%s\n M_Oy=%s\n M_Oz=%s\n\n',...
    char(M_Ox), char(M_Oy),char(M_Oz));
```

In MATLAB the data are introduced using:

```
items_s = {l,d_b,d_up,u,h_RL,h_BM,h_SK,...
    FK,FL,FM,alpha,d_R,l_R,u_S,ML};
items_t = {3,1,0.75,1,0.3,0.1,0.1,400,500,...
    600,35*pi/180,0.25,0.25,0.25,200};
```

and the numerical results are:

```
F_Ox=subs(F_Ox,items_s,items_t);
F_Oy=subs(F_Oy,items_s,items_t);
F_Oz=subs(F_Oz,items_s,items_t);
M_Ox=subs(M_Ox,items_s,items_t);
M_Oy=subs(M_Oy,items_s,items_t);
M_Oz=subs(M_Oz,items_s,items_t);
fprintf('The reaction forces and the ');
fprintf('reaction moment at the support O are \n')
fprintf(' F_Ox=%g (N)\n F_Oy=%g (N)\n F_Oz=%g (N)\n',...
```

```
    F_Ox, F_Oy, F_Oz);
fprintf(' M_Ox=%g (N m)\n M_Oy=%g (N m)\n M_Oz=%g (N m)\n',...
    M_Ox, M_Oy, M_Oz);
```

The numerical results (equivalent resultant force $\mathbf{F}_O$ and couple moment $\mathbf{M}_O$) are obtained in MATLAB as:

```
F_Ox=55.8541 (N)
F_Oy=-491.491 (N)
F_Oz=-500 (N)
M_Ox=-75.8509 (N m)
M_Oy=1440.59 (N m)
M_Oz=-1370.33 (N m)
```

Example 2.7
The L assembly is shown in Figure 2.18. The following dimensions are given: the distance from O to A is $l_1 = OA = 2$ m, the distance from A to point 2 is $l_2 = 2$ m, and the distance from point 2 to point 1 is $l_3 = 3$ m. The segment $O2$ is perpendicular to 12 and the assembly is on the xy plane. The weight of the assembly is neglected. At the point 1 acts a spatial force $\mathbf{F}_1 = F_{1x}\,\mathbf{1} + F_{1y}\,\mathbf{J} + F_{1z}\,\mathbf{k}$, where $F_{1x} = 0, F_{1y} = -1.5$ kN, $F_{1z} = 2$ kN and at the point 2 acts another spatial force $\mathbf{F}_2 = F_{2x}\,\mathbf{1} + F_{2y}\,\mathbf{J} + F_{2z}\,\mathbf{k}$, where $F_{2x} = 1$ kN, $F_{2y} = 0, F_{2z} = -1$ kN. Find the components of the internal loading at a section through the point A.

Solution
The origin of the xyz reference frame is at O. The position vectors of the points A, 2, and 1 are

$$\mathbf{r}_A = l_1\,\mathbf{J},$$
$$\mathbf{r}_2 = (l_1 + l_2)\,\mathbf{J},$$
$$\mathbf{r}_1 = l_3\,\mathbf{1} + (l_1 + l_2)\,\mathbf{J}. \tag{2.66}$$

The input data in MATLAB are introduced by:

```
syms l1 l2 l3 F1x F1y F1z F2x F2y F2z


F1_ = [F1x F1y F1z];
F2_ = [F2x F2y F2z];


rA_ = [0 l1 0];
r2_ = [0 l1+l2 0];
r1_ = [l3 l1+l2 0];


sl = {l1, l2, l3, F1x F1y F1z F2x F2y F2z};
nl = {2, 2, 3, 0, -1.5, 2, 1, 0, -1};
```

The MATLAB figure for the input data is obtain with:

```
xA = subs(rA_(1), sl, nl);
yA = subs(rA_(2), sl, nl);
```

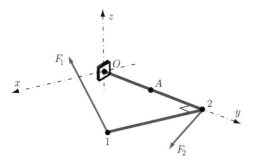

Figure 2.18 A shaped L assembly.

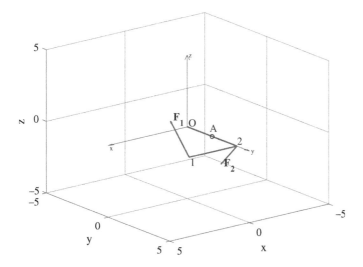

Figure 2.19 The MATLAB graphical representation of the forces F_1 and F_2 acting on the shaped L assembly.

```
zA  = subs(rA_(3), sl, nl);
x1  = subs(r1_(1), sl, nl);
y1  = subs(r1_(2), sl, nl);
z1  = subs(r1_(3), sl, nl);
x2  = subs(r2_(1), sl, nl);
y2  = subs(r2_(2), sl, nl);
z2  = subs(r2_(3), sl, nl);
F1n_ = subs(F1_, sl, nl);
F2n_ = subs(F2_, sl, nl);
```

The MATLAB representation is shown in Figure 2.19 The internal loading at a section through the point A are the internal force $\mathbf{F}_A = F_{Ax}\,\mathbf{i} + F_{Ay}\,\mathbf{j} + F_{Az}\,\mathbf{k}$ and the internal moment $\mathbf{M}_A = M_{Ax}\,\mathbf{i} + M_{Ay}\,\mathbf{j} + M_{Az}\,\mathbf{k}$. The internal loadings are calculated from the force equilibrium equation and the sum of the moments with respect to point A,

$$\sum \mathbf{F} = \mathbf{F}_A + \mathbf{F}_1 + \mathbf{F}_2,$$
$$\sum \mathbf{M} = \mathbf{M}_A + (\mathbf{r}_1 - \mathbf{r}_A) \times \mathbf{F}_1 + (\mathbf{r}_2 - \mathbf{r}_A) \times \mathbf{F}_2, \tag{2.67}$$

or with MATLAB:

```
syms FAx FAy FAz
syms MAx MAy MAz

FA_ = [FAx FAy FAz];
MA_ = [MAx MAy MAz];

SF_ = FA_ + F1_ + F2_;
SFx = SF_(1);
SFy = SF_(2);
SFz = SF_(3);

SMA_ = MA_ + cross(r1_-rA_, F1_) + cross(r2_-rA_, F2_);
SMx = SMA_(1);
SMy = SMA_(2);
SMz = SMA_(3);
```

The unknowns are determined with the `solve` command:

```
solC = solve(SFx,SFy,SFz,SMx,SMy,SMz, ...
       FAx, FAy, FAz, MAx, MAy, MAz);

FAxs = solC.FAx;
FAys = solC.FAy;
FAzs = solC.FAz;
MAxs = solC.MAx;
MAys = solC.MAy;
MAzs = solC.MAz;

FAxn = subs(FAxs, sl, nl);
FAyn = subs(FAys, sl, nl);
FAzn = subs(FAzs, sl, nl);
FAn_ = [FAxn FAyn FAxn];

MAxn = subs(MAxs, sl, nl);
MAyn = subs(MAys, sl, nl);
MAzn = subs(MAzs, sl, nl);
```

and the results are:

```
% FAx = - F1x - F2x = -1.000 (kN)
% FAy = - F1y - F2y =  1.500 (kN)
% FAz = - F1z - F2z = -1.000 (kN)
% MAx = - F1z*12 - F2z*12 = -2.000 (kN m)
% MAy = F1z*13 =  6.000 (kN m)
% MAz = F1x*12 + F2x*12 - F1y*13 =  6.500 (kN m)
```

The graphical representation of $\mathbf{F}_A$ and $\mathbf{M}_A$ are shown in Figure 2.20.

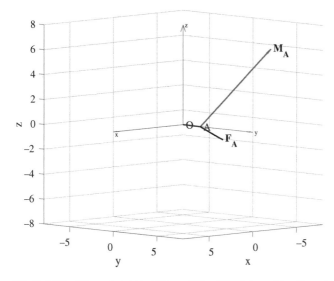

Figure 2.20 The components $\mathbf{M_A}$ and $\mathbf{F_A}$ of the internal loading at a section through the point A.

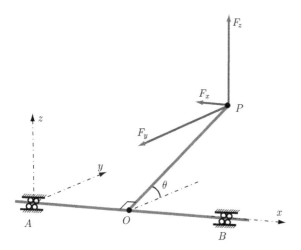

Figure 2.21 Graphical representation of a T shaped structure under the action of the several forces.

Example 2.8

Figure 2.21 shows a T structure $AOBP$. The segment PO is perpendicular to the segment AB at O. The T structure has the dimensions $AO = l_1 = 0.15$ m, $OB = l_2 = 0.15$ m, and $OP = l_3 = 0.10$ m. The angle between OP and the y axis is $\theta = \pi/3$ as shown in Figure 2.21. The forces acting at the end P of the link are $F_{Px} = F_x = -1$ kN, $F_{Py} = F_y = -1$ kN, and $F_{Pz} = F_z = 3$ kN. The directions of the forces are represented in Figure 2.21. The bearing at A can take axial (thrust) forces. Find the joint bearing reactions at A and B.

Solution

The origin of the *xyz* reference is at the axial bearing A. The position vectors of the radial bearing B, of the point O, and of the application point P are

$$\mathbf{r}_B = (l_1 + l_2)\, \mathbf{\imath},$$
$$\mathbf{r}_O = l_1\, \mathbf{J},$$
$$\mathbf{r}_P = l_1\, \mathbf{\imath} + l_3\, \cos\theta\, \mathbf{J} + l_3\, \sin\theta\, \mathbf{k}, \tag{2.68}$$

or in MATLAB the input data are:

```
Fx = -1.000;  %  (kN)
Fy = -1.000;  %  (kN)
Fz =  3.000;  %  (kN)
l1 = 0.15;    %  (m)
l2 = 0.15;    %  (m)
l3 = 0.10;    %  (m)


FP_ = [Fx,Fy,Fz] ;  % force vector at P


theta = pi/3;
rA_  = [0, 0, 0];
rB_  = [l1+l2, 0,0];
rO_  = [l1, 0,0];
% position vector of forces at P
rP_  = [l1, l3*cos(theta), l3*sin(theta)];
```

The unknowns are the reaction forces at the axial bearing A, $\mathbf{F}_A = F_{Ax}\, \mathbf{\imath} + F_{Ay}\, \mathbf{J} + F_{Ak}\, \mathbf{k}$, and at the radial bearing B, $\mathbf{F}_B = F_{By}\, \mathbf{J} + F_{Bz}\, \mathbf{k}$:

```
% reaction forces
syms FAx FAy FAz FBx FBy FBz
FA_  = [FAx,  FAy,  FAz];  % at A
FB_  = [ 0,  FBy,  FBz];  % at B
```

The joint reactions at the bearing B are determined using the sum of the moments about A:

```
% sum MA_ = rP_ x FP_ + rB_ x FB_  =>
MA_ = cross(rP_, FP_) + cross(rB_,FB_);
```

with the solution:

```
solFB = solve(MA_(2),MA_(3));
FBys=eval(solFB.FBy);
FBzs=eval(solFB.FBz);
FB_ = [0, FBys, FBzs];
FBr = sqrt(FB_(2)^2+FB_(3)^2);
% FB_ = [ 0  0.333 -1.789]  (kN)
% radial force FBr =  1.819  (kN)
```

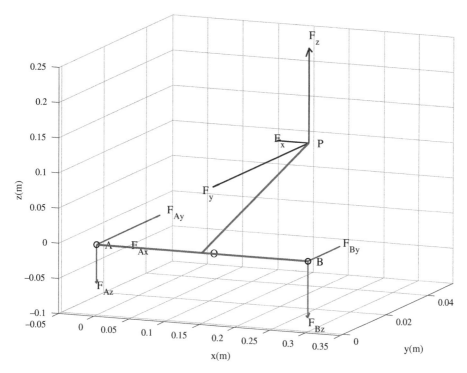

Figure 2.22 Graphical representation of the reaction forces.

The reaction forces at the axial bearing A are obtained with the sum of the forces for the system:

```
% sum F_  = FA_  + FB_  + FP_  =>
FA_  = -FP_ -FB_ ;
FAr = sqrt(FA_(2)^2+FA_(3)^2);
FAa = FA_(1);
% FA_  = [  1.000   0.667 -1.211]  (kN)
% radial force FAr =   1.383  (kN)
% axial force FAa =   1.000  (kN)
```

The graphical representation of the reaction forces are depicted in Figure 2.22.

3

Equilibrium of Structures

The mechanical engineering principles described in this chapter can be applied to framework structures to determine the internal forces in the members due to externally applied loads.

Frame structures are typically:

- Frames that are usually indeterminate and composed of straight members with hinges or rigid connections.
- Trusses that are usually arranged in a triangular fashion and composed of straight members pinned to the ground or to the rest of the frame at both end points (Figures 3.1 and 3.2).

When the members of a framework structure, e.g, frame or truss, are constructed in a plane, the framework structure is called a plane structure while a space structure is three dimensional. The members – considered to have a negligible weight – are assumed to carry loads only in the longitudinal direction.

A framework structure is called statically indeterminate if more links that are needed to prevent collapse are present. Additional links or supports that are not necessary for maintaining the equilibrium configuration are called redundant. Statically determinate plane structures do not have more supporting constraints than are necessary to maintain equilibrium.

3.1 Equilibrium Equations

A mechanical system (framework structure made of several members) is in equilibrium if and only if all of its subsystems (member) are in equilibrium, that is, each subsystem is stationary. For a system to remain stationary (be in equilibrium) it is necessary that all the forces and moments acting on it should be equivalent to the absence of forces and moments.

A body, acted upon by a system of forces and moments, is in equilibrium when the sum of all the forces acting on the body is zero and the sum of the moments about any point is zero, that is

$$\sum \mathbf{F} = 0 \text{ and } \sum \mathbf{M}_P = 0, \ \forall \ P. \tag{3.1}$$

Engineering Applications: Analytical and Numerical Calculation with MATLAB, First Edition.
Mihai Dupac and Dan B. Marghitu.

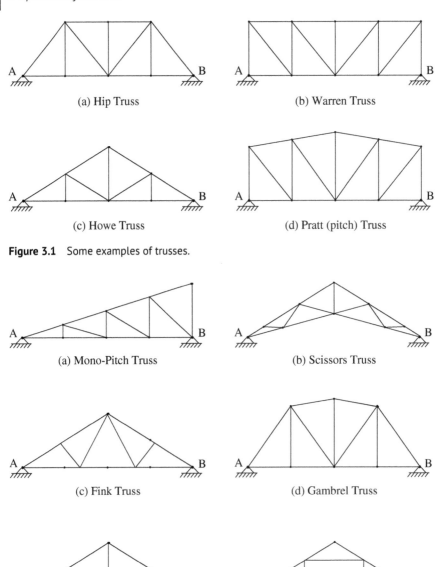

(a) Hip Truss

(b) Warren Truss

(c) Howe Truss

(d) Pratt (pitch) Truss

Figure 3.1 Some examples of trusses.

(a) Mono-Pitch Truss

(b) Scissors Truss

(c) Fink Truss

(d) Gambrel Truss

(e) King Post Truss

(f) Attic Truss

Figure 3.2 Some examples of trusses.

The state of equilibrium of a rigid body can be

- A stable equilibrium: when disturbed by a force the body returns to its initial position.
- An unstable equilibrium (when disturbed by a force the body moves away from its initial position).
- A neutral equilibrium (when the disturbed force is removed the body does not move).

Consider a rigid body subject to the forces $\mathbf{F}_i$, $i = 1, ..., n$, and couples $\mathbf{M}_j$, $j = 1, ..., m$. If the sum of the forces and the sum of the moments (about one point P) acting on the body is zero, it can be shown that the sum of the moments about every point is zero, that is

$$\sum_{i=1}^{n} \mathbf{F}_i = 0 \text{ and } \sum_{i=1}^{n} \mathbf{M}_P = \sum_{i=1}^{n} \mathbf{r}_{i_P} \times \mathbf{F}_i + \sum_{j=1}^{m} \mathbf{M}_j = 0.$$

Consider a body subjected to the concurrent forces $\mathbf{F}_1, \mathbf{F}_2, ..., \mathbf{F}_n$ and no couples. If the sum of the concurrent forces is zero, $\sum_{i=1}^{n} \mathbf{F}_i = 0$, the sum of the moments of the forces is also zero and so the sum of the moments about every point is zero. Therefore, for a set of concurrent forces the only condition imposed by equilibrium is that their sum is zero.

3.2 Supports

The members of a framework structure can be connected together in different ways so that it can support specific loading conditions. This involves the analysis of the forces that can be transferred or resisted by its support, where the resisted forces, e.g., the *reactions*, are forces and couples exerted by its supports.

A body has *redundant supports* when the body has more supports than the minimum number necessary to maintain it in equilibrium. Redundant supports are used whenever possible for strength and safety. Each support added to a body results in additional reactions. The difference between the number of reactions and the number of independent equilibrium equations is called the *degree of redundancy*.

A body has *improper supports* if it will not remain in equilibrium under the action of the loads exerted on it. The body with improper supports will move when the loads are applied.

Since the conditions at supports have a notably effect on the framework structure behaviour and therefore on the analysis, such connections should be analysed. Therefore, three of the most common connections namely roller, pinned and fixed will be analysed next. For the analysis the following force convention is defined: $\mathbf{F}_{ij}$ represents the force exerted by link i on link j.

Fixed Support
A fixed support, also known as rigid support, allows no relative translation or rotation, thus it can resist horizontal and vertical forces or moments. A fixed support or built-in support is shown in Figure 3.3(a), that is, the body (beam) is literally built into the wall. It can be seen that the fixed support denoted by A (Figure 3.3(a)) can exert two components of force and a couple on the support (Figure 3.3(b)), that is

$$\mathbf{F}_A = F_{Ax}\,\mathbf{i} + F_{Ay}\,\mathbf{j}, \text{ and } \mathbf{M}_A = M_{Az}\,\mathbf{k}.$$

Pin Support
A pinned support allows rotation but no relative translation, thus it cannot resist moments but only horizontal and vertical forces. Since a pinned connection allows rotation about one axis only it provides resistance to lateral movement, e.g, rotation in any other direction. The pin supports are used in mechanical devices that allow connected links to rotate relative to each other.

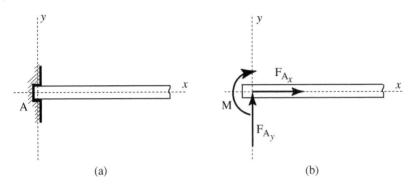

Figure 3.3 Fixed support.

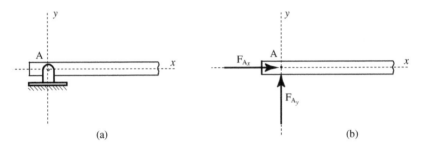

Figure 3.4 Pin joint.

A pinned support is shown in Figure 3.4(a), that is, the beam can rotate about the pin axis but cannot translate relative to the bracket due to the reactive force exerted by the support preventing this movement. The reaction forces at the pin support (Figure 3.4(b)) can be expressed by

$$\mathbf{F}_A = F_{Ax}\,\mathbf{1} + F_{Ay}\,\mathbf{J}$$

where F_{Ax} and F_{Ay} are the horizontal and vertical components of the reaction force.

Roller Support

A roller support allows rotation and relative translation along the surface upon which it rests, thus it cannot resist moments about the pin axis and forces parallel to the surface but only normal forces to the surface. The reaction force in this case is always normal and away from the surface upon the roller rests. Bridges and beams are usually roller supported at one end, so they are capable of contraction and expansion due to changes in temperature.

A roller support – which is a pin support mounted on rollers – is shown in Figure 3.5(a). The roller support can only exert a normal force (Figure 3.5(b)) to the rolling surface, that is

$$\mathbf{F}_A = F_{Ay}\,\mathbf{J}.$$

The roller support cannot exert a force parallel to the surface upon which it translates or a couple about the pin axis.

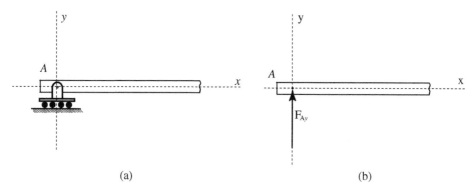

Figure 3.5 Roller support.

3.3 Free-Body Diagrams

A free-body diagram or force diagram is a graphical representation of a body along with all the loads acting on it. The body may consist of a single member (a block or a beam) or of multiple members (a truss) in which case a series of free-body diagrams may be necessary. Free-body diagrams are used to visualize and calculate the unknown forces and moments acting on bodies in equilibrium. The rule for drawing free-body diagrams is to depict all the forces that are acting on the body.

Two free-body diagrams are considered next, a block placed on a table (Figure 3.6(a)) and a simple supported beam (Figure 3.7(a)) with a pin support at the left and a roller support at the right.

For the block and the table shown in Figure 3.6(a) the only force acting on the table is the block weight (force due to gravity), while the table surface is exerting an upward force (named reaction force) on the block. The free-body diagram showing all the forces exerted on the block when the block is isolated from the table is shown in Figure 3.6(b).

The beam shown in Figure 3.7(a) has the length L and is loaded by a force F located at distance d_1 from the left end A and by a moment M at located at the distance d_2 from the same end A. The free-body diagram of the beam is obtained by depicting all the acting forces when the beam is isolated from its supports. In this case the acting forces including the loading force F, the moment M, the beam weight W and the reactions exerted on the beam by the supports are all shown on the free-body diagram, Figure 3.7(b). Once the free-body diagram is obtained one can apply the equilibrium equations.

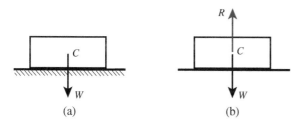

Figure 3.6 Free-body diagram of a block and table.

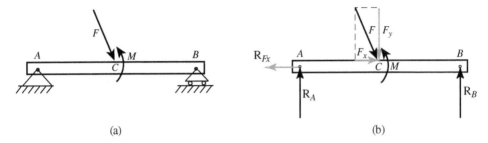

(a)   (b)

Figure 3.7 Free-body diagram of a simple supported beam.

The steps required to determine the reactions on bodies are

1) Draw the free-body diagram, isolating the body from its supports and showing the forces and the reactions
2) Apply the equilibrium equations to determine the reactions.

For two-dimensional systems, the forces and moments are related by three scalar equilibrium equations

$$\sum F_x = 0,$$ (3.2)

$$\sum F_y = 0,$$ (3.3)

$$\sum M_P = 0, \ \forall P.$$ (3.4)

One can obtain more than one equation from Equation (3.4) by evaluating the sum of the moments about more than one point. The additional equations will not be independent of Eqs. (3.2)-(3.4). One cannot obtain more than three independent equilibrium equations from a two-dimensional free-body diagram, which means one can solve for at most three unknown forces or couples.

3.4 Two-Force and Three-Force Members

A body is a *two-force member* if the system of forces and moments acting on the body is equivalent to two forces acting at different points.

For example a body is subjected to two forces, $\mathbf{F}_A$ and $\mathbf{F}_B$, at A and B. If the body is in equilibrium, the sum of the forces equals zero only if $\mathbf{F}_A = -\mathbf{F}_B$. Furthermore, the forces $\mathbf{F}_A$ and $-\mathbf{F}_B$ form a couple, so the sum of the moments is not zero unless the lines of action of the forces lie along the line through the points A and B. Thus for equilibrium the two forces are equal in magnitude, are opposite in direction, and have the same line of action. However, the magnitude cannot be calculated without additional information.

A body is a *three-force member* if the system of forces and moments acting on the body is equivalent to three forces acting at different points.

Theorem. If a three-force member is in equilibrium, the three forces are coplanar and the three forces are either parallel or concurrent.

The analysis of a body in equilibrium can often be simplified by recognizing the two-force or three-force member.

3.5 Plane Trusses

A framework structure composed of links joined at their ends to form a rigid structure is called a truss. Roof supports and bridges are common examples of trusses. When the links of the truss are in a single plane, the truss is called a plane truss. Three bars linked by pins joints at their ends form a rigid frame or non-collapsible frame. The basic element of a plane truss is the triangle, Figure 3.8. Four, five or more bars pin-connected to form a polygon of as many sides form a non-rigid frame. A non-rigid frame is made rigid, or stable, by adding a diagonal bars and forming triangles. Frameworks built using basic triangle are known as simple trusses.

A framework (or truss) can be externally indeterminate or internally indeterminate or both. If, in a frame, the number of external supports is more than necessary to ensure a stable equilibrium state, the extra supports are external redundancies and the frame as a whole is statically indeterminate. Additional links or supports that are not necessary for maintaining the equilibrium configuration are called redundant.

A plane truss is statically determinate internally if $n + 3 = 2c$, where n is number of its links and c is the number of its joints. If the number of internal members in a truss is larger than required to prevent collapse, the truss has internal redundancies and is said to be statically indeterminate. Therefore, if $n + 3 > 2c$, there are more members than equations and the frame is statically indeterminate with internal redundant members. If $n + 3 < 2c$ there are fewer members than independent equations and the truss is unstable and would collapse under load.

Several assumptions are made in the force analysis of simple trusses. First, each member of a truss is straight and has two nodes as points of application of the forces. To satisfy equilibrium, all the members are considered to be under two equal, opposite and collinear forces, in which case the members may be either in tension (T) or compression (C), as shown in Figure 3.9. A member in tension is called a tie and in compression is called a strut.

The weight of each member is considered small compared with the force it supports. If the weight of the link is not small, the weight W of the member is replaced by two forces, each $W/2$, one force acting at each end of the member. These weight forces are considered as external loads applied to the pin connections. The connection between the links are

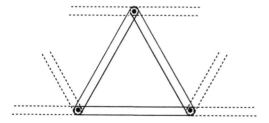

Figure 3.8 Basic element of a plane truss, a triangle.

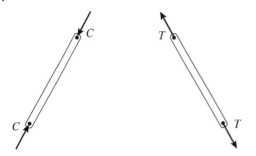

Figure 3.9 Link in tension (*T*) and compression (*C*).

assumed to be smooth pin joints. All the external forces are applied at the pin connections of the trusses.

3.6 Analysis of Simple Trusses

Two methods for the force analysis of simple trusses are described and demonstrated using a simple truss shown in Figure 3.10(a). The length of the truss members are $AB = BG = BC = CE = CD = a$. The external force at B is given and has the magnitude F. In the first stage of the analysis a free-body diagram of the whole truss is considered in order to find the external reactions at supports by previously applying the equilibrium equations. The reaction force of the ground on the truss at the pin support A is F_A and the reaction force of the ground on the truss at the roller support D is F_D. The two reaction forces can be calculated by

$$\sum F_y = 0 \Leftrightarrow F_A + F_D = F$$
$$\sum M_A = 0 \Leftrightarrow aF - 3aF_D = 0 \tag{3.5}$$

resulting in $F_A = \frac{2}{3}F$ and $F_D = \frac{1}{3}F$, that is, once the reactions are calculated the analysis of the internal forces may begin.

3.6.1 Method of Joints

The method of joints calculates the forces in the members of a truss by writing the conditions of equilibrium for each individual joint. The analysis applied for each joint separately involves the equilibrium of concurrent forces, that is, only two equilibrium equations, representing the sum of horizontal and sum of vertical forces, are involved. Therefore, in order to have in the equations two unknowns only, the joints should be solved in a certain order, that is, from the sides towards the centre of the truss.

The solution for each joint starts with the corresponding free-body diagram indicating all the external and internal forces acting on the joint. If a resultant member force acts away from the joint the member is in tension while if the force acts into the joint the member is in compression (principle of action and reaction). For this analysis it is initially assumed that each member is in tension (forces acting away from the joint), and the member status will change to compression if and only if the numerical force result is negative.

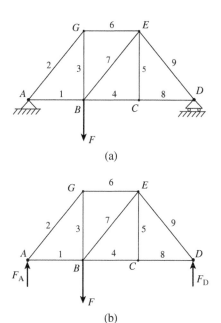

(a)

(b)

Figure 3.10 Simple truss.

For the truss shown in Figure 3.10(a) the analysis begins with the pin at A. The force in each link is designated by one letter defining the node and one number defining the member.

Joint A
The free-body diagram of joint A is shown in Figure 3.11. The forces acting at the joint A are:

- $\mathbf{F}_{1A}$ is the force of the member 1 (member AB) on the node A and is drawn acting away from the pin A.
- $\mathbf{F}_{2A}$ is the force of the member 2 (member AG) on the node A and is drawn acting away from the pin A.
- $\mathbf{F}_A$ is the external support reaction force of node A.

Figure 3.11 shows the free-body diagram of all the joints. The magnitudes of F_{1A} and F_{2A} are obtained from the conditions of equilibrium for the joint A (Figure 3.11) by

$$\sum \mathbf{F}^A = \mathbf{F}_A + \mathbf{F}_{1A} + \mathbf{F}_{2A} = \mathbf{0},$$

or equivalent

$$\sum F_x^A = 0 \iff F_{2A} \sqrt{2}/2 + F_{1A} = 0 \iff F_{1A} = \frac{2}{3}F$$

$$\sum F_y^A = 0 \iff F_A + F_{2A} \sqrt{2}/2 = 0 \iff F_{2A} = -\frac{2\sqrt{2}}{3}F. \tag{3.6}$$

Joint D is analysed next (Figure 3.11) since it contains only two unknowns, F_{8D} and F_{9D}. The magnitudes of the unknown forces are obtained from the conditions of equilibrium for

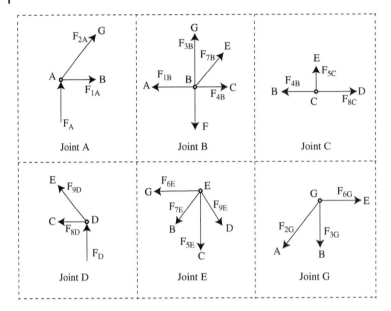

Figure 3.11 Free-body diagram of all joints.

the joint D by

$$\sum \mathbf{F}^D = \mathbf{F}_D + \mathbf{F}_{8D} + \mathbf{F}_{9D} = \mathbf{0},$$

or equivalent

$$\sum F_x^D = 0 \ \Leftrightarrow \ -F_{9D} \ \sqrt{2}/2 - F_{8D} = 0 \ \Leftrightarrow \ F_{8D} = \frac{1}{3}F$$

$$\sum F_y^D = 0 \ \Leftrightarrow \ F_D + F_{9D} \ \sqrt{2}/2 = 0 \ \Leftrightarrow \ F_{9D} = -\frac{\sqrt{2}}{3}F. \tag{3.7}$$

For joint C (Fig. 3.11) the force equilibrium conditions give

$$\sum \mathbf{F}^C = \mathbf{F}_{8C} + \mathbf{F}_{4C} + \mathbf{F}_{5C} = \mathbf{0},$$

or

$$\sum F_x^C = 0 \ \Leftrightarrow \ -F_{4C} + F_{8C} = 0 \Leftrightarrow F_{4C} = \frac{1}{3}F$$

$$\sum F_y^C = 0 \ \Leftrightarrow \ F_{5C} = 0 \tag{3.8}$$

where $F_{8C} = F_{8D} = \frac{1}{3}F$.

For joint E (Figure 3.11) the force equilibrium conditions give

$$\sum \mathbf{F}^E = \mathbf{F}_{5E} + \mathbf{F}_{6E} + \mathbf{F}_{7E} + \mathbf{F}_{9E} = \mathbf{0},$$

or

$$\sum F_x^E = 0 \ \Leftrightarrow -F_{6E} - F_{7E} \ \sqrt{2}/2 + F_{9E} \ \sqrt{2}/2 = 0$$

$$\sum F_y^E = 0 \ \Leftrightarrow -F_{5E} - F_{7E} \ \sqrt{2}/2 - F_{9E} \ \sqrt{2}/2 = 0. \tag{3.9}$$

Since $F_{5E} = F_{5C} = 0$ and $F_{9E} = F_{9D} = -\frac{\sqrt{2}}{3}F$ one can write

$$\sum F_x^E = 0 \Leftrightarrow F_{6E} + F_{7E}\frac{\sqrt{2}}{2} = F_{9E}\frac{\sqrt{2}}{2} \Leftrightarrow F_{6E} = -\frac{2}{3}F$$

$$\sum F_y^E = 0 \Leftrightarrow -F_{7E} - \frac{\sqrt{2}}{3}F = 0 \Leftrightarrow F_{7E} = \frac{\sqrt{2}}{3}F. \tag{3.10}$$

For joint G (Figure 3.11) the force equilibrium conditions give

$$\sum \mathbf{F}^G = \mathbf{F}_{2G} + \mathbf{F}_{3G} + \mathbf{F}_{6G} = \mathbf{0},$$

or

$$\sum F_x^G = 0 \Leftrightarrow F_{6G} - F_{2G}\sqrt{2}/2 = 0$$

$$\sum F_y^G = 0 \Leftrightarrow -F_{3G} - F_{2G}\sqrt{2}/2 = 0. \tag{3.11}$$

The correctness of the analysis is checked for $\sum F_x^G = 0$ with $F_{6G} = F_{6E}$ and $F_{2G} = F_{2A}$ that is

$$F_{6G} = F_{2G}\frac{\sqrt{2}}{2} \Leftrightarrow F_{6E} = F_{2A}\frac{\sqrt{2}}{2} \Leftrightarrow -\frac{2}{3}F = -\frac{2\sqrt{2}}{3}F\frac{\sqrt{2}}{2} \Leftrightarrow 1 = 1.$$

The force equilibrium condition for the y axis gives

$$\sum F_y^G = 0 \Leftrightarrow -F_{3G} - F_{2G}\frac{\sqrt{2}}{2} = 0$$

$$\Leftrightarrow F_{3G} = -F_{2G}\sqrt{2}/2$$

$$\Leftrightarrow F_{3G} = -F_{2A}\frac{\sqrt{2}}{2}$$

$$\Leftrightarrow F_{3G} = \frac{2\sqrt{2}}{3}F\frac{\sqrt{2}}{2}$$

$$\Leftrightarrow F_{3G} = \frac{2}{3}F \tag{3.12}$$

where $F_{2G} = F_{2A}$.

The method of joints for plane trusses uses only two of the three equilibrium equations because the method involves concurrent forces at each joint.

3.6.2 Method of Sections

The method of sections has the advantage that the force in almost any member may be found directly from an analysis of a section that has cut that link. Since there are only three independent equilibrium equations in plane not more than three members whose forces are unknown should be cut.

We can take advantage of the three equations by sectioning a frame and by selecting one section for a free-body diagram.

External forces acting in the direction of the member replace the cut members in the section. The main advantage of this method is that internal forces can be found directly without the need to carry out joint to joint calculations.

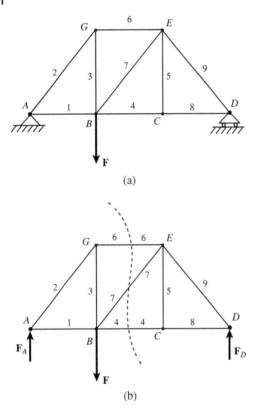

(a)

(b)

Figure 3.12 Method of sections.

When the method of sections is considered, the following rules should be followed:

- No more than three members to be cut.
- The external reaction forces are not to be removed.
- Section internal member forces are not considered in the free-body diagram analysis of the section.

In some cases the methods of sections and joints can be combined for an efficient solution.

The method of sections has the advantage that the force in almost any member may be found directly from an analysis of a section that has cut that link. Since there are only three independent equilibrium equations not more than three members whose forces are unknown should be cut. For the truss shown in Figure 3.12 the external reactions are first computed by considering the truss as a whole. The force in the member 4, 6 and 7 will be determined. An imaginary section, indicated by the dashed line, is passed through the truss, cutting it into two parts, Figure 3.12(b). This section has cut three links whose forces F_{B4}, F_{B6}, and F_{G7} are initially unknown. The left-hand section shown in Figure 3.13 is in equilibrium under the action of the external force F at B, the pin support reaction $F_A = \frac{2}{3}F$, and the three forces F_{G6}, F_{B6}, and F_{B4} exerted on the cut members by the right-hand section which has been removed. In general, the proper senses of the forces will result from the computation. To have equilibrium, the sum of all the forces acting on the left-hand section

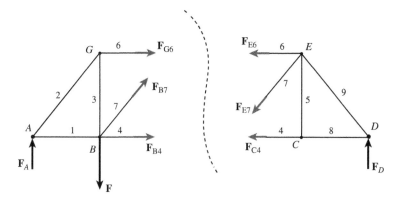

Figure 3.13 Method of sections.

is zero and the sum of the moments about any point is zero. The sum of forces acting on the left-hand section can be expressed by

$$\sum F_x = 0 \Leftrightarrow F_{G6} + F_{B4} + F_{B7} \sqrt{2}/2 = 0$$
$$\sum F_y = 0 \Leftrightarrow F_A - F + F_{B7} \sqrt{2}/2 = 0. \tag{3.13}$$

The sum of the moments about point B for the left-hand section (LHS) gives

$$\sum M_B^{LHS} = F_{G6} \, a + F_A \, a = 0 \text{ or } F_{G6} = F_A = \frac{2}{3}F.$$

3.7 Examples

Example 3.1

The uniform links AC and CE denoted by l_1 and l_2 are pin connected at C, link l_1 is roller supported at A, and link l_2 is fixed at E as shown in Figure 3.14. The distances d_{AB} from A to B, d_{BC} from B to C, d_{CD} from C to D, and d_{DE} from D to E are given: $d_{AB} = d_{11}, d_{BC} = d_{12}$, $d_{CD} = d_{22}$, and $d_{DE} = d_{23}$ respectively. The external forces F_1, F_2, F_3, and F_4 acts on the links at the points P_1, $P_2 = B$, P_3 and P_4, where $d_{AP_1} = d_{13}, d_{P_3 D} = d_{24}$ and $d_{EP_4} = d_{23}$. The angle of the force F_2 with the horizontal is α, the forces F_1 and F_3 act vertical and the force F_4 is a horizontal acting force. Find the reaction forces at the supports A and E, and at the joint C when the frame is in equilibrium. Numerical application: $d_{11} = 0.5$ m, $d_{12} = 0.2$ m, $d_{13} = 0.2$ m, $d_{21} = 0.3$ m, $d_{22} = 0.7$ m, $d_{23} = 0.15$ m, $d_{24} = 0.2$ m, $F_1 = 100$ N, $F_2 = 400$ N, $\alpha = -\pi/6$, $F_3 = -200$ N, and $F_4 = -150$ N.

Solution

For each of the two links l_1 and l_2 one can write the equilibrium equations, that is, the sum of the forces in the horizontal and vertical direction is zero and the sum of moments about any point is zero.

The free-body diagram for the links l_1 is shown in Figure 3.15. The sum of forces acting on the link l_1 can be expressed by

$$\sum F = 0 \Leftrightarrow F_A + F_1 + F_2 + F_{21} = 0 \tag{3.14}$$

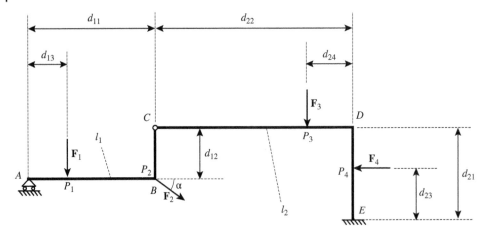

Figure 3.14 Uniform links *AC* and *CE* pin connected at *C*, link *AC* is roller supported at *A* and link *CE* is fixed at *E*.

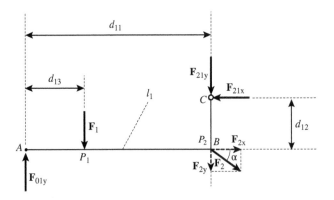

Figure 3.15 Free-body diagram of the link *AC*.

or equivalent

$$\sum F_{Oy} = 0 \Leftrightarrow -F_{2y} + F_{Ay} - F_{21y} - F_1 = 0 \Leftrightarrow -F_2 \sin(\alpha) + F_{Ay} - F_{21y} - F_1 = 0$$
$$\sum F_{Ox} = 0 \Leftrightarrow F_{2x} - F_{Ax} - F_{21x} = 0 \Leftrightarrow F_2 \cos(\alpha) - F_{21x} = 0 \tag{3.15}$$

where $F_{2x} = F_2 \cos(\alpha)$ and $F_{2y} = F_2 \sin(\alpha)$ are the magnitudes of the horizontal and vertical components of the force $\mathbf{F}_2$, F_{21x} and F_{21y} are the magnitudes of the horizontal and vertical reactions at *C*, and F_{Ay} is magnitude of the vertical reaction at *A*. There is no horizontal reaction at *A*, i.e. $F_{Ax} = 0$, since that link l_1 is roller supported at *A*.

The position vectors and all the forces components are input in MATLAB using:

```
clear all; clc; close all
syms d11 d12 d13 d21 d22 d23 d24
syms F1 F2 alpha F3 F4

rA_ = [0 0 0];      % origin at A
```

```
rP1_  =  [d13 0 0];   % position vector of P1
rP2_  =  [d11 0 0];   % position vector of P2
rC_  =  [d11 d12 0];  % position vector of C
rP3_  =  [d11+d22-d24, d12, 0];   % position vector of P3
rP4_  =  [d11+d22, d12-d21+d23, 0];   % position vector of P4
rE_  =  [d11+d22, d12-d21, 0];  % position vector of C
```

```
F1_  =  [0 -F1 0];
F2x  =  F2*cos(alpha);
F2y  =  -F2*sin(alpha);
F2_  =  [F2x F2y 0];
F3_  =  [0 -F3 0];
F4_  =  [-F4 0 0];
```

The sum of forces acting on the link l_1 can be expressed in MATLAB with:

```
syms FAy F21x F21y
FA_  =  [0 FAy 0];   % simple support reaction at A
F21_  =  [-F21x -F21y 0]; % joint reaction at C
```

```
% link 1
% sumF1_  = FA_+F1_+F2_+F21_  = 0_
SF1_  = FA_+F1_+F2_+F21_;
SF1x = SF1_(1);
SF1y = SF1_(2);
```

Taking moments about C one can write

$$\sum M^C = 0 \iff F_{Ay}d_{11} - F_1(d_{11} - d_{13}) - F_{2x}d_{12} = 0$$
$$\iff F_{Ay}d_{11} - F_1(d_{11} - d_{13}) - F_2\cos(\alpha)d_{12} = 0. \tag{3.16}$$

The moment about A is calculated and printed in MATLAB with:

```
% sumM1A_  = rP1_xF1_+rP2_xF2_+rC_xF21_  = 0_
SM1A_ = ...
cross(rP1_,F1_)+cross(rP2_,F2_)+cross(rC_,F21_);
SM1Az = SM1A_(3);
```

From Equations (3.15) and (3.16) one can write

$$\begin{cases} F_2\cos(\alpha) - F_{21x} = 0 \\ -F_2\sin(\alpha) + F_{Ay} - F_{21y} - F_1 = 0 \\ F_{Ay}d_{11} - F_1(d_{11} - d_{13}) - F_2\cos(\alpha)d_{12} = 0 \end{cases} \tag{3.17}$$

and in MATLAB:

```
fprintf('sumF1x = %s \n',char(SF1x))
fprintf('sumF1y = %s \n',char(SF1y))
fprintf('sumM1Az = %s \n\n',char(SM1Az))
```

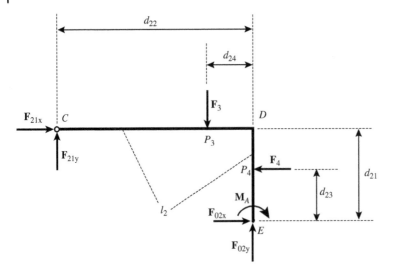

Figure 3.16 Free-body diagram of the link CE.

Solving Equation (3.17) one can obtain

$$F_{21x} = F_2 \cos(\alpha),$$

$$F_{21y} = -F_2 \sin(\alpha) + \frac{F_2 \cos(\alpha)d_{12} - F_1 d_{13}}{d_{11}},$$

$$F_{Ay} = F_1 + \frac{F_2 \cos(\alpha)d_{12} - F_1 d_{13}}{d_{11}}. \tag{3.18}$$

In MATLAB the results are calculated and printed using:

```
sol1=solve(SF1x,SF1y,SM1Az,...
    'FAy','F21x','F21y');
FAys  = sol1.FAy;
F21xs = sol1.F21x;
F21ys = sol1.F21y;


FAs_  = [0 FAys 0];
F21s_ = [F21xs F21ys 0];


fprintf('FAy = %s \n',char(FAys))
fprintf('F21x = %s \n',char(F21xs))
fprintf('F21y = %s \n\n',char(F21ys))
```

The free-body diagram for the links l_2 is shown in Figure 3.16. The equilibrium equations for the link l_2 can be written as

$$\sum F = 0 \Leftrightarrow \mathbf{F}_{12} + \mathbf{F}_3 + \mathbf{F}_4 + \mathbf{F}_E = 0 \tag{3.19}$$

or equivalent

$$\sum F_{Ox} = 0 \Leftrightarrow -F_4 + F_{Ex} + F_{21x} = 0$$

$$\sum F_{Oy} = 0 \Leftrightarrow -F_3 + F_{21y} + F_{Ey} = 0, \tag{3.20}$$

where F_{Ex} and F_{Ey} are the horizontal and vertical reactions at E.

The equilibrium equations for the link l_2 are written and solved in MATLAB using

```
% link 2
% sumF2_  = F02_+F3_+F4_+F21_  = 0_
FE_  = - (F3_+F4_+F21s_);
fprintf('FEx = %s \n',char(FE_(1)))
fprintf('FEy = %s \n\n',char(FE_(2)))
```

Taking moments about E one can write

$$\sum M^E = 0 \Leftrightarrow M_{Ez} + F_{21y}d_{22} - F_3 d_{24} + F_{21x}d_{21} - F_4 d_{23} = 0. \tag{3.21}$$

The moment is calculated in MATLAB with:

```
% sumM2E_ =
% rEC_x(-F21_)+rEP3_xF3_+rEP4_xF4_+M02_  = 0
ME_ = ...
cross(rC_-rE_,-F21s_)-cross(rP3_-rE_,F3_)-cross(rP4_-rE_,F4_);
fprintf('MEz = %s \n\n',char(ME_(3)))
```

From Equations (3.20) and (3.21) one can write

$$\begin{cases} -F_4 + F_{Ex} + F_{21x} = 0 \\ -F_3 + F_{21y} + F_{Ey} = 0 \\ M_{Ez} + F_{21y}d_{22} - F_3 d_{24} + F_{21x}d_{21} - F_4 d_{23} = 0. \end{cases} \tag{3.22}$$

Replacing F_{21x} and F_{21y} in Equation (3.22) by the values in Equation (3.18) one can write

$$\begin{cases} -F_4 + F_{Ex} + F_2\cos(\alpha) = 0 \\ F_{Ey} - F_3 - F_2\sin(\alpha) + \dfrac{F_2\cos(\alpha)d_{12} - F_1 d_{13}}{d_{11}} = 0 \\ M_{Ez} - \left(-F_2\sin(\alpha) + \dfrac{F_2\cos(\alpha)d_{12} - F_1 d_{13}}{d_{11}}\right)d_{22} - \\ \quad -F_3 d_{24} + F_2 d_{21}\cos(\alpha) - F_4 d_{23} = 0. \end{cases} \tag{3.23}$$

Solving Equation (3.23) one can obtain

$$\begin{aligned} F_{Ex} &= F_4 - F_2\cos(\alpha) \\ F_{Ey} &= F_3 + F_2\frac{\sin(\alpha)d_{11} - \cos(\alpha)d_{12}}{d_{11}} + F_1\frac{d_{13}}{d_{11}} \\ M_{Ez} &= \left(-F_2\sin(\alpha) + \frac{F_2\cos(\alpha)d_{12} - F_1 d_{13}}{d_{11}}\right)d_{22} \\ &\quad - F_4 d_{23} - F_3 d_{24} + F_2 d_{21}\cos(\alpha). \end{aligned} \tag{3.24}$$

The input numerical data are introduced in MATLAB with:

```
% numerical results
lists = {d11,d12,d13,d21,d22,d23,d24,...
```

```
            F1,F2,alpha,F3,F4};
listn = {0.5,0.2,0.2,0.3,0.7,0.15,0.2,...
            100,400,-pi/6,200,150};
```

The numerical results are calculated and printed in MATLAB using:

```
FAn_  = eval(subs(FAs_,lists,listn));
F21n_ = eval(subs(F21s_,lists,listn));
FEn_  = eval(subs(FE_,lists,listn));
MEn_  = eval(subs(ME_,lists,listn));

fprintf('\n')
fprintf('FA_  = [%d,%6.3f,%d]  (N)\n',FAn_)
fprintf('F21_ = [%6.3f,%6.3f,%d]  (N)\n',F21n_)
fprintf('\n')
fprintf('FE_  = [%6.3f,%6.3f,%d]  (N)\n',FEn_)
fprintf('ME_  = [%d,%d,%6.3f]  (N m)\n',MEn_)
```

The numerical results for the forces and moments obtained in MATLAB are:

```
FA_  = [0,198.564,0]  (N)
F21_ = [346.410,298.564,0]  (N)

FE_  = [-196.410,-98.564,0]  (N)
ME_  = [0,0,250.418]  (N m)
```

Example 3.2

Two uniform links AB (link l_1) and BD (link l_2) shown in Figure 3.17 are pin connected at B, and pin supported at A and D. The distances d_{AB} from A to B, d_{BC} from B to C, and d_{CD} from C to D, are given by $d_{AB} = d_{11}$, $d_{BC} = d_{21}$, and $d_{CD} = d_{22}$. An external vertical force $\mathbf{F}_1$ act at B and an external force $\mathbf{F}_2$ making an angle α with the horizontal direction act at C. The link l_1 makes an angle β with the horizontal, the segment BC makes and angle θ with the horizontal and the segment CD makes and angle γ with the horizontal as shown in Figure 3.17. Determine the reaction forces at the supports A and D (Figure 3.17). Numerical application: $d_{11} = 0.5$ m, $d_{21} = 0.3$ m, $d_{22} = 0.2$ m, $F_1 = 100$ N, $F_2 = 300$ N, $\alpha = \dfrac{\pi}{4}$, $\beta = \dfrac{\pi}{3}$, $\theta = \dfrac{\pi}{6}$ and $\gamma = \dfrac{\pi}{3}$.

Solution

The free-body diagram for the frame is shown in Figure 3.18. The sum of forces acting on the frame can be expressed by

$$\sum F = 0 \Leftrightarrow \mathbf{F}_A + \mathbf{F}_1 + \mathbf{F}_2 + \mathbf{F}_D = 0 \tag{3.25}$$

or equivalent

$$\sum F_{Ox} = 0 \Leftrightarrow F_{Ax} - F_{2x} + F_{D_x} = 0$$

$$\sum F_{Oy} = 0 \Leftrightarrow F_{Ay} - F_1 - F_{2y} + F_{D_y} = 0 \tag{3.26}$$

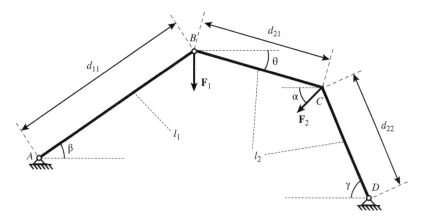

Figure 3.17 Uniform links AB and BD pin connected at B, link AB is pin supported at A and link BD is pin supported at D.

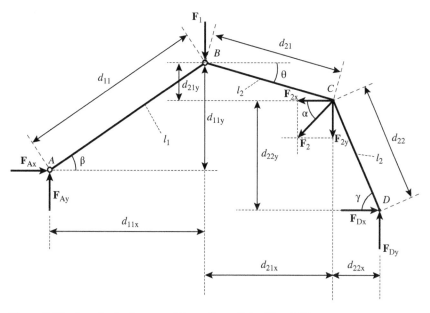

Figure 3.18 Free-body diagram of the uniform links AB and BD.

where F_1 is the magnitude of the external vertical force acting at B, $F_{2x} = F_2 \cos(\alpha)$ and $F_{2y} = F_2 \sin(\alpha)$ are the magnitudes of the horizontal and vertical components of the force $\mathbf{F}_2$, F_{A_x} and F_{A_y} are the Ox and Oy magnitude reactions at A, and F_{D_x} and F_{D_y} are the magnitudes of the Ox and Oy reactions at D.

All the angles, distances and force components are input in MATLAB using:

```
clear all; clc; close all
syms beta F1 F_Ax F_Bx F_Ay F_By d11
syms F2 F_Dx F_Dy  M_A d21x d22x d21 d22
syms alpha theta gamma
```

```
d11x=d11*cos(beta);
d11y=d11*sin(beta);
d21x=d21*cos(theta);
d21y=d21*sin(theta);
d22x=d22*cos(gamma);
d22y=d22*sin(gamma);
F_2x = F2*cos(alpha);
F_2y = F2*sin(alpha);
```

The sum of forces acting on the two link system can be expressed in MATLAB with:

```
sumFx = F_Ax + F_Dx - F_2x;
sumFy = F_Ay - F1 - F_2y + F_Dy;
fprintf('sumFx = %s \n',char(sumFx))
fprintf('sumFy = %s \n\n',char(sumFy))
```

Taking moments about the joint B to the left one can write

$$\sum M_B^l = 0 \Leftrightarrow -F_{Ax}d_{11y} + F_{Ay}d_{11x} = 0.$$

(3.27)

Taking moments about the joint B to the right one can write

$$\sum M_B^r = 0 \Leftrightarrow -F_{D_y}(d_{21x} + d_{22x}) + F_{2y}d_{21x} - F_{D_x}(d_{21y} + d_{22y}) + F_{2x}d_{21y} = 0. \quad (3.28)$$

Calculation of moments are performed in MATLAB with:

```
sumM_Bl = -F_Ax*d11y + F_Ay*d11x;
sumM_Br = - F_Dy*(d21x + d22x) ...
    + F_2y*d21x - F_Dx*(d21y + d22y) + F_2x*d21y;
fprintf('sumM_Bl = %s \n',char(sumM_Bl))
fprintf('sumM_Br = %s \n\n',char(sumM_Br))
```

Using Equations (3.26)–(3.28) one can write

$$\begin{cases} F_{Ax} - F_{2_x} + F_{D_x} = 0 \\ F_{Ay} - F_1 - F_{2_y} + F_{D_y} = 0 \\ -F_{Ax}d_{11y} + F_{Ay}d_{11x} = 0 \\ -F_{D_y}(d_{21x} + d_{22x}) + F_{2y}d_{21x} - F_{D_x}(d_{21y} + d_{22y}) + F_{2x}d_{21y} = 0. \end{cases}$$

(3.29)

The system of equations is solved in MATLAB with:

```
sol=solve(sumFx,sumFy,sumM_Bl,sumM_Br,...
    'F_Ax','F_Ay','F_Dx','F_Dy');
F_Ax = simplify(sol.F_Ax);
F_Ay = simplify(sol.F_Ay);
F_Dx = simplify(sol.F_Dx);
F_Dy = simplify(sol.F_Dy);
fprintf('F_Ax = %s \n',F_Ax)
fprintf('F_Ay = %s \n',F_Ay)
```

```
fprintf('F_Dx = %s \n',F_Dx)
fprintf('F_Dy = %s \n\n',F_Dy)
```

The input numerical data are introduced in MATLAB with:

```
% numerical results
lists = {d11,d21,d22,...
         F1,F2,alpha,beta,theta,gamma};
listn = {0.5,0.3,0.2,...
         100,300,pi/4,pi/6,pi/9,pi/3};
```

The numerical results are calculated in MATLAB using:

```
FAx = eval(subs(F_Ax,lists,listn));
FAy = eval(subs(F_Ay,lists,listn));
FDx = eval(subs(F_Dx,lists,listn));
FDy = eval(subs(F_Dy,lists,listn));
```

The numerical results for the supports reaction are printed in MATLAB with:

```
fprintf('FAx = [%f] (N)\n',FAx)
fprintf('FAy = [%f] (N)\n\n',FAy)
fprintf('FDx = [%f] (N)\n',FDx)
fprintf('FDy = [%f] (N)\n\n',FDy)
```

The obtained MATLAB results are:

```
FAx = [193.724011] (N)
FAy = [111.846610] (N)


FDx = [18.408024] (N)
FDy = [200.285425] (N)
```

Example 3.3

The frame shown in Figure 3.19 carries two external forces $\mathbf{F}_1$ and $\mathbf{F}_2$ located at C and L. The angle of the force $\mathbf{F}_2$ with the horizontal is α and $\mathbf{F}_1$ is a vertical force. The member AD of the frame is making an angle β with the member AB, and the distances d_{AB} from A to B, d_{BC} from B to C, and d_{CL} from C to L are given: $d_{AB} = d_1, d_{BC} = d_2$ and $d_{CL} = d_3$ respectively. The frame is pin supported at A and roller supported at B. Determine the reactions at the supports A and B. Numerical application: $d_1 = 0.5$ m, $d_2 = 0.4$ m, $d_3 = 1.2$ m, $F_1 = 100$ N, $F_2 = 200$ N, $\alpha = \dfrac{\pi}{4}, \beta = \dfrac{\pi}{3}$.

Solution

The free-body diagram of the frame is shown in Figure 3.20. The sum of forces acting on the frame can be expressed by

$$\sum F = 0 \Leftrightarrow \mathbf{F}_A + \mathbf{F}_1 + \mathbf{F}_2 + \mathbf{F}_B = 0 \tag{3.30}$$

or equivalent

$$\sum F_{Ox} = 0 \Leftrightarrow F_{Ax} + F_{B_x} - F_{2x} = 0 \Leftrightarrow F_{Ax} + F_{B_x} - F_2 \cos(\alpha) = 0$$

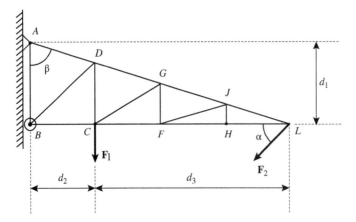

Figure 3.19 Frame pin supported at A and roller supported at B.

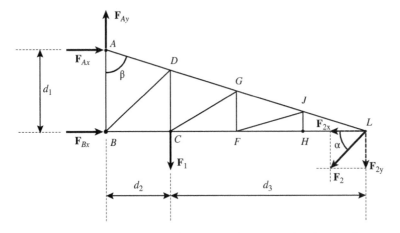

Figure 3.20 Free-body diagram of the frame pin supported at A and roller supported at B.

$$\sum F_{Oy} = 0 \Leftrightarrow F_{Ay} - F_1 - F_{2y} = 0 \Leftrightarrow F_{Ay} - F_1 - F_2 \sin(\alpha) = 0, \tag{3.31}$$

where $\mathbf{F}_A$ is the reaction force at A, $\mathbf{F}_B$ is the reaction force at B, $F_{2x} = F_2 \cos(\alpha)$ and $F_{2y} = F_2 \sin(\alpha)$ are the magnitudes of the horizontal and vertical components of the force $\mathbf{F}_2$, F_{A_x} and F_{A_y} are the magnitudes of the Ox and Oy components of the reaction $\mathbf{F}_A$ at A, and F_{B_x} is the magnitude of the Ox component of the reaction $\mathbf{F}_B$ at B. All the angles, distances and force components are input in MATLAB using:

```
clear all; clc; close all
syms F1 F2 alpha F_Ax F_Bx F_Ay d1 d2 d3
F2x = F2*cos(alpha);
F2y = F2*sin(alpha);
```

The sum of forces acting on the frame can be expressed in MATLAB with:

```
sumFOx = F_Ax + F_Bx - F2x;
sumFOy = F_Ay - F1 - F2y;
```

```
fprintf('F2x = %s \n',char(F2x))
fprintf('F2y = %s \n\n',char(F2y))
fprintf('sumFOx = %s \n',char(sumFOx))
fprintf('sumFOy = %s \n\n',char(sumFOy))
```

Taking moments about B one can write

$$\sum M_B = 0 \Leftrightarrow F_{A_x} d_1 + F_1 d_2 + F_{2y}(d_2 + d_3) = 0,$$
$$\Leftrightarrow F_{A_x} d_1 + F_1 d_2 + F_2 \sin(\alpha)(d_2 + d_3) = 0. \tag{3.32}$$

In MATLAB the calculation is performed using:

```
sumM_B = F_Ax*d1 + F1*d2 ...
    + F2y*(d2+d3);
fprintf('sumM_B = %s \n\n',char(sumM_B))
```

Using Equations (3.30)–(3.32) one can write

$$\begin{cases} F_{Ax} + F_{B_x} - F_2 \cos(\alpha) = 0 \\ F_{Ay} - F_1 - F_2 \sin(\alpha) = 0 \\ F_{A_x} d_1 + F_1 d_2 + F_2 \sin(\alpha)(d_2 + d_3) = 0 \end{cases} \tag{3.33}$$

Solving Equation (3.33) one can obtain the reactions at the supports A and B

$$F_{A_x} = -F_1 \frac{d_2}{d_1} - F_2 \sin(\alpha) \frac{d_2 + d_3}{d_1}$$
$$F_{A_y} = F_1 + F_2 \sin(\alpha)$$
$$F_{B_x} = F_1 \frac{d_2}{d_1} + F_2 \cos(\alpha) + F_2 \sin(\alpha) \frac{d_2 + d_3}{d_1}. \tag{3.34}$$

The system of equations is solved in MATLAB with:

```
sol=solve(sumFOx,sumFOy,sumM_B,...
    'F_Ax','F_Ay','F_Bx');
F_Ax = sol.F_Ax;
F_Ay = sol.F_Ay;
F_Bx = sol.F_Bx;
fprintf('F_Ax = %s \n',F_Ax)
fprintf('F_Ay = %s \n',F_Ay)
fprintf('F_Bx = %s \n\n',F_Bx)
```

The obtained MATLAB solution is:

```
F_Ax = -(F1*d2 + F2*d2*sin(alpha) + F2*d3*sin(alpha))/d1
F_Ay = F1 + F2*sin(alpha)
F_Bx = (F1*d2 + F2*d1*cos(alpha) + F2*d2*sin(alpha)
    + F2*d3*sin(alpha))/d1
```

The input numerical data are introduced in MATLAB with:

```
% numerical results
lists = {d1,d2,d3,F1,F2,alpha};
listn = {0.5,0.4,1.2,100,200,pi/4};
```

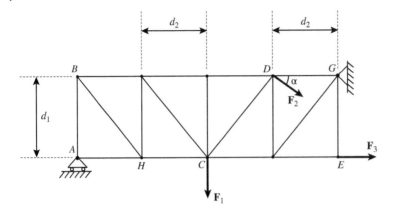

Figure 3.21 Frame pin supported at G and roller supported at A.

The numerical results are calculated in MATLAB using:

```
FAx = eval(subs(F_Ax,lists,listn));
FAy = eval(subs(F_Ay,lists,listn));
FBx = eval(subs(F_Bx,lists,listn));
```

The numerical results for the supports reactions are printed in MATLAB with:

```
fprintf('FAx = [%d] (N)\n',FAx);
fprintf('FAy = [%d] (N)\n',FAy);
fprintf('FBx = [%d] (N)\n',FBx);
```

The obtained MATLAB results are:

```
FAx = [-5.325483e+02] (N)
FAy = [2.414214e+02] (N)
FBx = [6.739697e+02] (N)
```

Example 3.4

The frame shown in Figure 3.21 carries three external forces F_1, F_2 and F_3 located at B, C and D. The angle of the force F_2 with the horizontal is α, the F_1 is a vertical force and F_3 is a acting along the horizontal direction. The distances d_{AB} from A to B, d_{DG} from D to G are given: $d_{AB} = d_1$ and $d_{DG} = d_2$. The distance d_{AC} is equal to d_{CE} and twice the distance d_{DG}, that is, $d_{AC} = d_{CE} = 2d_{DG} = 2d_2$. The frame is pin supported at G and roller supported at A. Determine the reactions at the supports A and G. Numerical application: $d_1 = 0.5$ m, $d_2 = 0.4$ m, $F_1 = 250$ N, $F_2 = 150$ N, $F_3 = 100$ N, $\alpha = \dfrac{\pi}{6}$.

Solution

The free-body diagram of the frame is shown in Figure 3.22. One can determine the reactions at the supports A and G by solving the equations for the sum of all the forces in the horizontal and vertical direction and the sum of moments about any considered point. Equilibrium forces acting on the frame can be expressed by

$$\sum F = 0 \Leftrightarrow \mathbf{F}_A + \mathbf{F}_1 + \mathbf{F}_2 + \mathbf{F}_3 + \mathbf{F}_G = 0 \tag{3.35}$$

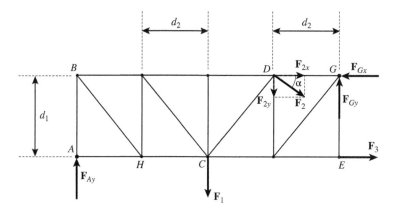

Figure 3.22 Free-body diagram of the frame pin supported at G and roller supported at A.

or equivalent

$$\sum F_{Oy} = 0 \Leftrightarrow F_{Ay} - F_1 - F_{2y} + F_{Gy} = 0 \Leftrightarrow F_{Ay} - F_1 - F_2\sin(\alpha) + F_{Gy} = 0$$

$$\sum F_{Ox} = 0 \Leftrightarrow F_{2x} - F_{Gx} + F_3 = 0 \Leftrightarrow F_2\cos(\alpha) - F_{Gx} + F_3 = 0 \quad (3.36)$$

where $\mathbf{F}_A$ is the reaction force at A, $\mathbf{F}_B$ is the reaction force at B, $F_{2x} = F_2\cos(\alpha)$ and $F_{2y} = F_2\sin(\alpha)$ are magnitudes of the horizontal and vertical components of the force $\mathbf{F}_2$, F_{G_x} and F_{G_y} are magnitudes of the Ox and Oy reactions at G, and F_{A_y} is the magnitude of the Oy reaction at A.

```
clear all; clc; close all
syms F1 F2 F3 alpha F_Ay F_Gx F_Gy  d1 d2
F2x = F2*cos(alpha);
F2y = F2*sin(alpha);

sumFOx = F2x -  F_Gx + F3;
sumFOy = F_Ay - F1 - F2y + F_Gy;
fprintf('F2x = %s \n',F2x);
fprintf('F2y = %s \n\n',F2y);
fprintf('sumFOx = %s \n',char(sumFOx));
fprintf('sumFOy = %s \n',char(sumFOy));
```

Taking moments about A one can write

$$\sum M_A = 0 \Leftrightarrow F_1 2d_2 + F_{2y}3d_2 + F_{2x}d_1 - F_{Gy}4d_2 - F_{Gx}d_1 = 0,$$

$$\Leftrightarrow (2F_1 + 3F_2\sin(\alpha) - 4F_{Gy})d_2 + (F_2\cos(\alpha) - F_{Gx})d_1 = 0. \quad (3.37)$$

```
sumM_A = F1*2*d2 + F2y*3*d2 - F_Gy*4*d2 ...
    + F2x*d1 - F_Gx*d1;
fprintf('sumM_A = %s \n\n',char(sumM_A));
```

Using Equations (3.35)–(3.37) one can write

$$\begin{cases} F_2\cos(\alpha) - F_{Gx} + F_3 = 0 \\ F_{Ay} - F_1 - F_2\sin(\alpha) + F_{Gy} = 0 \\ (2F_1 + 3F_2\sin(\alpha) - 4F_{Gy})d_2 + (F_2\cos(\alpha) - F_{Gx})d_1 = 0. \end{cases} \quad (3.38)$$

Solving Equation (3.38) one can obtain the reactions at the supports A and B

$$F_{A_y} = \frac{1}{2}F_1 + \frac{1}{4}F_2 \sin(\alpha) + \frac{1}{4}\frac{d_1}{d_2}F_3$$

$$F_{G_x} = F_2 \cos(\alpha) + F_3$$

$$F_{G_y} = \frac{1}{2}F_1 + \frac{3}{4}F_2 \sin(\alpha) - \frac{1}{4}\frac{d_1}{d_2}F_3. \qquad (3.39)$$

The obtained MATLAB solution, i.e. the reactions at the supports A and B, is:

```
F_Gxnew=solve(sumFOx,'F_Gx');
fprintf('F_Gx = %s \n',F_Gxnew);
sumFOy=subs(sumFOy, F_Gx, F_Gxnew);
sumM_A=subs(sumM_A, F_Gx, F_Gxnew);

sol=solve(sumFOy,sumM_A,F_Ay,F_Gy);
F_Ay = simplify(sol.F_Ay);
F_Gy = simplify(sol.F_Gy);
fprintf('F_Ay = %s \n',F_Ay)
fprintf('F_Gy = %s \n\n',F_Gy)
```

The input numerical data are introduced in MATLAB with:

```
% numerical results
lists = {d1,d2,F1,F2,F3,alpha};
listn = {0.5,0.4,250,150,100,pi/6};
```

The numerical results are calculated in MATLAB using:

```
FAy = eval(subs(F_Ay,lists,listn));
FGxnew = eval(subs(F_Gxnew,lists,listn));
FGy = eval(subs(F_Gy,lists,listn));
```

The numerical results for the support reactions are printed in MATLAB with:

```
fprintf('FAy = [%d]  (N)\n',FAy);
fprintf('FGx = [%d]  (N)\n',FGxnew);
fprintf('FGy = [%d]  (N)\n',FGy);
```

The obtained MATLAB results are:

```
FAy = [175]  (N)
FGx = [2.299038e+02]  (N)
FGy = [150]  (N)
```

Example 3.5

The frame shown in Figure 3.23 carries two external forces F_1 and F_2 located at C and E respectively. The angle of the force F_1 with the horizontal is θ and the force F_2 is acting along a horizontal direction. All the horizontal members of the frame are equal $d_{AC} = d_{CD} = d_{DH} = d_{HJ} = d_{BE} = d_{EG}$ and are given by $d_{AC} = d_1$, and all the vertical members are

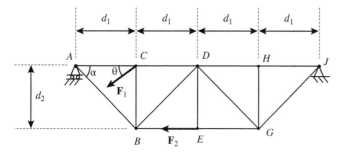

Figure 3.23 Frame roller supported at A and pin supported at J.

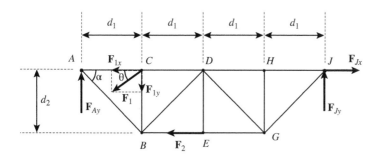

Figure 3.24 Free-body diagram of the frame roller supported at A and pin supported at J.

equal $d_{CB} = d_{DE} = d_{HG}$ and are given by $d_{CB} = d_2$. The frame is pin supported at J and roller supported at A. Using the joints method, determine all the internal forces and directions (tension or compression) in the frame. Numerical application: $d_1 = 0.5$ m, $d_2 = 0.4$ m, $F_1 = 100$ N, $F_2 = 200$ N, $\theta = \dfrac{\pi}{6}$.

Solution

The free-body diagram of the external forces and reaction forces at the supports acting on the frame is shown in Figure 3.24. One can determine the reactions at the supports by solving the equilibrium equations, i.e. the sum of all the forces and the sum of moments about any considered point is zero,

$$\sum F_x = 0 \Leftrightarrow -F_{1x} - F_2 + F_{Jx} = 0$$
$$\sum F_y = 0 \Leftrightarrow F_{Ay} - F_{1y} + F_{Jy} = 0$$
$$\sum M_J = 0 \Leftrightarrow 4d_1 F_{Ay} - 3d_1 F_{1y} + d_2 F_2 \tag{3.40}$$

where $F_{1x} = F_1 \cos\theta$ and $F_{1y} = F_1 \sin\theta$ are the magnitudes of the components of the force $\mathbf{F}_1$ on Ox and Oy.

Solving Equation (3.40) one can obtain

$$F_{Ay} = -\frac{F_2 d_2 - 3d_1 F_1 \sin\theta}{4d_1}$$
$$F_{Jx} = F_2 + F_1 \cos\theta$$
$$F_{Jy} = \frac{F_2 d_2 + F_1 d_1 \sin\theta}{4d_1}. \tag{3.41}$$

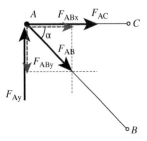

Figure 3.25 Free-body diagram of the joint A of frame.

The reactions at the support are obtained in MATLAB with:

```
clear all; clc; close all
syms d1 d2 F_Ay F_Jx F_Jy theta F1 F2 alpha
F1x = F1*cos(theta);
F1y = F1*sin(theta);

% Reactions at supports
sumFy=F_Ay-F1y-F_Jy;
sumFx=-F1x-F2+F_Jx;
SMJ=F_Ay*4*d1-3*F1y+F2*d2;
solR=solve(sumFx,sumFy,SMJ,F_Ay,F_Jx,F_Jy);
fprintf('Reactions at supports \n')
fprintf('F_Ay = %s \n',solR.F_Ay)
fprintf('F_Jx = %s \n',solR.F_Jx)
fprintf('F_Jy = %s \n\n',solR.F_Jy)
```

To calculate the forces in each member of the frame, the analysis is applied for each joint separately where two equilibrium equations, $\sum F_x = 0$ and $\sum F_y = 0$, must be satisfied. The solution for each pin or joint starts with a free-body diagram of the joint indicating all the external and internal forces acting on it.

The free-body diagram of the joint A indicating all the external and internal forces acting on the joint is shown in Figure 3.25. There, the unknown forces at joint A are denoted by F_{AB} and F_{AC}. It was assumed that each member is in tension, that is, the forces are acting away from the joint. One can write

$$\sum F_{Ax} = 0 \Leftrightarrow F_{AC} + F_{ABx} = 0 \Leftrightarrow F_{AC} + F_{AB} \cos\alpha = 0$$
$$\sum F_{Ay} = 0 \Leftrightarrow F_{Ay} - F_{ABy} = 0 \Leftrightarrow F_{Ay} - F_{AB} \sin\alpha = 0 \qquad (3.42)$$

where $F_{ABx} = F_{AB} \cos\alpha$ and $F_{ABy} = F_{AB} \sin\alpha$ are the Ox and Oy components of F_{AB}, and the angle between $\alpha = \arctan \dfrac{d_2}{d_1}$ is the angle between the direction of the member AB and member AC (Figure 3.24). One can calculate

$$\sin\alpha = \frac{d_2}{d_1\sqrt{\dfrac{d_2^2}{d_1^2}+1}} \quad \text{and} \quad \cos\alpha = \frac{1}{\sqrt{\dfrac{d_2^2}{d_1^2}+1}}. \qquad (3.43)$$

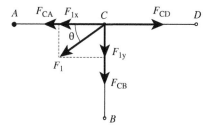

Figure 3.26 Free-body diagram of the joint C of frame.

Solving Equation (3.42) one can obtain

$$F_{AC} = -F_{Ay} \frac{\cos \alpha}{\sin \alpha}$$

$$F_{AB} = \frac{F_{Ay}}{\sin \alpha}.$$

$$(3.44)$$

The unknown forces F_{AB} and F_{AC} at joint A are calculated and printed in MATLAB with:

```
% joint A
syms F_AB F_AC alpha F_Ay
F_ABx = F_AB*cos(alpha);
F_ABy = F_AB*sin(alpha);
sumFAx = F_AC + F_ABx;
sumFAy = F_Ay - F_ABy;
fprintf('JOINT A - Sum of forces \n')
fprintf('sumFAx = %s \n',sumFAx)
fprintf('sumFAy = %s \n\n',sumFAy)

solA=solve(sumFAx,sumFAy,F_AB,F_AC);
F_AB = solA.F_AB;
F_AC = solA.F_AC;
fprintf('JOINT A - Results \n')
fprintf('F_AB = %s \n',F_AB)
fprintf('F_AC = %s \n\n',simplify(F_AC))
```

The free-body diagram of the joint C is shown in Figure 3.26. The unknown forces at joint C are F_{CB} and F_{CD}. One can write

$$\sum F_{Cx} = 0 \Leftrightarrow F_{CD} - F_{CA} - F_{1x} = 0 \Leftrightarrow F_{CD} - F_{CA} - F_1 \cos \theta = 0,$$

$$\sum F_{Cy} = 0 \Leftrightarrow -F_{CB} - F_{1y} = 0 \Leftrightarrow -F_{CB} - F_1 \sin \theta = 0 \qquad (3.45)$$

where $F_{1x} = F_1 \cos \theta$ and $F_{1y} = F_1 \sin \theta$ are the magnitudes of the Ox and Oy components of $\mathbf{F}_1$. Solving Equation (3.45) one can obtain

$$F_{CD} = F_{CA} + F_1 \cos \theta = F_{AC} + F_1 \cos \theta = F_1 \cos \theta - F_{Ay} \cot \alpha,$$

$$F_{CB} = -F_1 \sin \theta. \qquad (3.46)$$

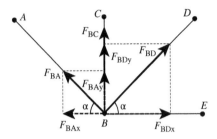

Figure 3.27 Free-body diagram of the joint B of frame.

The unknown forces F_{CB} and F_{CD} at joint C are calculated and printed in MATLAB with:

```
% joint C
syms F_CD F_CA F_CB
F_CA=F_AC;
sumFCx = F_CD - F_CA -F1x;
sumFCy = -F_CB - F1y;
fprintf('JOINT C - Sum of forces \n')
fprintf('sumFCx = %s \n',sumFCx)
fprintf('sumFCy = %s \n\n',sumFCy)

solC=solve(sumFCx,sumFCy,F_CB,F_CD);
F_CB = solC.F_CB;
F_CD = solC.F_CD;
fprintf('JOINT C - Results \n')
fprintf('F_CB = %s \n',F_CB)
fprintf('F_CD = %s \n\n',simplify(F_CD))
```

The free-body diagram of the joint B is shown in Figure 3.27. There, the unknown forces at joint B are F_{BE} and F_{BC}. One can write

$$\sum F_{Bx} = 0 \Leftrightarrow F_{BDx} - F_{BAx} = 0$$

$$\sum F_{By} = 0 \Leftrightarrow F_{BC} + F_{BAy} + F_{BDy} + F_{BE} = 0 \qquad (3.47)$$

or equivalent

$$F_{BD}\sin(\frac{\pi}{2} - \alpha) - F_{BA}\sin(\frac{\pi}{2} - \alpha) = 0$$

$$F_{BC} + F_{BE} + F_{BA}\cos(\frac{\pi}{2} - \alpha) + F_{BD}\cos(\frac{\pi}{2} - \alpha) = 0 \qquad (3.48)$$

where $F_{BDx} = F_{BD}\sin(\frac{\pi}{2} - \alpha)$, $F_{BAx} = F_{BA}\sin(\frac{\pi}{2} - \alpha)$, $F_{BDy} = F_{BD}\cos(\frac{\pi}{2} - \alpha)$, and $F_{BAy} = F_{BA}\cos(\frac{\pi}{2} - \alpha)$ are the Ox and Oy components of F_{BD} and F_{BA}. Using Equations (3.48) and (3.44) one can calculate

$$F_{BD} = F_{BA} = F_{AB} = \frac{F_{Ay}}{\sin \alpha},$$

$$F_{BE} = -F_{BC} - 2F_{BA}\cos(\frac{\pi}{2} - \alpha)$$

$$= -2F_{AB}\sin \alpha - F_{BC}$$

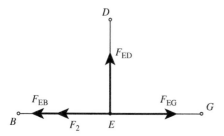

Figure 3.28 Free-body diagram of the joint E of frame.

$$= -2F_{Ay}\frac{\sin\alpha}{\sin\alpha} - F_{CB}$$
$$= -2F_{Ay} + F_1\sin\theta \tag{3.49}$$

where $\cos(\frac{\pi}{2} - \alpha) = \sin(\alpha)$. The forces F_{BE} and F_{BC} at joint B are calculated and printed in MATLAB with:

```
% joint B
syms F_BD F_BA F_BE pi
F_BA=F_AB;
F_BDx = F_BD*sin(pi/2 - alpha);
F_BAx = F_BA*sin(pi/2 - alpha);
sumFBx = F_BDx - F_BAx;
F_BC = F_CB;
F_BDy = F_BD*cos(pi/2 - alpha);
F_BAy = F_BA*cos(pi/2 - alpha);
sumFBy = F_BC + F_BAy + F_BDy + F_BE;
fprintf('JOINT B - Sum of forces \n')
fprintf('sumFBx = %s \n',sumFBx)
fprintf('sumFBy = %s \n\n',sumFBy)

solB=solve(sumFBx,sumFBy,F_BD,F_BE);
F_BD = solB.F_BD;
F_BE = solB.F_BE;
fprintf('JOINT B - Results \n')
fprintf('F_BD = %s \n',F_BD)
fprintf('F_BE = %s \n\n',simplify(F_BE))
```

The free-body diagram of the joint E is shown in Figure 3.27. There, the unknown forces at joint B are F_{ED} and F_{EG}. One can write

$$\sum F_{Ex} = 0 \Leftrightarrow F_{EG} - F_{EB} - F_2 = 0,$$
$$\sum F_{Ey} = 0 \Leftrightarrow F_{ED} = 0. \tag{3.50}$$

Using Equations (3.50) and (3.49) one can write

$$F_{EG} = F_{EB} + F_2 = -2F_{Ay} + F_1\sin\theta + F_2,$$

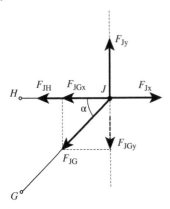

Figure 3.29 Free-body diagram of the joint J of frame.

$$F_{ED} = 0. \tag{3.51}$$

The forces F_{ED} and F_{EG} at joint E are calculated and printed in MATLAB with:

```
% joint E
syms F_EG F_EB F_ED
F_EB=F_BE;
sumFEx = F_EG - F_EB - F2;
sumFEy = F_ED;
fprintf('JOINT E - Sum of forces \n')
fprintf('sumFEx = %s \n',sumFEx)
fprintf('sumFEy = %s \n\n',sumFEy)


F_ED=solve(sumFEy,F_ED);
F_EG=solve(sumFEx,F_EG);
fprintf('JOINT E - Results \n')
fprintf('F_ED = %s \n',F_ED)
fprintf('F_EG = %s \n\n',simplify(F_EG))
```

The free-body diagram of the joint J is shown in Figure 3.29. The unknown forces at joint J are F_{JH} and F_{JG}. One can write

$$\sum F_{Jx} = 0 \Leftrightarrow F_{Jx} - F_{JH} - F_{JGx} = 0 \Leftrightarrow F_{Jx} - F_{JH} - F_{JG}\cos\alpha = 0,$$
$$\sum F_{Jy} = 0 \Leftrightarrow F_{Jy} - F_{JGy} = 0 \Leftrightarrow F_{Jy} - F_{JG}\sin\alpha = 0 \tag{3.52}$$

where $F_{JGx} = F_{JG}\cos\alpha$ and $F_{JGy} = F_{JG}\sin\alpha$ are the Ox and Oy components of F_{JG}. This results in

$$F_{JH} = F_{Jx} - F_{JG}\cos\alpha = F_{Jx} - F_{Jy}\cot\alpha,$$
$$F_{JG} = \frac{F_{Jy}}{\sin\alpha}. \tag{3.53}$$

The forces F_{JH} and F_{JG} at joint J are calculated and printed in MATLAB with:

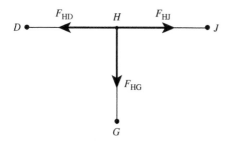

Figure 3.30 Free-body diagram of the joint *H* of frame.

```
% joint J
syms F_Jx F_Jy F_JH F_JG
F_JGx = F_JG*cos(alpha);
sumFJx = F_Jx - F_JH - F_JGx;
F_JGy = F_JG*sin(alpha);
sumFJy = F_Jy - F_JGy;
fprintf('JOINT J - Sum of forces \n')
fprintf('sumFJx = %s \n',sumFJx)
fprintf('sumFJy = %s \n\n',sumFJy)

solJ=solve(sumFJx,sumFJy,F_JH,F_JG);
F_JH = solJ.F_JH;
F_JG = solJ.F_JG;
fprintf('JOINT J - Results \n')
fprintf('F_JH = %s \n',simplify(F_JH))
fprintf('F_JG = %s \n\n',F_JG)
```

The free-body diagram of the joint *H* is shown in Figure 3.30. There, the unknown forces at joint *H* are F_{HD} and F_{HG}. One can write

$$\sum F_{Hx} = 0 \Leftrightarrow F_{HJ} - F_{HD} = 0 \Leftrightarrow F_{HD} = F_{HJ},$$
$$\sum F_{Hy} = 0 \Leftrightarrow F_{HG} = 0. \tag{3.54}$$

Using Equations (3.54) and (3.53) one can write

$$F_{HD} = F_{HJ} = F_{Jx} - F_{JG} \cos \alpha = F_{Jx} - F_{Jy} \cot \alpha,$$
$$F_{HG} = 0. \tag{3.55}$$

The forces F_{HD} and F_{HG} at joint *H* are calculated and printed in MATLAB with:

```
% joint H
syms F_HD F_HG
F_HJ = F_JH;
sumFHx = F_HJ - F_HD;
sumFHy = F_HG;
fprintf('JOINT H - Sum of forces \n')
fprintf('sumFHx = %s \n',sumFHx)
```

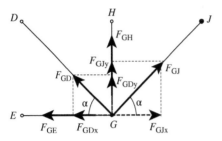

Figure 3.31 Free-body diagram of the joint G of frame.

```
fprintf('sumFHy = %s \n\n',sumFHy)
```

```
F_HD=solve(sumFHx,F_HD);
F_HG=solve(sumFHy,F_HG);
fprintf('JOINT H - Results \n')
fprintf('F_HD = %s \n',simplify(F_HD))
fprintf('F_HG = %s \n\n',F_HG)
```

The free-body diagram of the joint G is shown in Figure 3.31. The unknown forces at joint G are F_{GD} and F_{GE}. One can write

$$\sum F_{Gx} = 0 \Leftrightarrow F_{GJx} - F_{GE} - F_{GDx} = 0$$
$$\Leftrightarrow F_{GE} = F_{GJx} - F_{GDx}$$
$$\Leftrightarrow F_{GE} = F_{GJ}\cos\alpha - F_{GD}\cos\alpha,$$
$$\sum F_{Gy} = 0 \Leftrightarrow F_{GH} + F_{GDy} + F_{GJy} = 0$$
$$\Leftrightarrow F_{GD}\sin\alpha = -F_{GH} - F_{GJy}$$
$$\Leftrightarrow F_{GD}\sin\alpha = -F_{GH} - F_{GJ}\sin\alpha$$
$$\Leftrightarrow F_{GD} = \frac{-F_{GH} - F_{GJ}\sin\alpha}{\sin\alpha} \tag{3.56}$$

where $F_{GDx} = F_{GD}\cos\alpha$, $F_{GDy} = F_{GD}\sin\alpha$, and $F_{GJx} = F_{GJ}\cos\alpha$, $F_{GJy} = F_{GJ}\sin\alpha$ are the Ox and Oy components of F_{GD} and F_{GJ} respectively. Using Equations (3.56), (3.55) and (3.53) one can write

$$F_{GD} = -F_{GJ} = -\frac{F_{Jy}}{\sin\alpha},$$
$$F_{GE} = F_{GJ}\cos\alpha - F_{GD}\cos\alpha = 2F_{GJ}\cos\alpha = 2F_{Jy}\cot\alpha. \tag{3.57}$$

The forces F_{GD} and F_{GE} at joint G are calculated and printed in MATLAB with:

```
% joint G
syms F_GD F_GE F_GH
F_GJ=F_JG;
F_GJx = F_GJ*cos(alpha);
F_GDx = F_GD*cos(alpha);
sumFGx = F_GJx - F_GE - F_GDx;
F_GJy = F_JG*sin(alpha);
```

```
F_GDy = F_GD*sin(alpha);
F_GH=F_HG;
sumFGy = F_GH + F_GDy + F_GJy;
fprintf('JOINT G - Sum of forces \n')
fprintf('sumFGx = %s \n',sumFGx)
fprintf('sumFGy = %s \n\n',sumFGy)


solG=solve(sumFGx,sumFGy,F_GD,F_GE);
F_GD = solG.F_GD;
F_GE = solG.F_GE;
fprintf('JOINT G - Results \n')
fprintf('F_GD = %s \n',F_GD)
fprintf('F_GE = %s \n\n',simplify(F_GE))
```

The input numerical data are introduced in MATLAB with:

```
% numerical results
alpha=atan(d2/d1);
lists = {d1,d2,F1,F2,theta};
listn = {0.5,0.4,300,50,pi/6};
```

The numerical results – forces at each joint – are calculated in MATLAB using:

```
F_Ay = eval(subs(solR.F_Ay,lists,listn));
F_Jx = eval(subs(solR.F_Jx,lists,listn));
F_Jy = eval(subs(solR.F_Jy,lists,listn));
alpha = eval(subs(alpha,lists,listn));
F_AC = eval(subs(F_AC,lists,listn));
F_AB = eval(subs(F_AB,lists,listn));
F_CD = eval(subs(F_CD,lists,listn));
F_CB = eval(subs(F_CB,lists,listn));
F_BD = eval(subs(F_BD,lists,listn));
F_BE = eval(subs(F_BE,lists,listn));
F_EG = eval(subs(F_EG,lists,listn));
F_ED = eval(subs(F_ED,lists,listn));
F_JH = eval(subs(F_JH,lists,listn));
F_JG = eval(subs(F_JG,lists,listn));
F_HD = eval(subs(F_HD,lists,listn));
F_HG = eval(subs(F_HG,lists,listn));
F_GD = eval(subs(F_GD,lists,listn));
F_GE = eval(subs(F_GE,lists,listn));
```

The forces at each joint are printed in MATLAB using:

```
fprintf('F_Ay = [%f] (N)\n',F_Ay);
fprintf('F_Jx = [%f] (N)\n',F_Jx);
fprintf('F_Jy = [%f] (N)\n',F_Jy);
fprintf('F_AC = [%f] (N)\n',F_AC);
```

```
fprintf('F_AB = [%f] (N)\n',F_AB);
fprintf('F_CD = [%f] (N)\n',F_CD);
fprintf('F_CB = [%f] (N)\n',F_CB);
fprintf('F_BD = [%f] (N)\n',F_BD);
fprintf('F_BE = [%f] (N)\n',F_BE);
fprintf('F_EG = [%f] (N)\n',F_EG);
fprintf('F_ED = [%f] (N)\n',F_ED);
fprintf('F_JH = [%f] (N)\n',F_JH);
fprintf('F_JG = [%f] (N)\n',F_JG);
fprintf('F_HD = [%f] (N)\n',F_HD);
fprintf('F_HG = [%f] (N)\n',F_HG);
fprintf('F_GD = [%f] (N)\n',F_GD);
fprintf('F_GE = [%f] (N)\n',F_GE);
```

The numerical results for the forces are:

```
F_Ay = [102.500000] (N)
F_Jx = [309.807621] (N)
F_Jy = [47.500000] (N)
F_AC = [-128.125000] (N)
F_AB = [164.080059] (N)
F_CD = [131.682621] (N)
F_CB = [-150.000000] (N)
F_BD = [164.080059] (N)
F_BE = [-55.000000] (N)
F_EG = [-5.000000] (N)
F_ED = [0.000000] (N)
F_JH = [250.432621] (N)
F_JG = [76.037100] (N)
F_HD = [250.432621] (N)
F_HG = [0.000000] (N)
F_GD = [-76.037100] (N)
F_GE = [118.750000] (N)
```

Example 3.6

The frame shown in Figure 3.32 carries two external forces F_1 and F_2 located at B and D respectively. The force F_1 is a vertical force and the force F_2 is acting along the horizontal direction. The angle between the members AB and AC, CD and DB, EG and GD, and DC and CE is α, and $AC \perp CB$, $BC \perp CD$, $CD \perp DE$ and $GE \perp ED$ as shown in Figure 3.32. The distances d_{AC}, d_{CD} and d_{EG} are equal and are given by $d_{AC} = d_{CD} = d_{EG} = d$. The frame is pin supported at A and roller supported at G. Using the joints method, determine all the internal forces and directions (tension or compression) in the frame. Numerical application: $d = 0.5$ m, $F_1 = 300$ N, $F_2 = 50$ N, $\alpha = \dfrac{\pi}{6}$.

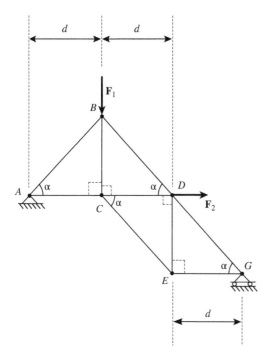

Figure 3.32 Frame pin supported at A and roller supported at G.

Solution

The free-body diagram of the external forces and reaction forces at the supports acting on the frame is shown in Figure 3.33. One can determine the reactions at the supports by solving the equilibrium equations, i.e. the sum of all the forces and the sum of moments about any considered point is zero,

$$\sum F_x = 0 \Leftrightarrow -F_{Ax} + F_2 = 0$$
$$\sum F_y = 0 \Leftrightarrow F_{Ay} - F_1 + F_{Gy} = 0$$
$$\sum M_D = 0 \Leftrightarrow -dF_{Gy} - dF_1 + 2dF_{Ay} = 0. \tag{3.58}$$

Solving Equation (3.58) one can obtain

$$F_{Ay} = F_2$$
$$F_{Jx} = \frac{2F_1}{3}$$
$$F_{Jy} = \frac{F_1}{3}. \tag{3.59}$$

The reaction forces F_{Ax}, F_{Ay} and F_{Gy} at the supports A and G are calculated and printed in MATLAB with:

```
clear all; clc; close all
syms d F_Ax F_Ay F_Gy F1 F2 alpha
```

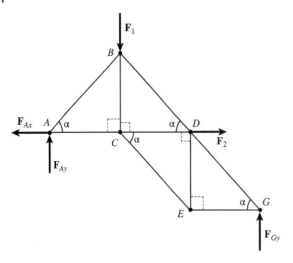

Figure 3.33 Free-body diagram of the frame pin supported at A and roller supported at G.

```
% Reactions at supports
sumFx = - F_Ax + F2;
sumFy = F_Ay - F1 + F_Gy;
SMD = - d*F_Gy - d*F1 + 2*d*F_Ay;
%SMA = - 3*d*F_Gy + d*F1;
solR=solve(sumFx,sumFy,SMD,F_Ax,F_Ay,F_Gy);
fprintf('Reactions at supports \n')
fprintf('F_Ax = %s \n',solR.F_Ax)
fprintf('F_Ay = %s \n',solR.F_Ay)
fprintf('F_Gy = %s \n\n',solR.F_Gy)
```

The free-body diagram of the joint A is shown in Figure 3.34. There, the unknown forces at joint A are denoted by F_{AB} and F_{AC}. It was assumed that each member is in tension, that is, the forces are acting away from the joint. Considering the joint equilibrium one can write

$$\sum F_{Ax} = 0 \Leftrightarrow F_{AC} + F_{ABx} - F_{Ax} = 0 \Leftrightarrow F_{AC} + F_{AB} \cos \alpha - F_{Ax} = 0,$$
$$\sum F_{Ay} = 0 \Leftrightarrow F_{Ay} + F_{ABy} = 0 \Leftrightarrow F_{Ay} + F_{AB} \sin \alpha = 0 \tag{3.60}$$

where $F_{ABx} = F_{AB} \cos \alpha$ and $F_{ABy} = F_{AB} \sin \alpha$ are the Ox and Oy components of F_{AB}.
This results in

$$F_{AC} = F_{Ax} + F_{Ay} \frac{\cos \alpha}{\sin \alpha}$$
$$F_{AB} = -\frac{F_{Ay}}{\sin \alpha}. \tag{3.61}$$

The forces F_{AC} and F_{AB} at joint A are calculated and printed in MATLAB with:

```
% joint A
syms F_AB F_AC
```

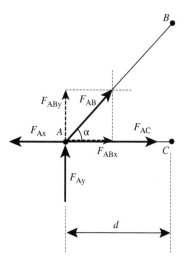

Figure 3.34 Free-body diagram of the joint A of frame.

```
F_ABx = F_AB*cos(alpha);
sumFAx = F_AC + F_ABx - F_Ax;
F_ABy = F_AB*sin(alpha);
sumFAy = F_Ay + F_ABy;

fprintf('JOINT A - Sum of forces \n')
fprintf('sumFAx = %s \n',char(sumFAx))
fprintf('sumFAy = %s \n\n',char(sumFAy))

solA=solve(sumFAx,sumFAy,F_AB,F_AC);
F_AB = solA.F_AB;
F_AC = solA.F_AC;
fprintf('JOINT A - Results \n')
fprintf('F_AB = %s \n',F_AB)
fprintf('F_AC = %s \n\n',F_AC)
```

The free-body diagram of the joint G is shown in Figure 3.35. The unknown forces at joint G are F_{GD} and F_{GE}. One can write

$$\sum F_{Gx} = 0 \Leftrightarrow -F_{GE} - F_{GDx} = 0 \Leftrightarrow F_{GE} + F_{GD}\cos\alpha = 0$$
$$\sum F_{Gy} = 0 \Leftrightarrow F_{GDy} + F_{Gy} = 0 \Leftrightarrow F_{Gy} + F_{GD}\sin\alpha = 0 \tag{3.62}$$

where $F_{GDx} = F_{GD}\cos\alpha, F_{GDy} = F_{GD}\sin\alpha$ are the Ox and Oy components of F_{GD}. This results in

$$F_{GE} = -F_{GD}\cos\alpha = F_{Gy}\cot\alpha,$$
$$F_{GD} = -\frac{F_{Gy}}{\sin\alpha}. \tag{3.63}$$

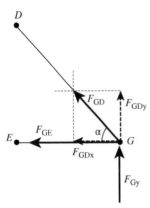

Figure 3.35 Free-body diagram of the joint G of frame.

The forces F_{GE} and F_{GD} at joint G are calculated and printed in MATLAB with:

```
% joint G
syms F_GD F_GE F_Gy
F_GDx = F_GD*cos(alpha);
sumFGx = - F_GE - F_GDx;
F_GDy = F_GD*sin(alpha);
sumFGy = F_GDy + F_Gy;

fprintf('JOINT G - Sum of forces \n')
fprintf('sumFGx = %s \n',sumFGx)
fprintf('sumFGy = %s \n\n',sumFGy)


solG=solve(sumFGx,sumFGy,F_GD,F_GE);
F_GD = solG.F_GD;
F_GE = solG.F_GE;
fprintf('JOINT G - Results \n')
fprintf('F_GD = %s \n',F_GD)
fprintf('F_GE = %s \n\n',F_GE)
```

The free-body diagram of the joint E is shown in Figure 3.36. The unknown forces at joint E are F_{ED} and F_{EC}. One can write

$$\sum F_{Ex} = 0 \Leftrightarrow F_{EG} - F_{ECx} = 0 \Leftrightarrow F_{EG} - F_{EC} \cos \alpha = 0$$
$$\sum F_{Ey} = 0 \Leftrightarrow F_{ED} + F_{ECy} = 0 \Leftrightarrow F_{ED} + F_{EC} \sin \alpha = 0 \qquad (3.64)$$

where $F_{ECx} = F_{EC} \cos \alpha$ and $F_{ECy} = F_{EC} \sin \alpha$ are the Ox and Oy components of F_{EC}. This results in

$$F_{EC} = \frac{F_{EG}}{\cos \alpha} = \frac{F_{GE}}{\cos \alpha} = \frac{F_{Gy} \cot \alpha}{\cos \alpha} = \frac{F_{Gy}}{\sin \alpha},$$
$$F_{ED} = -F_{EC} \sin \alpha = -\frac{F_{Gy}}{\sin \alpha} \sin \alpha = -F_{Gy}. \qquad (3.65)$$

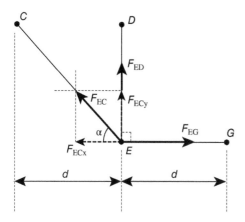

Figure 3.36 Free-body diagram of the joint E of frame.

The forces F_{ED} and F_{EC} at joint E are calculated and printed in MATLAB with:

```
% joint E
syms F_ED F_EC
F_ECx = F_EC*cos(alpha);
F_EG=F_GE;
sumFEx = F_EG - F_ECx;
F_ECy = F_EC*sin(alpha);
sumFEy = F_ED + F_ECy;

fprintf('JOINT E - Sum of forces \n')
fprintf('sumFEx = %s \n',sumFEx)
fprintf('sumFEy = %s \n\n',sumFEy)

solE=solve(sumFEx,sumFEy,F_ED,F_EC);
F_ED = solE.F_ED;
F_EC = solE.F_EC;
fprintf('JOINT E - Results \n')
fprintf('F_ED = %s \n',F_ED)
fprintf('F_EC = %s \n\n',F_EC)
```

The free-body diagram of the joint D is shown in Figure 3.37. The unknown forces at joint D are F_{DB} and F_{DC}. One can write

$$\sum F_{Dx} = 0 \Leftrightarrow F_2 + F_{DGx} - F_{DC} - F_{DBx} = 0,$$
$$\Leftrightarrow F_2 + F_{DG} \cos \alpha - F_{DC} - F_{DB} \cos \alpha = 0$$
$$\sum F_{Dy} = 0 \Leftrightarrow F_{DBy} - F_{DE} - F_{DGy} = 0$$
$$\Leftrightarrow F_{DB} \sin \alpha - F_{DE} - F_{DG} \sin \alpha = 0 \tag{3.66}$$

where $F_{DGx} = F_{DG} \cos(\alpha)$, $F_{DBx} = F_{DB} \cos(\alpha)$, $F_{DGy} = F_{DG} \sin(\alpha)$, and $F_{DBy} = F_{DB} \sin(\alpha)$ are the Ox and Oy components of F_{DG} and F_{DB}.

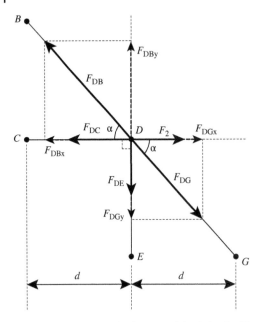

Figure 3.37 Free-body diagram of the joint *D* of frame.

Using Equations (3.66), (3.65) and (3.63) one can write

$$F_{DB} = \frac{F_{DE} + F_{DG} \sin \alpha}{\sin \alpha}$$

$$= \frac{F_{ED}}{\sin \alpha} + F_{GD}$$

$$= -\frac{F_{Gy}}{\sin \alpha} - \frac{F_{Gy}}{\sin \alpha},$$

$$= -2\frac{F_{Gy}}{\sin \alpha}$$

$$F_{DC} = F_2 + F_{DG} \cos \alpha - F_{DB} \cos \alpha$$

$$= F_2 + F_{GD} \cos \alpha + 2\frac{F_{Gy}}{\sin \alpha} \cos \alpha$$

$$= F_2 - \frac{F_{Gy}}{\sin \alpha} \cos \alpha + 2\frac{F_{Gy}}{\sin \alpha} \cos \alpha$$

$$= F_2 + F_{Gy} \cot \alpha. \tag{3.67}$$

The forces F_{DB} and F_{DC} at joint *D* are calculated and printed in MATLAB with:

```
% joint D
syms F_DC F2 F_DE F_DB
F_DG=F_GD; F_DE=F_ED;
F_DGx = F_DG*cos(alpha);
F_DBx = F_DB*cos(alpha);
F_DBy = F_DB*sin(alpha);
sumFDx = F2 + F_DGx - F_DC - F_DBx;
```

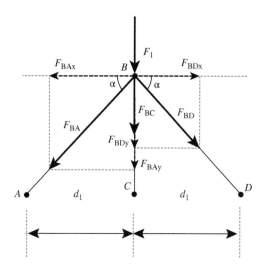

Figure 3.38 Free-body diagram of the joint B of frame.

```
F_DGy = F_DG*sin(alpha);
sumFDy = F_DBy - F_DE - F_DGy;

fprintf('JOINT D - Sum of forces \n')
fprintf('sumFDx = %s \n', sumFDx)
fprintf('sumFDy = %s \n\n', sumFDy)

solD=solve(sumFDx, sumFDy, F_DB, F_DC);
fprintf('JOINT D - Results \n')
F_DB = solD.F_DB;
F_DC = solD.F_DC;
fprintf('F_DB = %s \n', F_DB)
fprintf('F_DC = %s \n\n', F_DC)
```

The free-body diagram of the joint B is shown in Figure 3.38. There, the unknown forces at joint B are F_{BC} and F_{BD}. One can write

$$\sum F_{Bx} = 0 \Leftrightarrow F_{BDx} - F_{BAx} = 0 \Leftrightarrow F_{BD}\cos(\alpha) - F_{BA}\cos(\alpha) = 0$$
$$\sum F_{By} = 0 \Leftrightarrow -F_1 - F_{BC} - F_{BAy} - F_{BDy} = 0$$
$$\Leftrightarrow -F_1 - F_{BC} - F_{BA}\sin(\alpha) - F_{BD}\sin(\alpha) = 0 \tag{3.68}$$

where $F_{BDx} = F_{BD}\cos(\alpha)$, $F_{BAx} = F_{BA}\cos(\alpha)$, $F_{BDy} = F_{BD}\sin(\alpha)$, and $F_{BAy} = F_{BA}\sin(\alpha)$ are the Ox and Oy components of F_{BD} and F_{BA}. Using Equations (3.68) and (3.61) one can write

$$F_{BD} = F_{BA} \Leftrightarrow F_{BD} = F_{BA},$$
$$F_{BC} = -F_1 - F_{BA}\sin(\alpha) - F_{BD}\sin(\alpha)$$
$$= -F_1 - 2F_{AB}\sin(\alpha)$$

$$= -F_1 - 2\frac{F_{Ay}}{\sin\alpha}\sin(\alpha)$$
$$= -F_1 - 2F_{Ay}. \tag{3.69}$$

The forces F_{BC} and F_{BD} at joint B are calculated and printed in MATLAB with:

```
% joint B
syms F_BD F_BC F1
F_BA=F_AB;
F_BDx = F_BD*cos(alpha);
F_BAx = F_BA*cos(alpha);
sumFBx = F_BDx - F_BAx;
F_BDy = F_BD*sin(alpha);
F_BAy = F_BA*sin(alpha);
sumFBy = -F1 - F_BC - F_BAy - F_BDy;

fprintf('JOINT B - Sum of forces \n')
fprintf('sumFBx = %s \n',sumFBx)
fprintf('sumFBy = %s \n',sumFBy)
fprintf('\n')

solB=solve(sumFBx,sumFBy,F_BD,F_BC);
fprintf('JOINT B - Results \n')
F_BD = solB.F_BD;
F_BC = solB.F_BC
fprintf('F_BD = %s \n',F_BD)
fprintf('F_BC = %s \n',F_BC)
fprintf('\n')
```

The input numerical data are introduced in MATLAB with:

```
% numerical results
lists = {d,F1,F2,alpha};
listn = {0.5,300,50,pi/6};
```

The numerical results, i.e. forces at each joint, are calculated in MATLAB using:

```
F_Ax = eval(subs(solR.F_Ax,lists,listn));
F_Ay = eval(subs(solR.F_Ay,lists,listn));
F_Gy = eval(subs(solR.F_Gy,lists,listn));
F_AC = eval(subs(F_AC,lists,listn));
F_AB = eval(subs(F_AB,lists,listn));
F_GE = eval(subs(F_GE,lists,listn));
F_GD = eval(subs(F_GD,lists,listn));
F_EC = eval(subs(F_EC,lists,listn));
F_ED = eval(subs(F_ED,lists,listn));
F_DB = eval(subs(F_DB,lists,listn));
F_DC = eval(subs(F_DC,lists,listn));
```

```
F_BC = eval(subs(F_BC,lists,listn));
F_BD = eval(subs(F_BD,lists,listn));
```

The forces at each joint are printed in MATLAB using:

```
fprintf('F_Ax = [%f] (N)\n',F_Ax);
fprintf('F_Ay = [%f] (N)\n',F_Ay);
fprintf('F_Gy = [%f] (N)\n',F_Gy);
fprintf('F_AC = [%f] (N)\n',F_AC);
fprintf('F_AB = [%f] (N)\n',F_AB);
fprintf('F_GE = [%f] (N)\n',F_GE);
fprintf('F_GD = [%f] (N)\n',F_GD);
fprintf('F_EC = [%f] (N)\n',F_EC);
fprintf('F_ED = [%f] (N)\n',F_ED);
fprintf('F_DB = [%f] (N)\n',F_DB);
fprintf('F_DC = [%f] (N)\n',F_DC);
fprintf('F_BC = [%f] (N)\n',F_BC);
fprintf('F_BD = [%f] (N)\n',F_BD);
```

The numerical results for the forces are:

```
F_Ax = [50.000000]   (N)
F_Ay = [200.000000]  (N)
F_Gy = [100.000000]  (N)
F_AC = [396.410162]  (N)
F_AB = [-400.000000] (N)
F_GE = [173.205081]  (N)
F_GD = [-200.000000] (N)
F_EC = [200.000000]  (N)
F_ED = [-100.000000] (N)
F_DB = [-400.000000] (N)
F_DC = [223.205081]  (N)
F_BC = [100.000000]  (N)
F_BD = [-400.000000] (N)
```

Example 3.7
The frame shown in Figure 3.39 carries three external forces F_1, F_2 and F_3 located at A, D and E respectively. The force F_3 is a vertical force, F_1 is acting along the horizontal direction, and the force F_2 is making an angle β with the vertical direction. The angle between the members AB and BC, and CD and DE are equal and are denoted by α, and $CA \perp AB$, $CD \perp DB$, and $EC \perp CD$ as shown in Figure 3.39. The distances d_{AB} and d_{CD} are equal and are given by $d_{AB} = d_{CD} = d$. The frame is roller supported at A and pin supported at B. Using the joints method, determine all the internal forces and directions (tension or compression) in the frame when the externally applied loads act as shown. Numerical application: $d = 0.5$ m, $F_1 = 100$ N, $F_2 = 100$ N, $F_3 = 100$ N, $\alpha = \dfrac{\pi}{6}$, $\beta = \dfrac{\pi}{4}$.

Figure 3.39 Frame roller supported at A and pin supported at B.

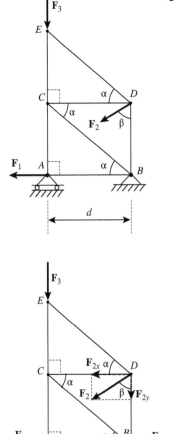

Figure 3.40 Free-body diagram of the frame roller supported at A and pin supported at B.

Solution
The free-body diagram of the external forces and reaction forces at the supports is shown in Figure 3.40. One can determine the reactions at the supports by solving the equilibrium equations, i.e. the sum of all the forces and the sum of moments about any considered point is zero,

$$\sum F_x = 0 \Leftrightarrow -F_1 - F_{2x} - F_{Bx} = 0$$
$$\sum F_y = 0 \Leftrightarrow F_{Ay} + F_{By} - F_{2y} - F_3 = 0$$
$$\sum M_A = 0 \Leftrightarrow -dF_{By} + dF_{2y} - d_{new}F_{2x} = 0 \tag{3.70}$$

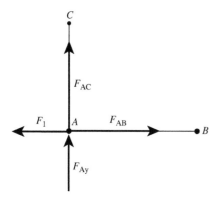

Figure 3.41 Free-body diagram of the joint A of frame.

where $d_{new} = d \tan \alpha$, $F_{2x} = F_2 \sin \beta$ and $F_{2y} = F_2 \cos \beta$ are the magnitudes of the Ox and Oy components of the force $\mathbf{F}_2$. Solving Equation (3.70) one can obtain

$$F_{Ay} = F_3 + F_2 \tan \alpha \sin \beta$$
$$F_{Bx} = -F_1 - F_2 \sin \beta$$
$$F_{By} = F_2 \cos \beta - F_2 \tan \alpha \sin \beta. \qquad (3.71)$$

The reaction forces F_{Ay}, F_{Bx} and F_{By} at the supports A and B are calculated and printed in MATLAB with:

```
clear all; clc; close all
syms d F_Ay F_Bx F_By F1 F2 F3 alpha beta

% Reactions at supports
F2x = F2*sin(beta);
F2y = F2*cos(beta);
dnew = d*tan(alpha);
sumFx = - F1 - F2x - F_Bx;
sumFy = F_Ay + F_By - F2y - F3;
SMB = d*F_Ay - dnew*F2x - d*F3;
%SMA = -d*F_By + d*F2y - dnew*F2x;
solR=solve(sumFx,sumFy,SMB,F_Ay,F_Bx,F_By);
fprintf('Reactions at supports \n')
fprintf('F_Ay = %s \n',solR.F_Ay)
fprintf('F_Bx = %s \n',solR.F_Bx)
fprintf('F_By = %s \n\n',solR.F_By)
```

The free-body diagram of the joint A indicating all the external and internal forces acting on the joint is shown in Figure 3.41. There, the unknown forces at joint A are denoted by F_{AB} and F_{AC}. Assuming that each member is in tension one can write

$$\sum F_{Ax} = 0 \Leftrightarrow F_{AB} - F_1 = 0,$$
$$\sum F_{Ay} = 0 \Leftrightarrow F_{AC} + F_{Ay} = 0. \qquad (3.72)$$

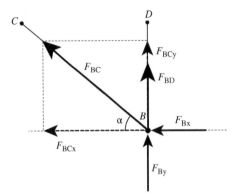

Figure 3.42 Free-body diagram of the joint B of frame.

This results in

$$F_{AC} = -F_{Ay},$$
$$F_{AB} = F_1. \tag{3.73}$$

The forces F_{AB} and F_{AC} at joint A are calculated and printed in MATLAB with:

```
% joint A
syms F_AB F_AC F_Ay
sumFAx = F_AB - F1;
sumFAy = F_AC + F_Ay;
fprintf('JOINT A - Sum of forces \n')
fprintf('sumFAx = %s \n',sumFAx)
fprintf('sumFAy = %s \n\n',sumFAy)

solA=solve(sumFAx,sumFAy,F_AB,F_AC);
F_AB = solA.F_AB;
F_AC = solA.F_AC;
fprintf('JOINT A - Results \n')
fprintf('F_AB = %s \n',F_AB)
fprintf('F_AC = %s \n\n',simplify(F_AC))
```

The free-body diagram of the joint B is shown in Figure 3.42. There, the unknown forces at joint B are F_{BC} and F_{BD}. One can write

$$\sum F_{Bx} = 0 \Leftrightarrow -F_{BCx} - F_{BA} - F_{Bx} = 0 \Leftrightarrow -F_{BC}\cos(\alpha) - F_{BA} - F_{Bx} = 0$$
$$\sum F_{By} = 0 \Leftrightarrow F_{BCy} + F_{BD} + F_{By} = 0 \Leftrightarrow F_{BC}\sin\alpha + F_{BD} + F_{By} = 0 \tag{3.74}$$

where $F_{BCx} = F_{BC}\cos(\alpha)$ and $F_{BCy} = F_{BC}\sin(\alpha)$ are the Ox and Oy components of F_{BC}. Using Equations (3.74) and (3.61) one can write

$$F_{BC} = -\frac{F_{BA} + F_{Bx}}{\cos\alpha} = -\frac{F_1 + F_{Bx}}{\cos\alpha}$$
$$F_{BD} = -F_{BC}\sin(\alpha) - F_{By}$$

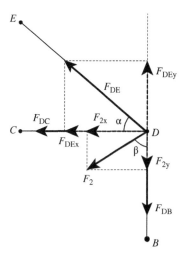

Figure 3.43 Free-body diagram of the joint D of frame.

$$= \frac{F_{BA} + F_{Bx}}{\cos \alpha} \sin(\alpha) - F_{By}$$
$$= (F_{BA} + F_{Bx}) \tan \alpha - F_{By}$$
$$= (F_1 + F_{Bx}) \tan \alpha - F_{By}. \tag{3.75}$$

The forces F_{BC} and F_{BD} at joint B are calculated and printed in MATLAB with:

```
% joint B
syms F_BD F_BC F_BA F_Bx F_By
F_BA=F_AB;
F_BCx = F_BC*cos(alpha);
F_BCy = F_BC*sin(alpha);
sumFBx = - F_BCx - F_BA - F_Bx;
sumFBy = F_BCy + F_BD + F_By;
fprintf('JOINT B - Sum of forces \n')
fprintf('sumFBx = %s \n', sumFBx)
fprintf('sumFBy = %s \n\n', sumFBy)

solB=solve(sumFBx, sumFBy, F_BD, F_BC);
F_BD = solB.F_BD;
F_BC = solB.F_BC;
fprintf('JOINT B - Results \n')
fprintf('F_BD = %s \n', simplify(F_BD))
fprintf('F_BC = %s \n\n', F_BC)
```

The free-body diagram of the joint D is shown in Figure 3.43. There, the unknown forces at the joint D are F_{DC} and F_{DE}. One can write

$$\sum F_{Dx} = 0 \Leftrightarrow -F_{DC} - F_{DEx} - F_{2x} = 0$$
$$\Leftrightarrow -F_{DC} - F_{DE} \cos \alpha - F_2 \sin \beta = 0$$

$$\sum F_{Dy} = 0 \Leftrightarrow F_{DEy} - F_{DB} - F_{2y} = 0$$
$$\Leftrightarrow F_{DE} \sin \alpha - F_{DB} - F_2 \cos \beta = 0 \tag{3.76}$$

where $F_{DEx} = F_{DE} \cos \alpha$, $F_{2x} = F_2 \sin \beta$, $F_{DEy} = F_{DE} \sin \alpha$, and $F_{2y} = F_2 \cos \beta$ are the Ox and Oy components of F_{DE} and F_2.

Using Equations (3.76) and (3.75) one can write

$$F_{DE} = \frac{F_{DB} + F_2 \cos \beta}{\sin \alpha},$$
$$= \frac{(F_1 + F_{Bx}) \tan \alpha - F_{By} + F_2 \cos \beta}{\sin \alpha}$$
$$= \frac{F_1 + F_{Bx}}{\cos \alpha} - \frac{F_{By} - F_2 \cos \beta}{\sin \alpha}$$
$$F_{DC} = -F_{DE} \cos \alpha - F_2 \sin \beta$$
$$= -\frac{(F_1 + F_{Bx}) \tan \alpha - F_{By} + F_2 \cos \beta}{\sin \alpha} - F_2 \sin \beta$$
$$= -(F_1 + F_{Bx}) + F_{By} \cot \alpha - F_2 \cos \beta \cot \alpha - F_2 \sin \beta. \tag{3.77}$$

The forces F_{DC} and F_{DE} at joint D are calculated and printed in MATLAB with:

```
% joint D
syms F_DC F_DE
F_DB=F_BD;
F_DEx = F_DE*cos(alpha);
F_DEy = F_DE*sin(alpha);
sumFDx = - F_DC - F2x - F_DEx;
sumFDy = F_DEy - F_DB - F2y;
fprintf('JOINT D - Sum of forces \n')
fprintf('sumFDx = %s \n',sumFDx);
fprintf('sumFDy = %s \n\n',sumFDy);

solD=solve(sumFDx,sumFDy,F_DE,F_DC);
F_DE = solD.F_DE;
F_DC = solD.F_DC;
fprintf('JOINT D - Results \n')
fprintf('F_DE = %s \n',simplify(F_DE));
fprintf('F_DC = %s \n\n',simplify(F_DC));
```

The free-body diagram of the joint E is shown in Figure 3.44. There, the unknown forces at joint E are F_{EC} and F_{ED}. One can write

$$\sum F_{Ex} = 0 \Leftrightarrow F_{EDx} = 0$$
$$\sum F_{Ey} = 0 \Leftrightarrow -F_{EC} - F_{EDy} - F_3 = 0 \tag{3.78}$$

where $F_{EDx} = F_{ED} \cos(\alpha)$ and $F_{EDy} = F_{ED} \sin(\alpha)$ are the Ox and Oy components of F_{ED}. Using Equation (3.78) one can write

$$F_{ED} = 0$$

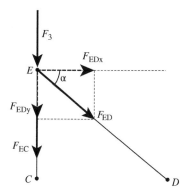

Figure 3.44 Free-body diagram of the joint E of frame.

$$F_{EC} = -F_{EDy} - F_3$$
$$= -F_{ED} \sin \alpha - F_3$$
$$= -F_3. \tag{3.79}$$

The forces F_{EC} and F_{ED} at joint E are calculated and printed in MATLAB with:

```
% joint E
syms F_EC F_ED
F_EDx = F_ED*cos(alpha);
F_EDy = F_ED*sin(alpha);
sumFEx = F_EDx;
sumFEy = - F_EC - F_EDy - F3;
fprintf('JOINT E - Sum of forces \n')
fprintf('sumFEx = %s \n',sumFEx)
fprintf('sumFEy = %s \n\n',sumFEy)

solE=solve(sumFEx,sumFEy,F_ED,F_EC);
F_ED = solE.F_ED;
F_EC = solE.F_EC;
fprintf('JOINT E - Results \n')
fprintf('F_ED = %s \n',simplify(F_ED))
fprintf('F_EC = %s \n\n',simplify(F_EC))
```

The input numerical data are introduced in MATLAB with:

```
% numerical results
lists = {d,F1,F2,F3,alpha,beta};
listn = {0.5,100,100,100,pi/6,pi/4};
```

The numerical results, i.e. forces at each joint, are calculated in MATLAB using:

```
F_Ay = eval(subs(solR.F_Ay,lists,listn));
F_Bx = eval(subs(solR.F_Bx,lists,listn));
F_By = eval(subs(solR.F_By,lists,listn));
```

```
F_AC = eval(subs(F_AC,lists,listn));
F_AB = eval(subs(F_AB,lists,listn));
F_BC = eval(subs(F_BC,lists,listn));
F_BD = eval(subs(F_BD,lists,listn));
F_DC = eval(subs(F_DC,lists,listn));
F_DE = eval(subs(F_DE,lists,listn));
F_ED = eval(subs(F_ED,lists,listn));
F_EC = eval(subs(F_EC,lists,listn));
```

The forces at each joint are printed in MATLAB using:

```
fprintf('F_Ay = [%f] (N)\n',F_Ay);
fprintf('F_Bx = [%f] (N)\n',F_Bx);
fprintf('F_By = [%f] (N)\n',F_By);
fprintf('F_AC = [%f] (N)\n',F_AC);
fprintf('F_AB = [%f] (N)\n',F_AB);
fprintf('F_BC = [%f] (N)\n',F_BC);
fprintf('F_BD = [%f] (N)\n',F_BD);
fprintf('F_DE = [%f] (N)\n',F_DE);
fprintf('F_DC = [%f] (N)\n',F_DC);
fprintf('F_ED = [%f] (N)\n',F_ED);
fprintf('F_EC = [%f] (N)\n',F_EC);
```

The numerical results for the forces are:

```
F_Ay = [140.824829] (N)
F_Bx = [-170.710678] (N)
F_By = [29.885849] (N)
F_AC = [-140.824829] (N)
F_AB = [100.000000] (N)
F_BC = [81.649658] (N)
F_BD = [-70.710678] (N)
F_DE = [-0.000000] (N)
F_DC = [-70.710678] (N)
F_ED = [0.000000] (N)
F_EC = [-100.000000] (N)
```

Example 3.8

The external forces $\mathbf{F}_1$ and $\mathbf{F}_2$ act on the frame shown in Figure 3.45 at the joints C and D. The force $\mathbf{F}_1$ is vertical and the force $\mathbf{F}_2$ is making the angle α with the vertical direction. The distances d_{AB}, d_{AC}, d_{CB} and d_{CD} are equal and are given by $d_{AC} = d_{AB} = d_{CB} = d_{CD} = d$. The frame is roller supported at A and pin supported at B. Using the section method, determine all the internal forces and directions (tension or compression) in the frame when the externally applied loads act as shown. Numerical application: $d = 1$ m, $F_1 = 300$ N, $F_2 = 25$ N, $\alpha = \dfrac{\pi}{6}$.

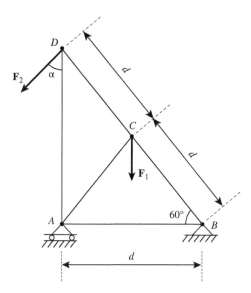

Figure 3.45 Frame roller supported at *A* and pin supported at *B*.

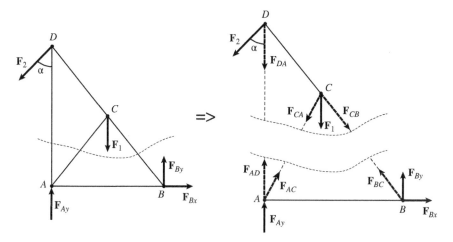

Figure 3.46 Method of sections illustrated by cutting the frame through the members *AD*, *AC* and *BC*.

Solution

The free-body diagram of the external forces and reaction forces at the supports is shown in Figure 3.46. One can determine the reactions at the supports by solving the equilibrium equations, i.e. the sum of all the forces and the sum of moments about any considered point is zero,

$$\sum F_x = 0 \iff -F_{2x} + F_{Bx} = 0$$
$$\sum F_y = 0 \iff F_{Ay} + F_{By} - F_{2y} - F_1 = 0$$
$$\sum M_D = 0 \iff \frac{d}{2}F_1 - d_{AD}F_{Bx} - dF_{By} = 0 \qquad (3.80)$$

where $d_{AD} = 2d\cos\dfrac{\pi}{6}$, $F_{2x} = F_2\sin\alpha$ and $F_{2y} = F_2\cos\alpha$ are the Ox and Oy components of the force F_2. Solving Equation (3.80) one can obtain

$$F_{Ay} = \frac{F_1}{2} + F_2\cos\alpha + \sqrt{3}F_2\sin\alpha$$
$$F_{Bx} = F_2\sin\alpha$$
$$F_{By} = \frac{F_1}{2} - \sqrt{3}F_2\sin\alpha. \tag{3.81}$$

The reaction forces F_{Ay}, F_{Bx} and F_{By} at the supports A and B are calculated and printed in MATLAB with:

```
clear all; clc; close all
syms d F_Ay F_Bx F_By F1 F2 alpha F2x F2y

% Reactions at supports
F2x = F2*sin(alpha);
F2y = F2*cos(alpha);
sumFx =   - F2x + F_Bx;
sumFy = F_Ay + F_By - F1 - F2y;
d_AD = 2*d*cos(pi/6);
%SMD = (d/2)*F1 - d_AD*F_Bx - d*F_By;
SMB = (d)*F_Ay - (1/2) *d*F1 - d*F2y - d_AD *F2x;
solR=solve(sumFx,sumFy,SMB,F_Ay,F_Bx,F_By);
fprintf('Reactions at supports \n')
fprintf('F_Ay = %s \n',solR.F_Ay)
fprintf('F_Bx = %s \n',solR.F_Bx)
fprintf('F_By = %s \n\n',solR.F_By)
```

For the method of sections, the external forces acting in the direction of the member replace the cut members in the section. The method of sections is illustrated for the frame in Figure 3.46, by cutting the frame through the members AD, AC and BC.

One of the resulting sections, the one that contains the member DC, is shown in Figure 3.47. For this section, one can resolve directly the member forces by solving the equilibrium equations, that is

$$\sum F_{Ox} = 0 \Leftrightarrow -F_{2x} - F_{CAx} + F_{CBx} = 0,$$
$$\Leftrightarrow -F_2\sin\alpha - F_{CA}\sin\frac{\pi}{6} + F_{CB}\sin\frac{\pi}{6} = 0$$
$$\sum F_{Oy} = 0 \Leftrightarrow -F_{DA} - F_{2y} - F_{CAy} - F_{CBy} - F_1 = 0,$$
$$\Leftrightarrow -F_{DA} - F_2\cos\alpha - F_{CA}\cos\frac{\pi}{6} - F_{CB}\cos\frac{\pi}{6} - F_1 = 0 \tag{3.82}$$

where $F_{CAx} = F_{CA}\sin\dfrac{\pi}{6}$ and $F_{CAy} = F_{CA}\cos\dfrac{\pi}{6}$, $F_{CBx} = F_{CB}\sin\dfrac{\pi}{6}$ and $F_{CBy} = F_{CB}\cos\dfrac{\pi}{6}$.

Taking moments about C one can write

$$\sum M_C = 0 \Leftrightarrow -F_{2x}\frac{d_{AD}}{2} - F_{DA}\frac{d}{2} - F_{2y}\frac{d}{2} = 0$$

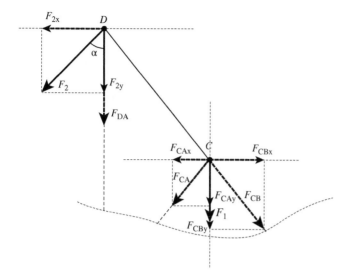

Figure 3.47 Resulting section containing the member DC.

$$\Leftrightarrow F_{DA}\frac{d}{2} = -F_{2x}\frac{d_{AD}}{2} - F_{2y}\frac{d}{2}$$

$$\Leftrightarrow F_{DA} = -F_{2x}\frac{d_{AD}}{d} - F_{2y}$$

$$\Leftrightarrow F_{DA} = -\sqrt{3}F_{2x} - F_{2y}$$

$$\Leftrightarrow F_{DA} = -\sqrt{3}F_2\sin\alpha - F_2\cos\alpha. \tag{3.83}$$

Using Equation (3.83) one can rewrite Equation (3.82) as

$$\begin{cases} -F_2\sin\alpha - \dfrac{F_{CA}}{2} + \dfrac{F_{CB}}{2} = 0 \\ F_2\sin\alpha - \dfrac{F_{CA}}{2} - \dfrac{F_{CB}}{2} - \dfrac{F_1}{\sqrt{3}} = 0 \end{cases} \tag{3.84}$$

Solving Equation (3.84) one can obtain

$$F_{CA} = -\frac{F_1}{\sqrt{3}}$$

$$F_{CB} = 2F_2\sin\alpha - \frac{F_1}{\sqrt{3}}. \tag{3.85}$$

The forces F_{DA}, F_{CA} and F_{CB} at the section that contains the member DC are calculated and printed in MATLAB with:

```
% section CD - cutting members AD, AC and BC
syms F_CA F_CB F_DA F1 F2 d_CD
F_CAx = F_CA*sin(pi/6);F_CAy = F_CA*cos(pi/6);
F_CBx = F_CB*sin(pi/6);F_CBy = F_CB*cos(pi/6);
sumFCDx = - F2x - F_CAx + F_CBx;
sumFCDy = - F_DA - F2y - F_CAy - F_CBy - F1;
```

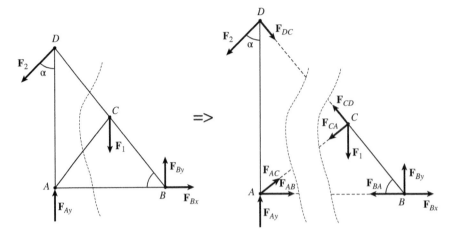

Figure 3.48 Method of sections illustrated by cutting the frame through the members *DC*, *AC* and *AB*.

```
sumMC =   - F2x*d_AD/2 - F_DA*d/2 -F2y*d/2;
fprintf('SECTION CD - Sum of forces \n')
fprintf('sumFCDx = %s \n', sumFCDx)
fprintf('sumFCDy = %s \n', sumFCDy)
fprintf('sumMC = %s \n\n', sumMC)


F_DAnew=solve(sumMC,F_DA);
fprintf('SECTION CD - Results \n')
fprintf('F_DA = %s \n',F_DAnew)
solCD=solve(sumFCDx,sumFCDy,F_CA,F_CB);
F_CA = solCD.F_CA;
F_CA = subs(F_CA,F_DA, F_DAnew);
F_CB = solCD.F_CB;
F_CB = subs(F_CB,F_DA, F_DAnew);
fprintf('F_CA = %s \n',simplify(F_CA))
fprintf('F_CB = %s \n\n',simplify(F_CB))
```

Since the only remaining unknown forces are F_{AB} and F_{DC} using the section that contain the member AB (Figure 3.46) after considering the one that contains the member DC, will not add any benefit (will not solve the problem). Therefore, in order to find forces F_{AB} and F_{DC} using the method of sections, the frame in Figure 3.46 should be cut through the members DC, AC and AB, as shown in Figure 3.48.

One of the resulting sections, the one that contains the member DA, is shown in Figure 3.49. For this section, one can resolve directly the member forces by solving the equilibrium equations, that is

$$\sum F_{Ox} = 0 \Leftrightarrow -F_{2x} + F_{DCx} + F_{ACx} + F_{AB} = 0,$$
$$\Leftrightarrow -F_2 \sin\alpha + F_{DC}\sin\frac{\pi}{6} + F_{AC}\sin\frac{\pi}{6} + F_{AB} = 0$$

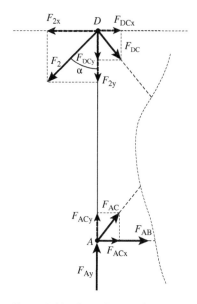

Figure 3.49 Resulting section containing the member *DA*.

$$\sum F_{Oy} = 0 \Leftrightarrow -F_{2y} - F_{DCy} + F_{ACy} + F_{Ay} = 0,$$
$$\Leftrightarrow -F_2 \cos \alpha - F_{DC} \cos \frac{\pi}{6} + F_{AC} \cos \frac{\pi}{6} + F_{Ay} = 0 \qquad (3.86)$$

where $F_{DCx} = F_{DC} \sin \frac{\pi}{6}$ and $F_{DCy} = F_{DC} \cos \frac{\pi}{6}$, $F_{ACx} = F_{AC} \sin \frac{\pi}{6}$ and $F_{ACy} = F_{AC} \cos \frac{\pi}{6}$, and $F_{2x} = F_2 \sin \alpha$ and $F_{2y} = F_2 \sin \alpha$ are the Ox and Oy components of F_{DC}, F_{AC} and F_2 respectively.

Taking moments about A one can write

$$\sum M_A = 0 \Leftrightarrow -F_{2x} d_{AB} + F_{DCx} d_{AB} = 0$$
$$\Leftrightarrow -F_2 \sin \alpha d_{AB} + F_{DC} \sin \frac{\pi}{6} = 0$$
$$\Leftrightarrow F_{DC} = 2F_2 \sin \alpha. \qquad (3.87)$$

Using Equation (3.87) one can rewrite Equation (3.86) as

$$\sum F_{Oy} = 0 \Leftrightarrow -F_2 \cos \alpha - F_{DC} \cos \frac{\pi}{6} + F_{AC} \cos \frac{\pi}{6} + F_{Ay} = 0$$
$$\Leftrightarrow -F_2 \cos \alpha - \sqrt{3} F_2 \sin \alpha + \frac{\sqrt{3}}{2} F_{AC} + \frac{F_1}{2} + F_2 \cos \alpha + \sqrt{3} F_2 \sin \alpha = 0$$
$$\Leftrightarrow F_{AC} = -\frac{F_1}{\sqrt{3}}. \qquad (3.88)$$

The forces F_{AC} and F_{DC} at the section that contains the member *DA* are calculated and printed in MATLAB with:

```
% section DA - cutting members DC, AC and AB
syms F_DC F_AC F_AB F_Ay
F_DCx = F_DC*sin(pi/6);F_DCy = F_DC*cos(pi/6);
```

```
F_ACx = F_AC*sin(pi/6);F_ACy = F_AC*cos(pi/6);
sumFDAx = - F2x + F_DCx + F_ACx + F_AB;
sumFDAy = - F2y - F_DCy + F_ACy + F_Ay;
sumMA = - F2x*d_AD + F_DCx*d_AD;
fprintf('section DA - Sum of forces \n')
fprintf('sumFDAx = %s \n',sumFDAx)
fprintf('sumFDAy = %s \n',sumFDAy)
fprintf('sumMA = %s \n\n',sumMA)


F_DCnew=solve(sumMA,F_DC);
fprintf('SECTION DA - Results \n')
fprintf('F_DC = %s \n',char(F_DCnew))
solDA=solve(sumFDAx,sumFDAy,F_AC,F_AB);
F_AC = solDA.F_AC;
F_AC = simplify(subs(F_AC,{F_DC,F_Ay},{F_DCnew,solR.F_Ay}));
F_AB = solDA.F_AB;
F_AB = subs(F_AB,{F_DC,F_Ay}, {F_DCnew,solR.F_Ay});
fprintf('F_AC = %s \n',simplify(F_AC))
fprintf('F_AB = %s \n\n',simplify(F_AB))
```

The input numerical data are introduced in MATLAB with:

```
% numerical results
lists = {d,F1,F2,alpha};
listn = {1,300,25,pi/6};
```

The numerical results – forces at each joint – are calculated in MATLAB using:

```
F2x = eval(subs(F2x,lists,listn));
F2y = eval(subs(F2y,lists,listn));
F_Ay = eval(subs(solR.F_Ay,lists,listn));
F_Bx = eval(subs(solR.F_Bx,lists,listn));
F_By = eval(subs(solR.F_By,lists,listn));
F_AC = eval(subs(F_AC,lists,listn));
F_CA = eval(subs(F_CA,lists,listn));
F_DA = eval(subs(F_DAnew,lists,listn));
F_CB = eval(subs(F_CB,lists,listn));
F_AB = eval(subs(F_AB,lists,listn));
F_DC = eval(subs(F_DCnew,lists,listn));
```

The forces at each joint are printed in MATLAB using:

```
fprintf('F_Ay = [%f] (N)\n',F_Ay);
fprintf('F_Bx = [%f] (N)\n',F_Bx);
fprintf('F_By = [%f] (N)\n',F_By);
fprintf('F_DA = [%f] (N)\n',F_DA);
fprintf('F_AC = [%f] (N)\n',F_AC);
fprintf('F_CA = [%f] (N)\n',F_CA);
```

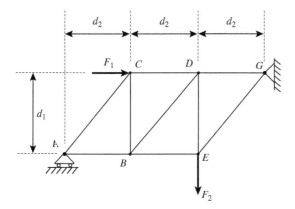

Figure 3.50 Frame roller supported at A and pin supported at G.

```
fprintf('F_CB = [%f] (N)\n',F_CB);
fprintf('F_AB = [%f] (N)\n',F_AB);
fprintf('F_DC = [%f] (N)\n',F_DC);
```

The numerical results for the forces are:

```
F_Ay = [193.301270]  (N)
F_Bx = [12.500000]   (N)
F_By = [128.349365]  (N)
F_DA = [-43.301270]  (N)
F_AC = [-173.205081] (N)
F_CA = [-173.205081] (N)
F_CB = [-148.205081] (N)
F_AB = [86.602540]   (N)
F_DC = [25.000000]   (N)
```

Example 3.9

The frame shown in Figure 3.50 carries two external forces, a horizontal force $\mathbf{F}_1$ acting at joint C and a vertical force $\mathbf{F}_2$ acting at joint E. The vertical members CB and DE have the lengths $d_{CB} = d_{DE} = d_1$, and the horizontal members AB, BE, CD and DG have all the same length, that is, $d_{AB} = d_{BE} = d_{CD} = d_{GB} = d_2$. The frame is roller supported at A and pin supported at G as shown in Figure 3.50. Using the section method, determine the magnitudes of the forces $\mathbf{F}_{AC}$, $\mathbf{F}_{CB}$, $\mathbf{F}_{CD}$, $\mathbf{F}_{BD}$, $\mathbf{F}_{BE}$ in the members AC, BC, CD, BD, and BE. Numerical application: $d_1 = 0.5$ m, $d_2 = 0.6$ m, $F_1 = 300$ N, $F_2 = 100$ N.

Solution

The free-body diagram of the external forces and reaction forces at the supports is shown in Figure 3.51. One can determine the reactions at the supports by solving the equilibrium equations, i.e. the sum of all the forces and the sum of moments about any considered point is zero,

$$\sum F_x = 0 \Leftrightarrow F_1 - F_{Gx} = 0$$

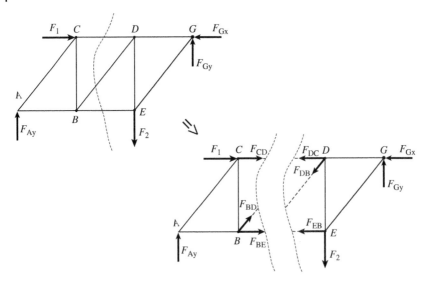

Figure 3.51 Method of sections illustrated by cutting the frame through the members *CD*, *BD* and *BE*.

$$\sum F_y = 0 \Leftrightarrow F_{Ay} - F_2 + F_{Gy} = 0$$
$$\sum M_G = 0 \Leftrightarrow 3d_2 F_{Ay} - d_2 F_2 = 0 \tag{3.89}$$

where F_{Gx} and F_{Gy} are magnitudes of the *Ox* and *Oy* reactions at the support *G* and F_{Ay} is the magnitude of the vertical reaction at support *A*. Solving Equation (3.89) one can obtain

$$F_{Ay} = \frac{F_2}{3}, \ F_{Gx} = F_1, \ F_{Gy} = \frac{2F_2}{3}. \tag{3.90}$$

The maginitudes of the reaction forces F_{Ay}, F_{Gx} and F_{Gy} at the supports *A* and *B* are calculated and printed in MATLAB with:

```
clear all; clc; close all
syms d F_Ay F_Gx F_Gy F1 F2 alpha d1 d2

% Reactions at supports
sumFx = F1 - F_Gx;
sumFy = F_Ay - F2 + F_Gy;
SMG = 3*d2*F_Ay - d2*F2;
solR=solve(sumFx,sumFy,SMG,F_Ay,F_Gx,F_Gy);
fprintf('Reactions at supports \n')
fprintf('F_Ay = %s \n',solR.F_Ay);
fprintf('F_Gx = %s \n',solR.F_Gx);
fprintf('F_Gy = %s \n\n',solR.F_Gy);
```

To determine the forces F_{CD}, F_{BD} and F_{BE}, the frame is cut through the members *CD*, *BD* and *BE*, as shown in Figure 3.51.

One of the resulting sections, the one that contains the members *AB*, *BC* and *AC*, is shown in Figure 3.52. For this section, one can resolve directly the member forces by solving the

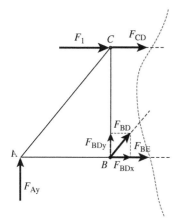

Figure 3.52 Resulting section containing the members *AB*, *AC* and *BC*.

equilibrium equations, that is

$$\sum F_{Ox} = 0 \Leftrightarrow F_1 + F_{CD} + F_{BE} + F_{BDx} = 0$$
$$\Leftrightarrow F_1 + F_{CD} + F_{BE} + F_{BD} \cos \alpha = 0$$
$$\sum F_{Oy} = 0 \Leftrightarrow F_{Ay} + F_{BDy} = 0$$
$$\Leftrightarrow F_{Ay} + F_{BD} \sin \alpha = 0$$
$$\Leftrightarrow F_{BD} = -\frac{F_{Ay}}{\sin \alpha} \tag{3.91}$$

where $\alpha = \arctan(d1/d2)$, and $F_{BDx} = F_{BD} \cos \alpha$ and $F_{BDy} = F_{BD} \sin \alpha$ are the magnitudes of the Ox and Oy components of $\mathbf{F}_{BD}$. Taking moments about B one can write

$$\sum M_B = 0 \Leftrightarrow F_{Ay} d_{AB} + F_1 d_{BC} + F_{CD} d_{CD} = 0$$
$$\Leftrightarrow F_{Ay} d_2 + F_1 d_1 + F_{CD} d_2 = 0$$
$$\Leftrightarrow F_{CD} = -F_{Ay} - F_1 \frac{d_1}{d_2}. \tag{3.92}$$

Using Equations (3.91) and (3.92) one can write

$$F_{BE} = -F_1 - F_{CD} - F_{BD} \cos \alpha$$
$$= -F_1 + F_{Ay} + F_1 \frac{d_1}{d_2} + F_{Ay} \cot \alpha$$
$$= F_1 \left(\frac{d_1}{d_2} - 1 \right) + F_{Ay}(1 + \cot \alpha). \tag{3.93}$$

The member forces magnitudes F_{CD}, F_{BD} and F_{BE} are calculated in MATLAB with:

```
syms F1 F_CD F_BE F_BD F_Ay d alpha
% section ABC - cutting members AB, BC and AC
alpha=atan(d1/d2);
F_BDx = F_BD*cos(alpha);
F_BDy = F_BD*sin(alpha);
```

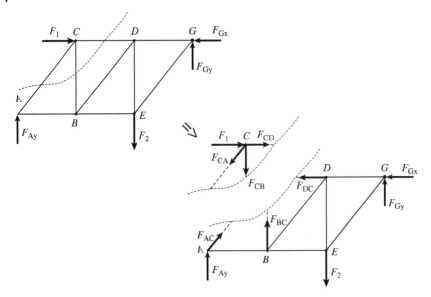

Figure 3.53 Method of sections illustrated by cutting the frame through the members *CD*, *CB* and *CA*.

```
d_AB=d2;d_BC=d1;
d_CD=d2;
sumFABCx = F1 + F_CD + F_BE + F_BDx;
sumFABCy = F_Ay + F_BDy;
sumMB =   F_Ay*d_AB + F1*d_BC + F_CD*d_CD;
fprintf('SECTION ABC - Sum of forces \n')
fprintf('sumFABCx = %s \n',sumFABCx)
fprintf('sumFABCy = %s \n',sumFABCy)
fprintf('sumMB = %s \n\n',sumMB)

fprintf('SECTION ABC - Results \n')
F_BDnew=solve(sumFABCy,F_BD);
fprintf('F_BD = %s \n',F_BDnew)
F_CDnew=solve(sumMB,F_CD);
f_CD=F_CDnew;
fprintf('F_CD = %s \n',simplify(F_CDnew))
F_BEnew=solve(sumFABCx,F_BE);
F_BEn = subs(F_BEnew,F_BD, F_BDnew);
F_BE = subs(F_BEn,F_CD, F_CDnew);
fprintf('F_BE = %s \n\n',simplify(F_BE))
```

To determine the forces F_{CD}, F_{BD} and F_{BE} the frame in Figure 3.53 is cut through the members *CD*, *BD* and *BE*, resulting in the section in Figure 3.54. The equilibrium equations for the section can be written as,

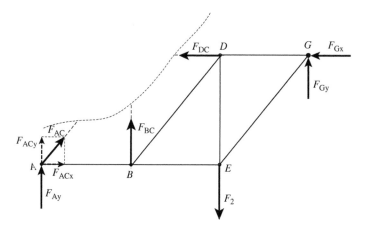

Figure 3.54 Resulting section containing the members *AB, BD, BE, ED, DG* and *EG*.

$$\sum F_{Ox} = 0 \Leftrightarrow F_{ACx} - F_{DC} - F_{Gx} = 0$$
$$\Leftrightarrow F_{AC} \cos \alpha - F_{CD} - F_{Gx} = 0$$
$$\Leftrightarrow F_{AC} \cos \alpha + F_{Ay} + F_1 \tan \alpha - F_{Gx} = 0$$
$$\Leftrightarrow F_{AC} = \frac{F_{Gx} - F_{Ay} - F_1 \tan \alpha}{\cos \alpha}$$

$$\sum F_{Oy} = 0 \Leftrightarrow F_{Ay} + F_{ACy} + F_{BC} - F_2 + F_{Gy} = 0$$
$$\Leftrightarrow F_{Ay} + F_{AC} \sin \alpha + F_{BC} - F_2 + F_{Gy} = 0$$
$$\Leftrightarrow F_{BC} = F_2 - F_{Gy} - F_{Ay} - F_{AC} \sin \alpha$$
$$\Leftrightarrow F_{BC} = F_2 - F_{Gy} - F_{Ay} - \left(\frac{F_{Gx} - F_{Ay} - F_1 \tan \alpha}{\cos \alpha} \right) \sin \alpha$$
$$\Leftrightarrow F_{BC} = F_2 - F_{Gy} - F_{Gx} \tan \alpha - F_{Ay}(1 - \tan \alpha) + F_1 \tan^2 \alpha \qquad (3.94)$$

where $F_{ACx} = F_{AC} \cos \alpha$ and $F_{ACy} = F_{AC} \sin \alpha$ are the magnitudes of the Ox and Oy components of $\mathbf{F}_{AC}$, and F_{Gx} is the magnitude of the Ox reaction at G.

The forces magnitude F_{CD}, F_{BD} and F_{BE} have been calculated in MATLAB with:

```
syms F_DC F_AC F_BC F_Gx F_Gy F2
% section AEG - cutting members DC, AC and AB
F_ACx = F_AC*cos(alpha);
F_ACy = F_AC*sin(alpha);
F_DC=F_CDnew;
sumFAEGx = F_ACx - F_DC - F_Gx;
sumFAEGy = F_Ay + F_ACy + F_BC - F2 + F_Gy;
fprintf('section AEG - Sum of forces \n')
fprintf('sumFAEGx = %s \n',sumFAEGx)
fprintf('sumFAEGy = %s \n\n',sumFAEGy)

fprintf('SECTION AEG - Results \n')
```

```
F_ACnew=solve(sumFAEGx,F_AC);
fprintf('F_AC = %s \n',simplify(F_ACnew))
F_BCnew=solve(sumFAEGy,F_BC);
F_BCn = subs(F_BCnew,F_AC, F_ACnew);
fprintf('F_BC = %s \n\n',simplify(F_BCn))
```

The input numerical data are introduced in MATLAB with:

```
% numerical results
lists = {d1,d2,F1,F2};
listn = {0.5,0.6,300,100};
```

The numerical results are calculated in MATLAB using:

```
F_Ay = eval(subs(solR.F_Ay,lists,listn));
F_Gx = eval(subs(solR.F_Gx,lists,listn));
F_Gy = eval(subs(solR.F_Gy,lists,listn));
F_BD = eval(subs(F_BDnew,lists,listn));
F_CD = eval(subs(F_CDnew,lists,listn));
F_BE = eval(subs(F_BEnew,lists,listn));
F_AC = eval(subs(F_ACnew,lists,listn));
F_BC = eval(subs(F_BCnew,lists,listn));
```

The forces are printed in MATLAB using:

```
fprintf('F_Ay = [%f] (N)\n',F_Ay);
fprintf('F_Gx = [%f] (N)\n',F_Gx);
fprintf('F_Gy = [%f] (N)\n',F_Gy);
fprintf('F_BD = [%f] (N)\n',F_BD);
fprintf('F_BE = [%f] (N)\n',F_BE);
fprintf('F_CD = [%f] (N)\n',F_CD);
fprintf('F_AC = [%f] (N)\n',F_AC);
fprintf('F_BC = [%f] (N)\n',F_BC);
```

The numerical results for the forces are:

```
F_Ay = [33.333333] (N)
F_Gx = [300.000000] (N)
F_Gy = [66.666667] (N)
F_BD = [-52.068331] (N)
F_BE = [23.333333] (N)
F_CD = [-283.333333] (N)
F_AC = [21.695138] (N)
F_BC = [-13.888889] (N)
```

Example 3.10

The frame shown in Figure 3.55 carries two external forces F_1 and F_2. The force F_1 acting at B is making an angle β with the horizontal direction and the force F_2 acting at C is a vertical force. The angle between the members EC and CG, DC and CE, DB and BC, and

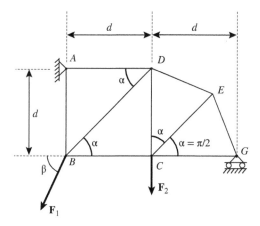

Figure 3.55 Frame pin supported at A and roller supported at G.

AB and BD is $\alpha = \frac{\pi}{4}$. The members CG, CE, CD, AD, AB and BC have the same length, that is, $CG = CE = CD = AD = AB = BC = d$. Combine the section method and the method of joints to determine all the internal forces and directions (tension or compression) in the frame. Numerical application: $d = 0.5$ m, $F_1 = 100$ N, $F_2 = 250$ N, $\beta = \frac{\pi}{4}$.

Solution

The free-body diagram of the external and reaction forces at the supports acting on the frame is shown in Figure 3.56. One can determine the reactions at the supports by solving the equilibrium equations, i.e. the sum of all the forces and the sum of moments about any considered point is zero,

$$\sum F_x = 0 \Leftrightarrow F_{Ax} - F_{1x} = 0$$
$$\sum F_y = 0 \Leftrightarrow F_{Ay} - F_{1y} - F_2 + F_{Gy} = 0$$
$$\sum M_B = 0 \Leftrightarrow dF_{Ax} + dF_2 - 2dF_{Gy} = 0 \tag{3.95}$$

where F_{Gy} is the magnitude of the Oy reaction at the support G, F_{Ax} and F_{Ay} are the magnitudes of the horizontal and vertical reactions at support A, and $F_{1x} = F_1 \cos \beta$ and $F_{1y} = F_1 \sin \beta$ are the magnitudes of the horizontal and vertical components of force $\mathbf{F}_1$. Solving Equation (3.95) one can obtain

$$F_{Ax} = F_1 \cos \beta$$
$$F_{Ay} = \frac{1}{2}F_2 - \frac{1}{2}F_1 \cos \beta + F_1 \sin \beta$$
$$F_{Gy} = \frac{1}{2}F_2 + \frac{1}{2}F_1 \cos \beta. \tag{3.96}$$

The reaction forces magnitudes F_{Ax}, F_{Ay} and F_{Gy} at the supports A and G are calculated and printed in MATLAB with:

```
% method of sections and method of joints
clear all; clc; close all
% Reactions at supports
```

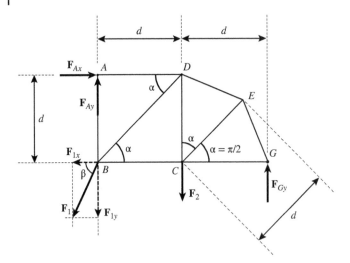

Figure 3.56 Free-body diagram of the frame pin supported at *A* and roller supported at *G*.

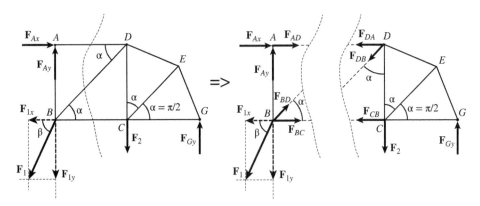

Figure 3.57 Method of sections illustrated by cutting the frame through the members *AD*, *BD* and *BC*.

```
syms d F_Ax F_Ay F_Gy F1 F2 beta
F1x = F1*cos(beta); F1y = F1*sin(beta);
sumFx = F_Ax - F1x;
sumFy = F_Ay - F1y - F2 + F_Gy;
SMB = d*F_Ax + d*F2 -2*d*F_Gy;
solR=solve(sumFx,sumFy,SMB,F_Ax,F_Ay,F_Gy);
fprintf('Reactions at supports \n');
fprintf('F_Ax = %s \n',solR.F_Ax);
fprintf('F_Ay = %s \n',solR.F_Ay);
fprintf('F_Gy = %s \n\n',solR.F_Gy);
```

To determine the forces magnitude F_{AD}, F_{BD} and F_{BC} the frame is cut through the members *AD*, *BD* and *BC*, as shown in Figure 3.57. One of the resulting sections is shown in Figure 3.58. For this section, one can resolve directly the member forces by solving the

Figure 3.58 Resulting section containing the member *AB*.

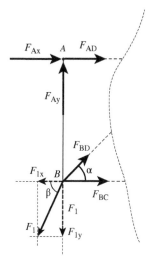

equilibrium equations, that is,

$$\sum F_{Ox} = 0 \Leftrightarrow F_{Ax} + F_{AD} - F_{1x} + F_{BC} + F_{BDx} = 0$$
$$\Leftrightarrow F_{Ax} + F_{AD} - F_1 \cos \beta + F_{BC} + F_{BD} \cos \alpha = 0$$
$$\sum F_{Oy} = 0 \Leftrightarrow F_{Ay} - F_{1y} + F_{BDy} = 0$$
$$\Leftrightarrow F_{Ay} - F_1 \sin \beta + F_{BD} \sin \alpha = 0$$
$$\Leftrightarrow F_{BD} = \frac{F_1 \sin \beta - F_{Ay}}{\sin \alpha}$$
$$\sum M_B = 0 \Leftrightarrow dF_{Ax} + dF_{AD} = 0$$
$$\Leftrightarrow F_{Ax} + F_{AD} = 0$$
$$\Leftrightarrow F_{AD} = -F_{Ax} = -F_1 \cos \beta \qquad (3.97)$$

where $F_{BDx} = F_{BD} \cos \alpha$ and $F_{BDy} = F_{BD} \sin \alpha$ are the Ox and Oy components of F_{BD}.
Using Equation (3.97) one can calculate

$$F_{BC} = F_1 \cos \beta - F_{Ax} - F_{AD} - F_{BD} \cos \alpha$$
$$= F_1 \cos \beta - F_{Ax} + F_{Ax} - F_{BD} \cos \alpha$$
$$= F_1 \cos \beta - F_{BD} \cos \alpha$$
$$= F_1 \cos \beta - (F_1 \sin \beta - F_{Ay}) \frac{\cos \alpha}{\sin \alpha}. \qquad (3.98)$$

The forces magnitude F_{AD}, F_{BC} and F_{BD} at section AB are calculated and printed in MATLAB
with:

```
% section AB - cutting members AD, BD and BC
syms F1 F_AD F_BD F_BC F_Ax F_Ay d alpha beta
F_BDx = F_BD*cos(alpha);  F_BDy = F_BD*sin(alpha);
sumFABx = F_Ax + F_AD - F1x + F_BC + F_BDx;
sumFABy = F_Ay - F1y + F_BDy;
d_BA=d;
```

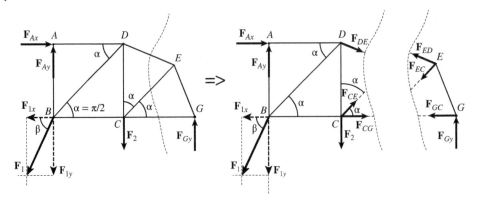

Figure 3.59 Method of sections illustrated by cutting the frame through the members *DE*, *CE* and *CG*.

```
sumMB = F_Ax*d_BA + F_AD*d_BA;
fprintf('SECTION AB - Sum of forces \n')
fprintf('sumFABx = %s \n',sumFABx)
fprintf('sumFABy = %s \n',sumFABy)
fprintf('sumMB = %s \n',sumMB)
fprintf('\n')


fprintf('SECTION AB - Results \n')
F_ADnew=solve(sumMB,F_AD);
fprintf('F_AD = %s \n',F_ADnew)
F_BDnew=solve(sumFABy,F_BD);
fprintf('F_BD = %s \n',simplify(F_BDnew))
F_BCnew=solve(sumFABx,F_BC);
F_BCn = subs(F_BCnew,F_AD, F_ADnew);
F_BC = subs(F_BCn,F_BD, F_BDnew);
fprintf('F_BC = %s \n',F_BC)
fprintf('\n')
```

To determine the forces magnitude F_{DE}, F_{CE} and F_{CG} the frame in Figure 3.59 is cut through the members *DE*, *CE* and *CG*, as shown in Figure 3.60. For the section shown in Figure 3.60, one can resolve directly the member forces by solving the equilibrium equations, that is

$$\sum F_{Ox} = 0 \Leftrightarrow -F_{ECx} - F_{EDx} - F_{GC} = 0$$

$$\Leftrightarrow -F_{EC}\cos 2\alpha - F_{ED}\cos\left(\frac{\alpha}{2}\right) - F_{GC} = 0$$

$$\sum F_{Oy} = 0 \Leftrightarrow F_{EDy} - F_{ECy} + F_{Gy} = 0$$

$$\Leftrightarrow F_{ED}\sin\left(\frac{\alpha}{2}\right) - F_{EC}\sin 2\alpha + F_{Gy} = 0 \tag{3.99}$$

where $F_{ECx} = F_{EC}\cos 2\alpha$ and $F_{ECy} = F_{EC}\sin 2\alpha$ are the *Ox* and *Oy* components of F_{EC}, $F_{EDx} = F_{ED}\cos\left(\frac{\alpha}{2}\right)$ and $F_{EDy} = F_{ED}\sin\left(\frac{\alpha}{2}\right)$ are the *Ox* and *Oy* components of F_{ED}.

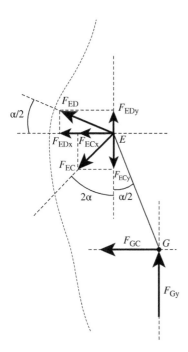

Figure 3.60 Resulting section containing the member *EG*.

Taking moments about E one can write

$$\sum M_E = 0 \Leftrightarrow F_{GC}d_{EGy} - F_{Gy}d_{EGx} = 0$$

$$\Leftrightarrow F_{GC}d_{EG}\cos\left(\frac{\alpha}{2}\right) - F_{Gy}d_{EG}\sin\left(\frac{\alpha}{2}\right) = 0$$

$$\Leftrightarrow F_{GC} = F_{Gy}\tan\left(\frac{\alpha}{2}\right). \tag{3.100}$$

Using Equations (3.99) and (3.100) one can write

$$\begin{cases} -F_{EC}\cos 2\alpha - F_{ED}\cos\left(\frac{\alpha}{2}\right) - F_{Gy}\tan\left(\frac{\alpha}{2}\right) = 0 \\ -F_{EC}\sin 2\alpha + F_{ED}\sin\left(\frac{\alpha}{2}\right) + F_{Gy} = 0 \end{cases} \tag{3.101}$$

Solving Equation (3.101) one can obtain

$$F_{EC} = 2F_{Gy}\frac{\cos\alpha}{\sin 2\alpha + \sin 3\alpha}$$

$$F_{ED} = -F_{Gy}\frac{2\cos\alpha - 1}{\sin\dfrac{5\alpha}{2}}. \tag{3.102}$$

The forces F_{GC}, F_{EC} and F_{ED} at section EG are calculated and printed in MATLAB with:

```
% section EG - cutting members DE, CE and BC
syms F_ED F_EC F_GC F_EG F_Gy F_Gy d_EG
F_ECx = F_EC*cos(2*alpha); F_ECy = F_EC*sin(2*alpha);
F_EDx = F_ED*cos(alpha/2); F_EDy = F_ED*sin(alpha/2);
```

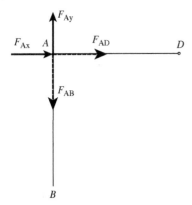

Figure 3.61 Free-body diagram of the joint *A* of frame.

```
sumFEGx = -F_ECx - F_EDx - F_GC;
sumFEGy = F_EDy - F_ECy + F_Gy;
d_EGx=d_EG*sin(alpha/2);
d_EGy=d_EG*cos(alpha/2);
sumME = F_GC*d_EGy - F_Gy*d_EGx;
fprintf('section EG - Sum of forces \n')
fprintf('sumFEGx = %s \n',sumFEGx)
fprintf('sumFEGy = %s \n',sumFEGy)
fprintf('sumME = %s \n\n',sumME)

fprintf('SECTION EG - Results \n')
F_GCnew=solve(sumME,F_GC);
fprintf('F_GC = %s \n',simplify(F_GCnew))
sol=solve(sumFEGx,sumFEGy,F_EC,F_ED);
F_EC = sol.F_EC;
F_EC = simplify(subs(F_EC,F_GC, F_GCnew));
F_ED = sol.F_ED;
F_ED = simplify(subs(F_ED,F_GC, F_GCnew));
fprintf('F_EC = %s \n',eval(F_EC))
fprintf('F_ED = %s \n\n',eval(F_ED))
```

The remaining unknown forces after applying the section method are F_{AB}, F_{CD} and F_{EG}; forces that can be found using the method of joints. The free-body diagram of the joint *A* indicating all the external and internal forces acting on the joint is shown in Figure 3.61. There, the unknown force at joint *A* is F_{AB}. One can write

$$\sum F_{Ay} = 0 \Leftrightarrow -F_{AB} + F_{Ay} = 0$$
$$\Leftrightarrow F_{AB} = F_{Ay}, \tag{3.103}$$

The force magnitude F_{AB} at joint *A* is calculated and printed in MATLAB with:

```
% joint A
syms F_AB F_Ay
```

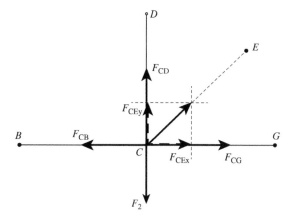

Figure 3.62 Free-body diagram of the joint C of frame.

```
sumFAy = F_Ay - F_AB;
fprintf('JOINT A - Sum of forces \n')
fprintf('sumFAy = %s \n\n',sumFAy)
F_AB=solve(sumFAy,F_AB);
fprintf('JOINT A - Results \n')
fprintf('F_AB = %s \n\n',F_AB)
```

The free-body diagram of the joint C, indicating all the external and internal forces acting on the joint, is shown in Figure 3.62. There, the unknown force at joint C is F_{CD}. One can write

$$\sum F_{Cy} = 0 \Leftrightarrow -F_2 + F_{CD} + F_{CEy} = 0$$
$$\Leftrightarrow -F_2 + F_{CD} + F_{CE} \sin \alpha = 0$$
$$\Leftrightarrow F_{CD} = F_2 - F_{CE} \sin \alpha \qquad (3.104)$$

where $F_{ECy} = F_{EC} \sin \alpha$ is the Oy component of F_{EC}.

The force magnitude F_{CD} at joint C is calculated and printed in MATLAB with:

```
% joint C
syms F_CD F2 F_CE
F_CEy=F_CE*sin(alpha);
sumFCy = - F2 + F_CD + F_CEy;
fprintf('JOINT C - Sum of forces \n')
fprintf('sumFCy = %s \n\n',sumFCy)
F_CD=solve(sumFCy,F_CD);
F_CD = subs(F_CD,F_CE, F_EC);
fprintf('JOINT C - Results \n')
fprintf('F_CD = %s \n\n',F_CD)
```

The free-body diagram of the joint G indicating all the external and internal forces acting on the joint is shown in Figure 3.63. There, the unknown force at joint G is F_{GE}. One can

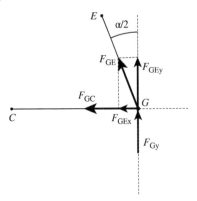

Figure 3.63 Free-body diagram of the joint G of frame.

write

$$\sum F_{Gy} = 0 \Leftrightarrow F_{Gy} + F_{GEy} = 0$$

$$\Leftrightarrow F_{Gy} + F_{GE} \cos \frac{\alpha}{2} = 0$$

$$\Leftrightarrow F_{GE} = \frac{F_{Gy}}{\cos \dfrac{\alpha}{2}} \tag{3.105}$$

where $F_{GEy} = F_{GE} \cos \dfrac{\alpha}{2}$.

The force magnitude F_{GE} at joint G is calculated and printed in MATLAB with:

```
% joint G
syms F_GE F_Gy
F_EGy=F_EG*cos(alpha/2);
sumFGy = F_Gy - F_GEy;
fprintf('JOINT G - Sum of forces \n')
fprintf('sumFGy = %s \n\n',sumGEy)
F_GEnew=solve(sumFGy,F_GE);
fprintf('JOINT G - Results \n')
fprintf('F_GE = %s \n\n',F_GEnew)
```

The input numerical data are introduced in MATLAB with:

```
% numerical results
lists = {d,F1,F2,beta,alpha};
listn = {0.5,100,250,pi/4,pi/4};
```

The numerical results are calculated in MATLAB using:

```
F_Ax = eval(subs(solR.F_Ax,lists,listn));
F_Ay = eval(subs(solR.F_Ay,lists,listn));
F_Gy = eval(subs(solR.F_Gy,lists,listn));
F_AD = eval(subs(F_ADnew,lists,listn));
F_BD = eval(subs(F_BDnew,lists,listn));
```

```
F_BC = eval(subs(F_BC,lists,listn));
F_GC = eval(subs(F_GCnew,lists,listn));
F_EC = eval(subs(F_EC,lists,listn));
F_AB = eval(subs(F_AB,lists,listn));
F_CD = eval(subs(F_CD,lists,listn));
F_EG = eval(subs(F_EGnew,lists,listn));
```

The forces are printed in MATLAB using:

```
fprintf('F_Ax = [%f] (N)\n',F_Ax);
fprintf('F_Ay = [%f] (N)\n',F_Ay);
fprintf('F_Gy = [%f] (N)\n',F_Gy);
fprintf('F_AD = [%f] (N)\n',F_AD);
fprintf('F_BD = [%f] (N)\n',F_BD);
fprintf('F_BC = [%f] (N)\n',F_BC);
fprintf('F_GC = [%f] (N)\n',F_GC);
fprintf('F_EC = [%f] (N)\n',F_EC);
fprintf('F_AB = [%f] (N)\n',F_AB);
fprintf('F_CD = [%f] (N)\n',F_CD);
fprintf('F_EG = [%f] (N)\n',F_EG);
```

The numerical results for the forces are:

```
F_Ax = [70.710678]  (N)
F_Ay = [160.355339]  (N)
F_Gy = [160.355339]  (N)
F_AD = [-70.710678]  (N)
F_BD = [-126.776695]  (N)
F_BC = [160.355339]  (N)
F_GC = [66.421356]  (N)
F_EC = [132.842712]  (N)
F_AB = [160.355339]  (N)
F_CD = [156.066017]  (N)
F_EG = [173.567368]  (N)
```

4

Centroids and Moments of Inertia

4.1 Centre of the Mass and Centroid

Consider the set of n particles S_k, $k = 1, 2, \ldots, n$ shown in Figure 4.1. For each particle S_k one can define its position by the vector $\mathbf{r}_{S_k}$ with respect to a chosen point O.

The position vector $\mathbf{r}_C$ (Figure 4.1) of the mass centre of the set of the particles S_k, $k = 1, 2, \ldots, n$, relative to the origin O of the reference frame $Oxyz$, is defined as

$$\mathbf{r}_C = \frac{\displaystyle\sum_{k=1}^{n} m_k \, \mathbf{r}_{S_k}}{\displaystyle\sum_{k=1}^{n} m_k}.$$

For the set of particles S_k, $k = 1, \ldots, n$, the coordinates x_C, y_C, z_C of the mass centre C can be expressed as

$$x_C = \frac{\displaystyle\sum_{k=1}^{n} m_k \, x_k}{\displaystyle\sum_{k=1}^{n} m_k}, \quad y_C = \frac{\displaystyle\sum_{i=1}^{n} m_i \, y_k}{\displaystyle\sum_{k=1}^{n} m_k}, \quad z_C = \frac{\displaystyle\sum_{k=1}^{n} m_k \, z_k}{\displaystyle\sum_{k=1}^{n} m_k}.$$

If the particles have equal masses, $m_1 = m_2 = \ldots = m_k = \ldots = m_n = m$ then the position vector $\mathbf{r}_C$ of the set of the particles S_k, $k = 1, 2, \ldots, n$ is calculated as

$$\mathbf{r}_C = \frac{\displaystyle\sum_{k=1}^{n} m \, \mathbf{r}_{S_k}}{\displaystyle\sum_{k=1}^{n} m} = \frac{1}{n} \sum_{k=1}^{n} \mathbf{r}_{S_k}$$

and represents the position of the centroid of the particles S_k, $k = 1, 2, \ldots, n$ relative to the fixed point O. The centroidal coordinates x_C, y_C, z_C of C can be expressed in the reference

Engineering Applications: Analytical and Numerical Calculation with MATLAB, First Edition.
Mihai Dupac and Dan B. Marghitu.
© 2021 John Wiley & Sons Ltd. Published 2021 by John Wiley & Sons Ltd.

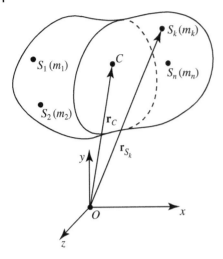

Figure 4.1 Centroid representation of a set of points.

frame *Oxyz* by

$$x_C = \frac{\sum\limits_{k=1}^{n} x_k}{n}, \quad y_C = \frac{\sum\limits_{k=1}^{n} y_k}{n}, \quad z_C = \frac{\sum\limits_{k=1}^{n} z_k}{n}.$$

The position of a centroid of a set of particles is uniquely determined and represents the geometric centre of the system, thus its location is independent of the origin and orientation of the used Cartesian frame.

One can define the *first moment* of a particle S_k with regard to the position of a reference point as a vector $\mathbf{M} = m_k \mathbf{r}_{S_k}$, where m_k is a scalar called the *strength* of S_k, e.g., the mass m_k of the particle located at S_k.

Using the definition of the centroid, one can alternatively define the *mass centre* of a number of particles S_k, $k = 1, 2, \ldots, n$ as the centroid of the particles location each particle having an associated strength equal to its own mass.

One can define the *first moment* of the set of the particles S_k, $k = 1, 2, \ldots, n$ relative to the fixed point O as a vector

$$\mathbf{M}_C = \sum_{k=1}^{n} m_k \mathbf{r}_{S_k}.$$

4.2 Centroid and Centre of the Mass of a Solid Region, Surface or Curve

The mass centre and centroid definitions presented above can be extended to define the mass centre and the centroid of a solid region, surface or a curve. One can define the vector position $\mathbf{r}_C$ of the mass centre C of a solid region, surface, or curve by

$$\mathbf{r}_C = \frac{\displaystyle\int_\sigma \mathbf{r}\, \rho\, d\sigma}{\displaystyle\int_\sigma \rho\, d\sigma} = \frac{1}{m}\int_\sigma \mathbf{r}\, \rho\, d\sigma, \tag{4.1}$$

where $\mathbf{r}$ is the vector position of a generic element σ, $d\sigma$ is the differential element for the volume, area or length, $m = \int_\sigma \rho\, d\sigma$ is the mass of the system, and ρ is the mass per unit volume, area or length when σ is a solid region, surface or curve respectively.

If the density ρ is the same (uniform) at all points of a solid region, surface or curve then the centre of the corresponding mass centre and centroid are identical. The centroid C of such solid region, surface, or curve can be determined using

$$\mathbf{r}_C = \frac{\displaystyle\int_\sigma \mathbf{r}\, d\sigma}{\displaystyle\int_\sigma d\sigma} = \frac{1}{\sigma}\int_\sigma \mathbf{r}\, d\sigma, \tag{4.2}$$

where $\mathbf{r}$ is the vector position of a generic element σ, $d\sigma$ is the differential element volume, area or length, and $\int_\sigma d\sigma$ represents the total volume, area or length when the element σ is a solid region, surface or curve.

When the density ρ is not constant the coordinates x_C, y_C and z_C of the mass centre of a solid region, surface or of a curve, with respect to the origin O of the frame $Oxyz$ can be defined by

$$x_C = \frac{\displaystyle\int_\sigma x\, \rho d\sigma}{\displaystyle\int_\sigma \rho d\sigma}, \quad y_C = \frac{\displaystyle\int_\sigma y\, \rho d\sigma}{\displaystyle\int_\sigma \rho d\sigma}, \quad z_C = \frac{\displaystyle\int_\sigma z\, \rho d\sigma}{\displaystyle\int_\sigma \rho d\sigma} \tag{4.3}$$

or equivalent by

$$x_C = \frac{\displaystyle\int_\sigma x\, \rho d\sigma}{\sigma}, \quad y_C = \frac{\displaystyle\int_\sigma y\, \rho d\sigma}{\sigma}, \quad z_C = \frac{\displaystyle\int_\sigma z\, \rho d\sigma}{\sigma} \tag{4.4}$$

where $\rho d\sigma$ is the mass of a differential element $d\sigma$.

The coordinates x_C, y_C and z_C for the centroid of a solid region, surface or curve can be defined by

$$x_C = \frac{\displaystyle\int_\sigma x\, d\sigma}{\displaystyle\int_\sigma d\sigma}, \quad y_C = \frac{\displaystyle\int_\sigma y\, d\sigma}{\displaystyle\int_\sigma d\sigma}, \quad z_C = \frac{\displaystyle\int_\sigma z\, d\sigma}{\displaystyle\int_\sigma d\sigma} \tag{4.5}$$

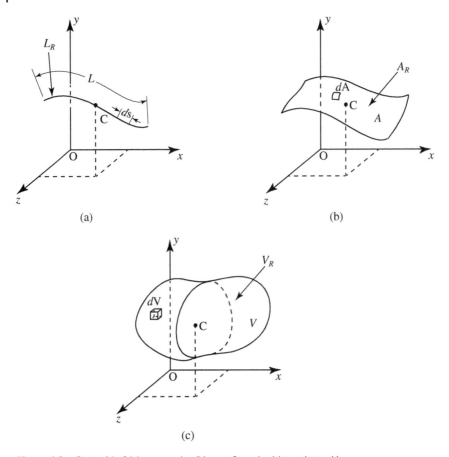

Figure 4.2 Centroid of (a) a curve L_R, (b) a surface A_R, (c) a volume V_R.

or equivalent by

$$x_C = \frac{1}{\sigma} \int_\sigma x \, d\sigma, \quad y_C = \frac{1}{\sigma} \int_\sigma y \, d\sigma, \quad z_C = \frac{1}{\sigma} \int_\sigma z \, d\sigma. \tag{4.6}$$

The coordinates x_C, y_C and z_C for the centroid of the curve L_R shown in Figure 4.2(a) can be determined by

$$x_C = \frac{\displaystyle\int_{L_R} x \, ds}{\displaystyle\int_{L_R} ds} = \frac{1}{L} \int_{L_R} x \, ds,$$

$$y_C = \frac{\displaystyle\int_{L_R} y \, ds}{\displaystyle\int_{L_R} ds} = \frac{1}{L} \int_{L_R} y \, ds,$$

$$z_C = \frac{\displaystyle\int_{L_R} z \, ds}{\displaystyle\int_{L_R} ds} = \frac{1}{L}\int_{L_R} z \, ds \tag{4.7}$$

where $L = \displaystyle\int_{L_R} ds$ is the curve length. It should be noted that the centroid may not be located throughout the length the curve.

The coordinates x_C, y_C and z_C of the centroid of the surface A_R shown Figure 4.2(b) can be calculated using

$$x_C = \frac{\displaystyle\int_{A_R} x \, dA}{\displaystyle\int_{A_R} dA} = \frac{1}{A}\int_{A_R} x \, dA,$$

$$y_C = \frac{\displaystyle\int_{A_R} y \, dA}{\displaystyle\int_{A_R} dA} = \frac{1}{A}\int_{A_R} y \, dA,$$

$$z_C = \frac{\displaystyle\int_{A_R} z \, dA}{\displaystyle\int_{A_R} dA} = \frac{1}{A}\int_{A_R} z \, dA, \tag{4.8}$$

where $A = \displaystyle\int_{A_R} dA$ is the area of the surface A_R.

The coordinates x_C, y_C and z_C of the centroid of a solid region V_R shown in Figure 4.2(c) are computed with:

$$x_C = \frac{\displaystyle\int_{V_R} x \, dV}{\displaystyle\int_{V_R} dV} = \frac{1}{V}\int_{V_R} x \, dV,$$

$$y_C = \frac{\displaystyle\int_{V_R} y \, dV}{\displaystyle\int_{V_R} dV} = \frac{1}{V}\int_{V_R} y \, dV,$$

$$z_C = \frac{\displaystyle\int_{V_R} z \, dV}{\displaystyle\int_{V_R} dV} = \frac{1}{V}\int_{V_R} z \, dV, \tag{4.9}$$

where $V = \displaystyle\int_{V_R} dV$ is the volume of the solid region V_R.

4.3 Method of Decomposition

The mass centre of a body can be determined using the *method of decomposition* considering the following steps:

- Decompose the body into simpler body shapes, e.g., solids regions, surfaces, or curves
- Consider holes as regions with negative size, e.g, negative mass.
- Calculate the x_{C_i}, y_{C_i}, z_{C_i} coordinates of each simple shape
- Calculate the mass centre of the whole body using

$$x_C = \frac{\sum\limits_{k=1}^{n}\int_\sigma x\, d\sigma}{\sum\limits_{k=1}^{n}\int_\sigma d\sigma}, \quad y_C = \frac{\sum\limits_{k=1}^{n}\int_\sigma y\, d\sigma}{\sum\limits_{k=1}^{n}\int_\sigma d\sigma}, \quad z_C = \frac{\sum\limits_{k=1}^{n}\int_\sigma z\, d\sigma}{\sum\limits_{k=1}^{n}\int_\sigma d\sigma}, \tag{4.10}$$

where σ is an length element, area element, or volume element as required.

The centroid can be determined using the steps of the *method of decomposition* above. The equations to calculate the centroid may be simplified as

$$x_C = \frac{\sum\limits_{k=1}^{n} x_{C_k}\,\sigma_k}{\sum\limits_{k=1}^{n}\sigma_k}, \quad y_C = \frac{\sum\limits_{k=1}^{n} y_{C_k}\,\sigma_k}{\sum\limits_{k=1}^{n}\sigma_k}, \quad z_C = \frac{\sum\limits_{k=1}^{n} z_{C_k}\,\sigma_k}{\sum\limits_{k=1}^{n}\sigma_k}, \tag{4.11}$$

where σ_i is the volume element, area element, or length element related to each shape.

4.4 First Moment of an Area

Consider the planar surface in Figure 4.3 and the associated Cartesian reference frame Oxy. One can define the first moment of area by taking the moment about a particular axis, i.e. about the Oy axis or respectively Ox axis, by

$$M_y = \int_A x\, dA, \quad M_x = \int_A y\, dA. \tag{4.12}$$

where x and y represents the normal distance measured from the Ox and respectively Oy axis to the differential area element dA.

One can consider x_C and y_C the centroidal coordinates of the planar surface shown in Figure 4.3, and the whole area A concentrated at the centroid $C(x_C, y_C)$ of the surface. The centroidal coordinates can be inferred by equating the moment of the area, and moment of the concentrated area about Ox and Oy, that is

$$y_C A = \int_A y\, dA, \quad x_C A = \int_A x\, dA. \tag{4.13}$$

From Equation (4.13) one can calculate

$$y_C = \frac{1}{A}\int_A y\, dA = \frac{M_x}{A}, \quad x_C = \frac{1}{A}\int_A x\, dA = \frac{M_y}{A}. \tag{4.14}$$

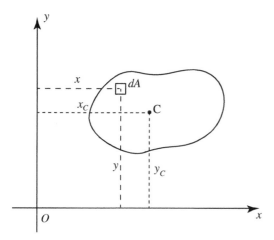

Figure 4.3 Centroidal coordinates and centroid.

If the centroid of the area is positioned at the origin of the frame Oxy, i.e. $O \equiv C$, then Ox and Oy are called *centroidal axes*.

In countless mechanical engineering applications, the area of interest can be made by simple shapes that can be subtracted or added together. The final shape made up from such simple shapes is named a *composite* shape. The centroid of a composite shape can be calculated using

$$x_C = \frac{\sum_k A_k x_{Ck}}{A} \quad \text{and} \quad y_C = \frac{\sum_k A_k y_{C_k}}{A}, \tag{4.15}$$

where x_{C_k} and y_{C_k} are the centroidal coordinates (with the correct signs), and where A_k is the area of each k-th shape.

4.5 The Centre of Gravity

The *centre of gravity* of a system of particles or a solid is a point where the distribution of the weight is equal in all directions. Correspondingly, the *centre of mass* of a solid or a set of particles is a point where the distribution of the mass is equal in all directions. The centre of gravity depends on the gravitational field, if the gravity field is uniform the mass centre and the centre of gravity are the same (identical).

The centre of gravity coordinates can be computed using

$$x_C = \frac{\int_V g \, \rho \, x \, dV}{\int_V g \, \rho \, dV}, \quad y_C = \frac{\int_V g \, \rho \, y \, dV}{\int_V g \, \rho \, dV}, \quad z_C = \frac{\int_V g \, \rho \, z \, dV}{\int_V g \, \rho \, dV} \tag{4.16}$$

where g is the gravitational acceleration, e.g., $g = 9.81 \text{ m s}^{-2}$.

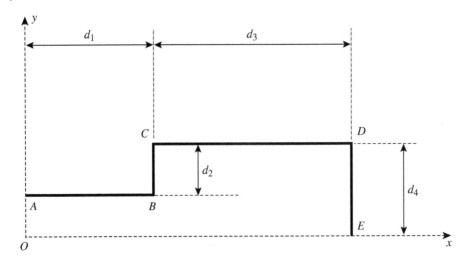

Figure 4.4 Uniform rod *ABCDE*.

4.6 Examples

Example 4.1

Determine the centroid of the uniform rod *ABCDE* shown in Figure 4.4 with respect to the *Oxy* Cartesian frame. The distances d_{AB} from *A* to *B*, d_{BC} from *B* to *C*, d_{CD} from *C* to *D*, and d_{DE} from *D* to *E* are given by: $d_{AB} = d_1$, $d_{BC} = d_2$, $d_{CD} = d_3$, and $d_{DE} = d_4$ respectively. The links *AB* and *CD* are horizontal and the links *BC* and *DE* are vertical, that is, $AB \perp BC$, $BC \perp CD$, $CD \perp DE$, where $\perp$ means *perpendicular to*.

Numerical application: $d_1 = 0.5$ m, $d_2 = 0.2$ m, $d_3 = 0.8$ m, $d_4 = 0.35$ m.

Solution

The coordinates *x* and *y* of the points denoted by *A*, *B*, *C*, *D*, and *E* of the road *ABCDE* can be calculated as

$$\text{Point } A \quad \Leftrightarrow \quad x_A = x_O = 0,$$
$$y_A = d_4 - d_2,$$
$$\text{Point } B \quad \Leftrightarrow \quad x_B = d_1,$$
$$y_B = d_4 - d_2,$$
$$\text{Point } C \quad \Leftrightarrow \quad x_C = d_1,$$
$$y_C = d_4,$$
$$\text{Point } D \quad \Leftrightarrow \quad x_D = d_1 + d_3,$$
$$y_D = d_4,$$
$$\text{Point } E \quad \Leftrightarrow \quad x_E = d_1 + d_3,$$
$$y_B = y_O = 0. \tag{4.17}$$

The coordinates x and y of A, B, C, D, and E are computed and printed in MATLAB with:

```
clear all; clc; close all
syms x0 y0 d1 d2 d3 d4

xA=x0;
yA=d4-d2;
xB=d1;
yB=d4-d2;
xC=d1;
yC=d4;
xD=d1+d3;
yD=d4;
xE=d1+d3;
yE=y0;
fprintf('The coordinates x and y of A, B, C, D and E are \n');
fprintf('xA = %s  \n',xA);
fprintf('yA = %s  \n',yA);
fprintf('xB = %s  \n',xB);
fprintf('yB = %s  \n',yB);
fprintf('xC = %s  \n',xC);
fprintf('yC = %s  \n',yC);
fprintf('xD = %s  \n',xD);
fprintf('yD = %s  \n',yD);
fprintf('xE = %s  \n',xE);
fprintf('yE = %s  \n\n',yE);
```

Considering Equation (4.17) and the free body diagram in Figure 4.4 one can calculate the centroid for each individual link/segment AB, BC, CD and DE by

$$\text{Link } AB \quad \Leftrightarrow \quad x_{C_{AB}} = \frac{x_A + x_B}{2} = \frac{d_1}{2},$$

$$y_{C_{AB}} = \frac{y_A + y_B}{2} = d_4 - d_2,$$

$$\text{Link } BC \quad \Leftrightarrow \quad x_{C_{BC}} = \frac{x_B + x_C}{2} = d_1,$$

$$y_{C_{BC}} = \frac{y_B + y_C}{2} = \frac{2d_4 - d_2}{2},$$

$$\text{Link } CD \quad \Leftrightarrow \quad x_{C_{CD}} = \frac{x_C + x_D}{2} = \frac{2d_1 + d_3}{2},$$

$$y_{C_{CD}} = \frac{y_C + y_D}{2} = d_4,$$

$$\text{Link } DE \quad \Leftrightarrow \quad x_{C_{DE}} = \frac{x_D + x_E}{2} = d_1 + d_3,$$

$$y_{C_{DE}} = \frac{y_D + y_E}{2} = \frac{d_4}{2}. \tag{4.18}$$

Table 4.1 Table for centroidal coordinates.

Link	Length L	x_C coordinate	y_C coordinate	$x_C L$	$y_C L$
AB	d_1	$\dfrac{d_1}{2}$	$d_4 - d_2$	$\dfrac{d_1^2}{2}$	$d_1(d_4 - d_2)$
BC	d_2	d_1	$\dfrac{2d_4 - d_2}{2}$	$d_1 d_2$	$d_2 \dfrac{2d_4 - d_2}{2}$
CD	d_3	$\dfrac{2d_1 + d_3}{2}$	d_4	$d_3 \dfrac{2d_1 + d_3}{2}$	$d_3 d_4$
DE	d_4	$d_1 + d_3$	$\dfrac{d_4}{2}$	$d_4(d_1 + d_3)$	$\dfrac{d_4^2}{2}$

The centroid of the links *AB*, *BC*, *CD* and *DE* are calculated and printed in MATLAB with:

```
xC_AB=(xA+xB)/2;
yC_AB=(yA+yB)/2;
xC_BC=(xB+xC)/2;
yC_BC=(yB+yC)/2;
xC_CD=(xC+xD)/2;
yC_CD=(yC+yD)/2;
xC_DE=(xD+xE)/2;
yC_DE=(yD+yE)/2;

fprintf('The centroidal coordinates of each link AB, BC, CD,and
          DE are \n');
fprintf('xC_AB = %s  \n',xC_AB);
fprintf('yC_AB = %s  \n',yC_AB);
fprintf('xC_BC = %s  \n',xC_BC);
fprintf('yC_BC = %s  \n',yC_BC);
fprintf('xC_CD = %s  \n',xC_CD);
fprintf('yC_CD = %s  \n',yC_CD);
fprintf('xC_DE = %s  \n',xC_DE);
fprintf('yC_DE = %s  \n\n',yC_DE);
```

To calculate the centroid of the road *ABCDE*, Table 4.1 is set.
From Table(4.1) one can calculate

$$\sum L = d_1 + d_2 + d_3 + d_4,$$

$$\sum x_C L = \frac{d_1^2}{2} + d_1 d_2 + d_3 \frac{2d_1 + d_3}{2} + d_4(d_1 + d_3),$$

$$\sum y_C L = d_1(d_4 - d_2) + d_2 \frac{2d_4 - d_2}{2} + d_3 d_4 + \frac{d_4^2}{2}, \tag{4.19}$$

and in MATLAB:

```
sum_L = d1 + d2 + d3 + d4
dAB=d1;
```

```
dBC=d2;
dCD=d3;
dDE=d4;
sum_xC_L = dAB*xC_AB + dBC*xC_BC + dCD*xC_CD + dDE*xC_DE;
sum_yC_L = dAB*yC_AB + dBC*yC_BC + dCD*yC_CD + dDE*yC_DE;
fprintf('The next sums are calculated with \n');
fprintf('sum_L = %s   \n',sum_L);
fprintf('sum_xC_L = %s   \n',sum_xC_L);
fprintf('sum_yC_L = %s   \n\n',sum_yC_L);
```

The centroidal coordinates $X_{C_{ABCDE}}$ and $Y_{C_{ABCDE}}$ of the road $ABCDE$ are calculated from Equation (4.19) using

$$X_{C_{ABCDE}} = \frac{\sum x_C L}{\sum L} = \frac{\dfrac{d_1^2}{2} + d_1 d_2 + d_3 \dfrac{2d_1 + d_3}{2} + d_4(d_1 + d_3)}{d_1 + d_2 + d_3 + d_4},$$

$$Y_{C_{ABCDE}} = \frac{\sum y_C L}{\sum L} = \frac{\sum d_1(d_4 - d_2) + d_2 \dfrac{2d_4 - d_2}{2} + d_3 d_4 + \dfrac{d_4^2}{2}}{d_1 + d_2 + d_3 + d_4}. \qquad (4.20)$$

The centroidal coordinates are calculated and printed in MATLAB using:

```
X_C_ABCDE=sum_xC_L/sum_L;
Y_C_ABCDE=sum_yC_L/sum_L;
fprintf('The centroidal coordinates of the link ABCDE are cal-
culated with \n');
fprintf('X_C_ABCDE = %s   \n',X_C_ABCDE);
fprintf('Y_C_ABCDE = %s   \n\n',Y_C_ABCDE);
```

In MATLAB, the following numerical data is considered:

```
% numerical results
lists = {x0,y0,d1,d2,d3,d4};
listn = {0,0,0.5,0.2,0.8,0.35};
```

The numerical results are calculated and printed in MATLAB using:

```
X_C_ABCDE = eval(subs(X_C_ABCDE,lists,listn));
Y_C_ABCDE = eval(subs(Y_C_ABCDE,lists,listn));
fprintf('The numerical values of the centroidal coordinates are  \n');
fprintf('X_C_ABCDE = %f [m] \n',X_C_ABCDE);
fprintf('Y_C_ABCDE = %f [m]\n',Y_C_ABCDE);
```

The MATLAB numerical results for centroidal coordinates are:

```
The numerical values of the centroidal coordinates are
X_C_ABCDE = 0.756757 [m]
Y_C_ABCDE = 0.252027 [m]
```

Example 4.2
Determine the centroid of the rod $ABCD$ (Figure 4.5) with respect to the Oxy Cartesian frame. The distances d_{AB} from A to B, d_{BC} from B to C, and d_{CD} from C to D, are given by

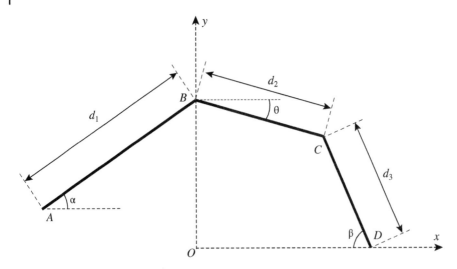

Figure 4.5 Uniform rod ABCD.

$d_{AB} = d_1$, $d_{BC} = d_2$, and $d_{CD} = d_3$. The link AB makes an angle α with the horizontal, the link BC makes and angle θ with the horizontal and the link CD makes and angle β with the horizontal as shown in Figure 4.5.

Numerical application: $d_1 = 0.7$ m, $d_2 = 0.4$ m, $d_3 = 0.4$ m, $\alpha = \dfrac{\pi}{4}$, $\theta = \dfrac{\pi}{6}$ and $\beta = \dfrac{\pi}{3}$.

Solution

From Figure 4.6 one can deduce

$$\sin \alpha = \frac{d_{BG}}{d_{AB}} \quad \Leftrightarrow \quad d_{BG} = d_{AB} \sin \alpha = d_1 \sin \alpha,$$

$$\cos \alpha = \frac{d_{AG}}{d_{AB}} \quad \Leftrightarrow \quad d_{AG} = d_{AB} \cos \alpha = d_1 \cos \alpha,$$

$$\sin \beta = \frac{d_{CE}}{d_{CD}} \quad \Leftrightarrow \quad d_{CE} = d_{CD} \sin \beta = d_3 \sin \beta,$$

$$\cos \beta = \frac{d_{DE}}{d_{CD}} \quad \Leftrightarrow \quad d_{DE} = d_{CD} \cos \beta = d_3 \cos \beta,$$

$$\sin \theta = \frac{d_{BF}}{d_{BC}} \quad \Leftrightarrow \quad d_{BF} = d_{BC} \sin \theta = d_2 \sin \theta,$$

$$\cos \theta = \frac{d_{CF}}{d_{BC}} \quad \Leftrightarrow \quad d_{CF} = d_{BC} \cos \theta = d_2 \cos \theta, \tag{4.21}$$

where $d_{AB} = d_1$, $d_{BC} = d_2$, $d_{CD} = d_3$, d_{BG} is the distance from B to G, d_{AG} is the distance from A to G, d_{CE} is the distance from C to E, d_{DE} is the distance from D to E, d_{BF} is the distance from B to F, and d_{CF} is the distance from C to F.

The lengths d_{BG}, d_{AG}, d_{CE}, d_{DE}, d_{BF} and d_{CF} are computed in MATLAB with:

```
clear all; clc; close all
syms x0 y0 d1 d2 d3 alpha beta theta
```

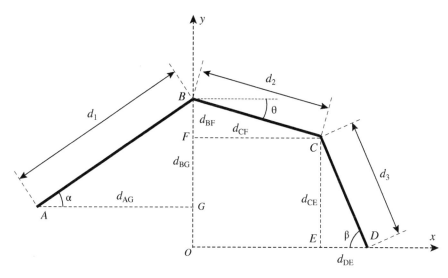

Figure 4.6 Dimensions of the uniform rod *ABCD*.

```
dAB=d1; dBC=d2; dCD=d3;
dBG=dAB*sin(alpha);
dAG=dAB*cos(alpha);
dCE=dCD*sin(beta);
dDE=dCD*cos(beta);
dBF=dBC*sin(theta);
dCF=dBC*cos(theta);
```

The coordinates x and y of A, B, C and D of the road $ABCD$ can be calculated as

$$\text{Point A} \quad \Leftrightarrow \quad x_A = -d_{AG} = -d_1 \cos\alpha,$$
$$y_A = d_{OG} = d_{OF} + d_{FB} - d_{GB} = d_{CE} + d_{BF} - d_{BG}$$
$$= d_3 \sin\beta + d_2 \sin\theta - d_1 \sin\alpha,$$
$$\text{Point B} \quad \Leftrightarrow \quad x_B = x_O = 0,$$
$$y_B = d_{OF} + d_{FB} = d_{CE} + d_{BF} = d_3 \sin\beta + d_2 \sin\theta,$$
$$\text{Point C} \quad \Leftrightarrow \quad x_C = d_{OE} = d_{CF} = d_2 \cos\theta,$$
$$y_C = d_{OF} = d_{CE} = d_3 \sin\beta,$$
$$\text{Point D} \quad \Leftrightarrow \quad x_D = d_{OE} + d_{ED} = d_{CF} + d_{DE} = d_2 \cos\theta + d_3 \cos\beta,$$
$$y_D = y_O = 0, \tag{4.22}$$

and in MATLAB with:

```
xO=0; yO=0;
dGB=dBG; dFB=dBF; dED=dDE; dOF=dCE;
dOG=dOF+dFB-dGB;
dOE=dCF;
xA=-dAG;
```

```
yA=dOG;
xB=xO;
yB=dOF+dFB;
xC=dOE;
yC=dOF;
xD=dOE+dED;
yD=yO;
fprintf('The coordinates x and y of A, B, C and D are \n');
fprintf('xA = %s  \n',xA);
fprintf('yA = %s  \n',yA);
fprintf('xB = %f  \n',xB);
fprintf('yB = %s  \n',yB);
fprintf('xC = %s  \n',xC);
fprintf('yC = %s  \n',yC);
fprintf('xD = %s  \n',xD);
fprintf('yD = %f  \n\n',yD);
```

Considering Equation (4.22) and Figure 4.6 one can calculate the centroid for each individual link (segment) *AB*, *BC* and *CD* by

$$\text{Link AB} \Leftrightarrow x_{C_{AB}} = \frac{x_A + x_B}{2} = \frac{-d_1 \cos\alpha}{2},$$

$$y_{C_{AB}} = \frac{y_A + y_B}{2} = \frac{2d_3 \sin\beta + 2d_2 \sin\theta - d_1 \sin\alpha}{2},$$

$$\text{Link BC} \Leftrightarrow x_{C_{BC}} = \frac{x_B + x_C}{2} = \frac{d_2 \cos\theta}{2},$$

$$y_{C_{BC}} = \frac{y_B + y_C}{2} = \frac{2d_3 \sin\beta + d_2 \sin\theta}{2},$$

$$\text{Link CD} \Leftrightarrow x_{C_{CD}} = \frac{x_C + x_D}{2} = \frac{2d_2 \cos\theta + d_3 \cos\beta}{2},$$

$$y_{C_{CD}} = \frac{y_C + y_D}{2} = \frac{d_3 \sin\beta}{2}. \tag{4.23}$$

The centroidal coordinates of the links *AB*, *CD* and *BC* are calculated and printed in MATLAB with:

```
xC_AB=(xB+xA)/2;
yC_AB=(yB+yA)/2;
xC_BC=(xC+xB)/2;
yC_BC=(yC+yB)/2;
xC_CD=(xD+xC)/2;
yC_CD=(yD+yC)/2;
fprintf('The centroidal coordinates of each link AB, BC, CD,and DE are \n');
fprintf('xC_AB = %s  \n',xC_AB);
fprintf('yC_AB = %s  \n',yC_AB);
fprintf('xC_BC = %s  \n',xC_BC);
fprintf('yC_BC = %s  \n',yC_BC);
fprintf('xC_CD = %s  \n',xC_CD);
fprintf('yC_CD = %s  \n',yC_CD);
```

To calculate the centroid of the road *ABCD* the Table 4.2 is set where *L* is the length of each corresponding link.

Table 4.2 Table for centroidal coordinates.

Link	L	x_C coordinate	y_C coordinate	$x_C L$	$y_C L$
AB	d_1	$\dfrac{-d_1 \cos\alpha}{2}$	$\dfrac{2d_3 \sin\beta + 2d_2 \sin\theta - d_1 \sin\alpha}{2}$	$d_1 \dfrac{-d_1 \cos\alpha}{2}$	$d_1 \dfrac{2d_3 \sin\beta + 2d_2 \sin\theta - d_1 \sin\alpha}{2}$
BC	d_2	$\dfrac{d_2 \cos\theta}{2}$	$\dfrac{2d_3 \sin\beta + d_2 \sin\theta}{2}$	$d_2 \dfrac{d_2 \cos\theta}{2}$	$d_2 \dfrac{2d_3 \sin\beta + d_2 \sin\theta}{2}$
CD	d_3	$\dfrac{2d_2 \cos\theta + d_3 \cos\beta}{2}$	$\dfrac{d_3 \sin\beta}{2}$	$d_3 \dfrac{2d_2 \cos\theta + d_3 \cos\beta}{2}$	$d_3 \dfrac{d_3 \sin\beta}{2}$

One can calculate

$$\sum L = d_1 + d_2 + d_3,$$

$$\sum x_C L = \frac{-d_1^2 \cos \alpha}{2} + \frac{d_2^2 \cos \theta}{2} + d_3 \frac{2d_2 \cos \theta + d_3 \cos \beta}{2},$$

$$= \frac{-d_1^2 \cos \alpha}{2} + \frac{d_2^2 \cos \theta}{2} + \frac{d_3^2 \cos \beta}{2} + d_3 d_2 \cos \theta,$$

$$\sum y_C L = d_1 \frac{2d_3 \sin \beta + 2d_2 \sin \theta - d_1 \sin \alpha}{2} + d_2 \frac{2d_3 \sin \beta + d_2 \sin \theta}{2} + \frac{d_3^2 \sin \beta}{2},$$

$$= d_1 d_3 \sin \beta + d_1 d_2 \sin \theta + d_2 d_3 \sin \beta - \frac{d_1^2 \sin \alpha}{2} + \frac{d_2^2 \sin \theta}{2} + \frac{d_3^2 \sin \beta}{2}$$

$$(4.24)$$

and in MATLAB:

```
sum_L = d1 + d2 + d3
sum_xC_L = dAB*xC_AB + dBC*xC_BC + dCD*xC_CD;
sum_yC_L = dAB*yC_AB + dBC*yC_BC + dCD*yC_CD;
fprintf('The next sums are calculated with \n');
fprintf('sum_L = %s   \n',sum_L);
fprintf('sum_xC_L = %s   \n',sum_xC_L);
fprintf('sum_yC_L = %s   \n\n',sum_yC_L);
```

The centroidal coordinates $X_{C_{ABCD}}$ and $Y_{C_{ABCD}}$ of the rod *ABCD* are calculated from Equation (4.24) using

$$X_{C_{ABCD}} = \frac{\sum x_C L}{\sum L} = \frac{\dfrac{-d_1^2 \cos \alpha}{2} + \dfrac{d_2^2 \cos \theta}{2} + \dfrac{d_3^2 \cos \beta}{2} + d_3 d_2 \cos \theta}{d_1 + d_2 + d_3},$$

$$Y_{C_{ABCD}} = \frac{\sum y_C L}{\sum L} = \frac{d_1 d_3 \sin \beta + d_1 d_2 \sin \theta + d_2 d_3 \sin \beta - \dfrac{d_1^2 \sin \alpha}{2} + \dfrac{d_2^2 \sin \theta}{2} + \dfrac{d_3^2 \sin \beta}{2}}{d_1 + d_2 + d_3},$$

$$(4.25)$$

```
X_C_ABCD=sum_xC_L/sum_L;
Y_C_ABCD=sum_yC_L/sum_L;
fprintf('The centroidal coordinates of the link ABCD ');
fprintf('with respect to the Oxy datum are calculated with \n');
fprintf('X_C_ABCD = %s   \n',X_C_ABCD);
fprintf('Y_C_ABCD = %s   \n\n',Y_C_ABCD);
```

The following numerical data is considered in MATLAB:

```
% numerical results
lists = {x0,y0,d1,d2,d3,alpha, beta, theta};
listn = {0,0,0.7,0.4,0.4,pi/4,pi/3,pi/6};
```

The numerical results of the centroidal coordinates of the link *ABCD* with respect to the *Oxy* Cartesian frame are calculated and printed in MATLAB using:

```
X_C_ABCD = eval(subs(X_C_ABCD,lists,listn));
Y_C_ABCD = eval(subs(Y_C_ABCD,lists,listn));
fprintf('The numerical values of the centroidal coordinates ');
fprintf('with respect to the Oxy datum are \n');
fprintf('X_C_ABCD = %s [m] \n',X_C_ABCD);
fprintf('Y_C_ABCD = %s [m] \n',Y_C_ABCD);
```

The MATLAB numerical results of the centroidal coordinates are:

```
X_C_ABCD = 4.973662e-02 [m]
Y_C_ABCD = 3.047280e-01 [m]
```

Example 4.3

Determine the location of the mass centre of the homogeneous wire *AC* (Figure 4.7) with respect to the *Oxy* cartesian reference frame. The wire defined by the equation $x = a\left(1 - \left(\frac{y}{c}\right)^2\right)$ has ends located at $A(0, a)$ and $C(b, 0)$.

Numerical application: $a = 2$ m and $c = 3$.

Solution

The centroidal coordinates for the wire *AC* defined by the equation $y = a\left(1 - \left(\frac{x}{c}\right)^3\right)$ can be calculated with

$$x_{C_{AC}} = \int \frac{x_{el}\mathrm{d}L}{\mathrm{d}L}$$

$$y_{C_{AC}} = \int \frac{y_{el}\mathrm{d}L}{\mathrm{d}L} \qquad (4.26)$$

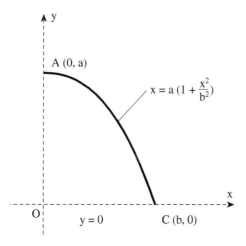

Figure 4.7 Homogeneous wire *AC*.

where dL is the differential length element evaluated using

$$dL = \sqrt{(dy)^2 + (dx)^2},$$

$$= \left(\sqrt{\left(\frac{dx}{dy}\right)^2 + 1}\ dy\right). \tag{4.27}$$

To evaluate the integrals in Equation (4.27) the next substitution is considered

$$x_{el} = x = a\left(1 - \frac{y^2}{c^2}\right),$$

$$dx = \left(-\frac{2a}{c^2}y\right)dy,$$

$$dL = \left(\sqrt{1 + \left(-\frac{2a}{c^2}y\right)^2}\ dy\right) \tag{4.28}$$

The equivalent substitution is considered in MAtlab:

```
clear all; clc; close all
syms a b c x y

dOB=b; dOC=c;dAO=a;

x_elAC=a*(1-y^2/c^2);
dx__dyAC = (-a^2*y/c^2);
dL_AC = sqrt(1+ dx__dyAC^2);
```

Therefore the integral $\int dL$ can be calculated by

$$\int dL = \int\left(\sqrt{1 + \left(-\frac{2a}{c^2}y\right)^2}\right)dy$$

$$= \int\left(\sqrt{1 + \frac{4a^2}{c^4}y^2}\right)dy$$

$$= \frac{1}{2}y\sqrt{1 + \frac{4a^2y^2}{c^4}} + \frac{c^2}{4a}\ln\left(\frac{2ay}{c^2} + \sqrt{1 + \frac{4a^2y^2}{c^4}}\right) + C \tag{4.29}$$

and in MATLAB with:

```
IdLAC=int(dL_AC,0,a);
fprintf('The integral IdlAC is calculated as\n');
fprintf('IdLAC = %s  \n\n',IdLAC);
```

One can calculate

$$\int x_{el}dL = \int a\left(1 - \frac{y^2}{c^2}\right)\left(\sqrt{1 + \left(-\frac{2a}{c^2}y\right)^2}\right)dy$$

$$= a\int\left[\sqrt{1 + \frac{4a^2y^2}{c^4}}\ dy - \frac{a}{c^2}\int y^2\left(\sqrt{1 + \frac{4a^2y^2}{c^4}}\right)dy\right]$$

$$= \frac{ay}{2}\sqrt{1 + \frac{4a^2y^2}{c^4}} + \frac{c^2}{4}\ln\left(\frac{2ay}{c^2} + \sqrt{1 + \frac{4a^2y^2}{c^4}}\right)$$

$$-\frac{c^2 y}{32a}\sqrt{1+\frac{4a^2y^2}{c^4}} - \frac{ay^3}{4c^2}\sqrt{1+\frac{4a^2y^2}{c^4}} + \frac{c^4}{64a^2}\ln\left(\frac{2ay}{c^2} + \sqrt{1+\frac{4a^2y^2}{c^4}}\right)$$

$$(4.30)$$

and

$$\int y_{el}dL = \int y\left(\sqrt{1+\left(-\frac{2a}{c^2}y\right)^2}\right)dy$$

$$= \int y\left(\sqrt{1+\frac{4a^2}{c^4}y^2}\right)dy$$

$$= \frac{c^4}{12a^2}\sqrt{\left(1+\frac{4a^2y^2}{c^4}\right)^3} + C. \qquad (4.31)$$

In MATLAB, the integrals $\int x_{el}dL$ and $\int y_{el}dL$ are calculated with:

```
IydLAC=int(x_elAC*dL_AC,0,a);
fprintf('The integral Iydl is calculated as\n');
fprintf('IydLAC = %s   \n\n',IydLAC);


IxdLAC=int(y*dL_AC,0,a);
fprintf('The integral IxdlAC is calculated as\n');
fprintf('IxdLAC = %s   \n\n',IxdLAC);
```

When $a = 2$ m, $b = 4$ $c = 3$ one can calculate

$$\int dL = \int_0^2\left(\sqrt{1+\frac{16}{81}y^2}\right)dy = 2.238$$

$$\int x_{el}dL = \int_0^2 2\left(1-\frac{y^2}{9}\right)\left(\sqrt{1+\frac{16}{81}y^2}\right)dy = 4.476 - 0.717 = 3.759$$

$$\int y_{el}dL = \int_0^2 y\left(\sqrt{1+\frac{16}{81}y^2}\right)dy = 2.235. \qquad (4.32)$$

In MATLAB, the next numerical data is considered:

```
% numerical results
lists = {a,b,c};
listn = {2,4,3};
```

The numerical results of the centroidal coordinates of the homogeneous wire and calculated and printed in MATLAB using:

```
IdLAC=eval(subs(int(dL_AC,0,a),lists,listn));
IydLAC=eval(subs(int(x_elAC*dL_AC,0,a),lists,listn));
IxdLAC=eval(subs(int(y*dL_AC,0,a),lists,listn));
fprintf('The integral IxdlAC is evaluated to \n');
fprintf('IxdLAC = %s   \n\n',IxdLAC);
fprintf('The integral IydlAC is evaluated to \n');
fprintf('IydLAC = %s   \n\n',IydLAC);
```

```
fprintf('The integral Iydl is evaluated to \n');
fprintf('IydLAC = %s  \n\n',IydLAC);
```

The centroidal coordinates for the curved part *AC* are calculated using

$$x_{C_{AC}} = \int \frac{x_{el} dL}{dL} = \frac{3.759}{2.238} = 1.679$$

$$y_{C_{AC}} = \int \frac{y_{el} dL}{dL} = \frac{2.235}{2.238} = 0.998 \tag{4.33}$$

and in MATLAB:

```
x_cAC=IxdLAC/IdLAC;
y_cAC=IydLAC/IdLAC;
fprintf('The centroidal coordinates for the wire AC are given by \n');
fprintf('x_cAC = %s  \n',x_cAC);
fprintf('y_cAC = %s  \n\n',y_cAC);
```

The MATLAB numerical results of the centroidal coordinates are:

```
The centroidal coordinates for the wire AC are given by
x_cAC = 1.051651e+00
y_cAC = 1.679428e+00
```

Example 4.4
Determine the location of the mass centre for a homogeneous planar plate *ANBEFG* shown in Figure 4.8. The distance d_{AG} from *A* to *G* is equal to the distance d_{BE} from *B* to *E*, that is, the rectangle 1 in Figure 4.8 has the dimensions $d_{AG} = d_{BE} = d_1$ and $d_{AB} = d_{GE} = 2d_2$. The radius of the semicircle 2 defined by *ANB* is d_2, that is $d_{PA} = d_{PB} = d_{PN} = d_2$. The circular sector 4 (quarter of a circle) defined by the points *LEF* and the circular sector 6 (quarter

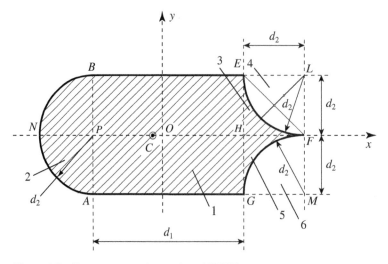

Figure 4.8 Homogeneous planar plate *ANBEFG*.

of a circle) defined by the points MFG have the same radius d_2, that is, $d_{EL} = d_{HF} = d_{GM} = d_{GH} = d_{HE} = d_{MF} = d_{FL} = d_2$. Numerical application: $d_1 = 0.7$ m, $d_2 = 0.4$ m.

Solution

One can decompose the plate in four elements namely the semicircular area ANB (sector 2), the rectangular area $ABEG$ (sector 1), the square $HELF$ (sector 3) with the corresponding quarter-circular area LFE (sector 4) to be subtracted, and the square $GHFM$ (sector 5) with the corresponding quarter-circular area MGF (sector 6) to be subtracted.

For the planar plate shown in Figure 4.8, the Ox represents an axis of symmetry and the origin $O = O(0, 0)$ of the reference frame is located at the centre of the rectangular area $ABEG$ (sector 1). Due to symmetry, the centroid (mass centre) y_i for the corresponding areas A_1 (sector 1), corresponding area A_2 (sector 2), and for the area $GHEF$ (combined sectors 4, 5, 6 and 7), are all located on the Ox axis, therefore the centroidal coordinate y_C of the homogeneous planar plate is also located on the Ox axis.

Considering the decomposition method one can calculate the centroidal coordinate (mass centre) x_i for each corresponding area A_i, and the first moment M_{x_i} for all four elements. The centroidal coordinates x_i, areas A_i and products $x_i A_i$ are given in Table 4.3. The centroidal coordinates x_i in Table 4.3 are computed in MATLAB with:

```
clear all; clc; close all
syms d1 d2

dAG=d1; dBE=d2; dAB=2*d2;
dNP=d2; dPA=d2; dPB=d2;
dHE=d2; dLE=d2; dFL=d2;
dHG=d2; dMG=d2; dFM=d2;
dHF=d2; dOP=d1/2; dOH=d1/2;
dxC_ANB=4*dNP/(3*pi);
dxC_HFLE=dHF/2; dxC_HFMG=dHF/2;
```

Table 4.3 Table for x_i centroidal coordinates.

Element number	x_i	A_i	$M_{y_i} = x_i A_i$
Sector 2 – semicircular area	$-\dfrac{d_1}{2} - \dfrac{4d_2}{3\pi}$	$\dfrac{\pi d_2^2}{2}$	$\dfrac{\pi d_2^2}{2}\left(-\dfrac{d_1}{2} - \dfrac{4d_2}{3\pi}\right)$
Sector 1 – rectangular area	0	$d_1 2d_2$	0
Sector 3 – square area	$\dfrac{d_1}{2} + \dfrac{d_2}{2}$	d_2^2	$d_2^2\left(\dfrac{d_1}{2} + \dfrac{d_2}{2}\right)$
Sector 4 – quarter-circular area	$\dfrac{d_1}{2} + d_2 - \dfrac{4d_2}{3\pi}$	$-\dfrac{\pi d_2^2}{4}$	$-\dfrac{\pi d_2^2}{4}\left(\dfrac{d_1}{2} + d_2 - \dfrac{4d_2}{3\pi}\right)$
Sector 5 – square area	$\dfrac{d_1}{2} + \dfrac{d_2}{2}$	d_2^2	$d_2^2\left(\dfrac{d_1}{2} + \dfrac{d_2}{2}\right)$
Sector 6 – quarter-circular area	$\dfrac{d_1}{2} + d_2 - \dfrac{4d_2}{3\pi}$	$-\dfrac{\pi d_2^2}{4}$	$-\dfrac{\pi d_2^2}{4}\left(\dfrac{d_1}{2} + d_2 - \dfrac{4d_2}{3\pi}\right)$
Σ	$-$	ΣA_i	$\Sigma M_{y_i} = \Sigma M_{y_i}$

```
dxC_LFE=4*dLE/(3*pi);
dxC_MFG=4*dMG/(3*pi);

x1=0;
x2=-dOP-dxC_ANB;
x3=dOH+dxC_HFLE;
x4=dOH+dHF-dxC_LFE;
x5=dOH+dxC_HFMG;
x6=dOH+dHF-dxC_MFG;
fprintf('The x_i coordinates of sectors 1 to 6 are \n');
fprintf('x1 = %f  \n',x1);
fprintf('x2 = %s  \n',x2);
fprintf('x3 = %s  \n',x3);
fprintf('x4 = %s  \n',x4);
fprintf('x5 = %s  \n',x5);
fprintf('x6 = %s  \n\n',x6);
```

The area A_i for each sector $i = \overline{1,6}$ shown in Figure 4.8 (and Table 4.3) are calculated and printed in MATLAB with:

```
A1=dAG*dAB;
A2=pi*dNP^2/2;
A3=dHF*dHE;
A4=-pi*dNP^2/4;
A5=dHF*dHG;
A6=-pi*dNP^2/4;
fprintf('The area of the sectors 1 to 6 are are \n');
fprintf('A1 = %s  \n',A1);
fprintf('A2 = %s  \n',A2);
fprintf('A3 = %s  \n',A3);
fprintf('A4 = %s  \n',A4);
fprintf('A5 = %s  \n',A5);
fprintf('A6 = %s  \n\n',A6);
```

In MATLAB, the product $x_i A_i$ for each sector $i = \overline{1,6}$ shown in Figure 4.8 is calculated and printed with:

```
M1=A1*x1;
M2=A2*x2;
M3=A3*x3;
M4=A4*x4;
M5=A5*x5;
M6=A6*x6;
fprintf('The area of the sectors 1 to 6 are are \n');
fprintf('M1 = %s  \n',M1);
fprintf('M2 = %s  \n',M2);
fprintf('M3 = %s  \n',M3);
```

```
fprintf ('M4 = %s   \n',M4);
fprintf ('M5 = %s   \n',M5);
fprintf ('M6 = %s   \n\n',M6);
```

One can calculate the sums

$$\sum A_i = \frac{\pi d_2^2}{2} + d_1 2d_2 + d_2^2 - \frac{\pi d_2^2}{4} + d_2^2 - \frac{\pi d_2^2}{4},$$

$$= 2d_1 d_2 + 2d_2^2,$$

$$= 2d_2(d_1 + d_2),$$

$$\sum x_i A_i = \frac{\pi d_2^2}{2}\left(-\frac{d_1}{2} - \frac{4d_2}{3\pi}\right) + d_2^2\left(\frac{d_1}{2} + \frac{d_2}{2}\right) - \frac{\pi d_2^2}{4}\left(\frac{d_1}{2} + d_2 - \frac{4d_2}{3\pi}\right),$$

$$+ d_2^2\left(\frac{d_1}{2} + \frac{d_2}{2}\right) - \frac{\pi d_2^2}{4}\left(\frac{d_1}{2} + d_2 - \frac{4d_2}{3\pi}\right),$$

$$= \frac{\pi d_2^2}{2}\left(-\frac{d_1}{2} - \frac{4d_2}{3\pi}\right) + d_2^2(d_1 + d_2) - \frac{\pi d_2^2}{2}\left(\frac{d_1}{2} + d_2 - \frac{4d_2}{3\pi}\right),$$

$$= \frac{\pi d_2^2}{2}\left(-\frac{d_1}{2} - \frac{4d_2}{3\pi} - \frac{d_1}{2} - d_2 + \frac{4d_2}{3\pi}\right) + d_2^2(d_1 + d_2),$$

$$= -\frac{\pi d_2^2}{2}(d_1 + d_2) + d_2^2(d_1 + d_2),$$

$$= d_2^2\left(1 - \frac{\pi}{2}\right)(d_1 + d_2), \tag{4.34}$$

and in MATLAB:

```
sum_Ai = A1+A2+A3+A4+A5+A6;
sum_Mi = M1+M2+M3+M4+M5+M6;
fprintf ('The next sums are calculated with \n');
fprintf ('sum_Ai = %s   \n',simplify(sum_Ai));
fprintf ('sum_Mi = %s   \n\n',simplify(sum_Mi));
```

The X_C and Y_C centroidal coordinates of the homogeneous planar plate *ANBEFG* are

$$X_C = \frac{\sum x_i A_i}{\sum A_i} = \frac{d_2^2\left(1 - \frac{\pi}{2}\right)(d_1 + d_2)}{2d_2(d_1 + d_2)} = d_2\left(\frac{1}{2} - \frac{\pi}{4}\right),$$

$$Y_C = 0.$$

In MATLAB, the centroidal coordinates X_C and Y_C of the homogeneous planar plate are computed with:

```
X_C=sum_Mi/sum_Ai;
Y_C=0;
fprintf ('The X_C centroidal coordinate ');
fprintf ('of the plate is \n');
fprintf ('X_C = %s   \n',simplify(X_C));
fprintf ('Due to symmetry ');
fprintf ('the Y_C centroidal coordinate ');
fprintf ('of the plate is \n');
```

```
fprintf('Y_C = %f   \n\n',Y_C);
```

The numerical data considered in MATLAB are:

```
% numerical results
lists = {d1,d2,};
listn = {0.7,0.4};
```

The numerical results of the centroidal coordinates of the plate are computed and printed in MATLAB with:

```
X_C = eval(subs(X_C,lists,listn));
Y_C = 0;
fprintf('The centroidal coordinates of the plate are');
fprintf('X_C = %s [m] \n',X_C);
fprintf('Y_C = %f [m]\n',Y_C);
```

The MATLAB numerical results of the centroidal coordinates of the plate are:

```
The centroidal coordinates of the plate are
X_C = -1.141593e-01 [m]
Y_C = 0.000000 [m]
```

Example 4.5
The homogeneous plate *AFBMGDEN* shown in Figure 4.9 has a hole of radius d_1 such that $d_{C_1C_3} = d_1$. The centre of the hole is located at C_3 and C1 is the centre (centroid) of the square *ABGE*. The plate has the dimensions $d_{AN} = d_{NE} = d_{BM} = d_{MG} = d_{AF} = d_{FB} = d_{FC_1} = d_2$. The radius of the semicircle defined by *EHGD* is d_2, that is $d_{HG} = d_{HE} = d_{HD} = d_2$.

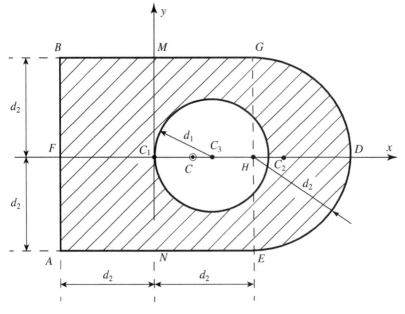

Figure 4.9 Homogeneous planar plate *AFBMGDEN* with hole.

(a) Calculate the centroidal position of the homogeneous plate delimitated by the hatched area (Figure 4.9) when d_1 and d_2 are given. Numerical application: $d_1 = 0.3$ m, $d_2 = 0.5$ m.

(b) Calculate the value of the radius d_1 such as the plate centre of the mass position C is located at the centre of the hole C_3.

Numerical application: $d_2 = 0.5$ m.

Solution

(a) Since the x axis (Figure 4.9) is a symmetry axis, the centroidal coordinate y_C of the homogeneous planar plate is located on the x axis, that is, $y_C = 0$.

Consider $C_1 = O(0,\ 0)$ the origin of the Oxy reference frame. The centroidal coordinates x_i, areas A_i and products $x_i A_i$ for sector 1 (rectangular area $ABGE$), sector 2 (semicircular area $EHGDE$) and sector 3 (removed circular area of radius d_1 centred at C_3) are calculated in Table 4.4.

The x_i centroidal coordinates for each sector 1, 2 and 3, areas A_i and products $x_i A_i$ in the Table 4.4 are computed in MATLAB with:

```
clear all; clc; close all
syms d1 d2 dHD

dAN=d2; dNE=d2; dAF=d2;
dFB=d2; dHD= d2;
dC2E=d2; dC1H=d2;
dC1C3=d1; dHC2=(4*dHD)/(3*pi);

x1=0;
x2=dC1H+dHC2;
x3=dC1C3;
fprintf('The x_i coordinates of sectors 1 to 6 are \n');
fprintf('x1 = %f   \n',x1);
fprintf('x2 = %s   \n',x2);
fprintf('x3 = %s   \n\n',x3);
```

Table 4.4   Table for x_i centroidal coordinates.

Element number	x_i	A_i	$M_{y_i} = x_i A_i$
Sector 1 – rectangular area	0	$4d_2^2$	0
Sector 2 – semicircular area	$d_2 + \dfrac{4d_2}{3\pi}$	$\dfrac{\pi d_2^2}{2}$	$\left(d_2 + \dfrac{4d_2}{3\pi}\right)\dfrac{\pi d_2^2}{2}$
Sector 3 – circular area	d_1	$-\pi d_1^2$	$-d_1 \pi d_1^2$
Σ	–	ΣA_i	$\Sigma M_{y_i} = \Sigma M_{y_i}$

```
A1=(dAN+dNE)*(dAF+dFB);
A2=pi*dHD^2/2;
A3=-pi*dC1C3^2;
fprintf('The area of the sectors 1 to 6 are are \n');
fprintf('A1 = %s   \n',A1);
fprintf('A2 = %s   \n',A2);
fprintf('A3 = %s   \n\n',A3);
```

```
M1=A1*x1;
M2=A2*x2;
M3=A3*x3;
fprintf('The area of the sectors 1 to 6 are are \n');
fprintf('M1 = %s   \n',M1);
fprintf('M2 = %s   \n',M2);
fprintf('M3 = %s   \n\n',M3);
```

One can calculate

$$\sum A_i = 4d_2^2 + \frac{\pi d_2^2}{2} - \pi d_1^2,$$
$$= \left(4 + \frac{\pi}{2}\right) d_2^2 - \pi d_1^2,$$
$$\sum x_i A_i = \left(d_2 + \frac{4d_2}{3\pi}\right) \frac{\pi d_2^2}{2} - d_1 \pi d_1^2$$
$$= \left(1 + \frac{4}{3\pi}\right) \frac{\pi d_2^3}{2} - \pi d_1^3$$
$$= \left(\frac{\pi}{2} + \frac{2}{3}\right) d_2^3 - \pi d_1^3 \qquad (4.35)$$

and in MATLAB:

```
sum_Ai = A1+A2+A3;
sum_Mi = M1+M2+M3;
fprintf('The next sums are calculated with \n');
fprintf('sum_Ai = %s   \n',simplify(sum_Ai));
fprintf('sum_Mi = %s   \n\n',simplify(sum_Mi));
```

The X_C and Y_C centroidal coordinates of the homogeneous planar plate *AFBMGDEN* with a hole of radius d_1 (such that $d_{C_1 C_3} = d_1$) are

$$X_C = \frac{\left(\frac{\pi}{2} + \frac{2}{3}\right) d_2^3 - \pi d_1^3}{\left(\frac{\pi}{2} + 4\right) d_2^2 - \pi d_1^2}, \qquad (4.36)$$

$$Y_C = 0. \qquad (4.37)$$

where $Y_C = 0$ due to the symmetry of the part about the $C_1 x$ axis (Figure 4.9).

The X_C and Y_C centroidal coordinates of the homogeneous planar plate are calculated in MATLAB with:

```
X_C=sum_Mi/sum_Ai;
Y_C=0;
fprintf('The X_C centroidal coordinate ');
fprintf('of the plate is \n');
fprintf('X_C = %s  \n',X_C);
fprintf('Due to symmetry ');
fprintf('the Y_C centroidal coordinate ');
fprintf('of the plate is \n');
fprintf('Y_C = %f  \n\n',Y_C);
```

(b) If the centroid of the plate is located at the centre of the hole C_3, one can write

$$X_C = x_{C_3}.\tag{4.38}$$

From Equations (4.36) and (4.38) one can deduce

$$\frac{\left(\frac{\pi}{2}+\frac{2}{3}\right)d_2^3 - \pi d_1^3}{\left(\frac{\pi}{2}+4\right)d_2^2 - \pi d_1^2} = d_1,$$

$$\Leftrightarrow \quad \left(\frac{\pi}{2}+\frac{2}{3}\right)d_2^3 - \pi d_1^3 = \left(\left(\frac{\pi}{2}+4\right)d_2^2 - \pi d_1^2\right)d_1,$$

$$\Leftrightarrow \quad \left(\frac{\pi}{2}+\frac{2}{3}\right)d_2^3 = \left(\frac{\pi}{2}+4\right)d_1 d_2^2,$$

$$\Leftrightarrow \quad d_1 = \frac{\frac{\pi}{2}+\frac{2}{3}}{\frac{\pi}{2}+4}d_2.\tag{4.39}$$

In MATLAB the value of the radius d_1 (when the mass centre C of the plate and the centre of the hole C_3 are the same) is calculated with:

```
d1new=solve(X_C-x3,d1);
fprintf('The centroid X_C is the same as x3 when\n');
fprintf('d1 = %s \n\n',d1new);
```

In MATLAB, the numerical data is:

```
% numerical results
lists = {d1,d2};
listn = {0.3,0.5};
```

The numerical results of the centroidal coordinates and distance d_1 are computed and printed in MATLAB with:

```
X_C = eval(subs(X_C,lists,listn));
Y_C = 0;
fprintf('The numerical values of the ');
fprintf('centroidal coordinates of the plate are\n');
```

```
fprintf('X_C = %s [m] \n',X_C);
fprintf('Y_C = %f [m]\n\n',Y_C);
```

```
d1new=eval(subs(d1new,lists,listn));
fprintf('The centroid X_C is the same as x3 when \n');
fprintf('d1 = %s [m] \n',d1new);
```

The MATLAB numerical results of the centroidal coordinates of the homogeneous plate *AFBMGDEN* are:

```
The numerical values of the centroidal
coordinates of the plate are
X_C = 1.755564e-01 [m]
Y_C = 0.000000 [m]
```

In MATLAB, the position of the centroid of the whole plate is located at the centre of the hole C_3 when the numerical value of the distance d_1 is

```
The centroid X_C is the same as x3 when
d1 = 2.008208e-01 [m]
```

Example 4.6
Calculate the position of the centroid of the homogeneous planar plate *ABC* (Figure 4.10) with respect to the *Oxy* Cartesian frame. The plate could be decomposed in two elements, the quarter of the circle *OBC* of radius *r* (that is $d_{OC} = r$) and the right angle triangle *AOB* with $d_{OB} = r$ and $d_{OA} = b$.
Numerical application: $r = 1$ m and $b = 4$.

Solution
The quarter of the circle *OBC* is bounded by the circular edge *BC*, by the segment *OC* and by the segment *OB*. The circular edge *BC* is defined by the equation $y = \sqrt{r^2 - x^2}$, the segment

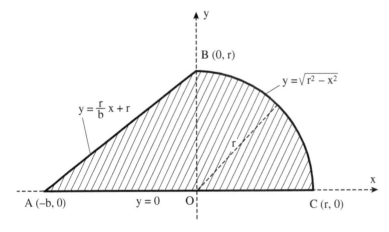

Figure 4.10 Homogeneous planar plate *ABC*.

OC is located on the Ox axis defined by $y = 0$ and segment OB is located on the Oy axis defined by the equation $x = 0$. The area of the quarter of a circle OBC can be obtained by integration using

$$M_{xQC} = \int_0^r \sqrt{r^2 - x^2}dx. \tag{4.40}$$

To evaluate the integral in Equation (4.40) the substitution

$$x = r\sin\theta,$$
$$dx = r\cos\theta d\theta,$$
$$\sqrt{r^2 - x^2} = \sqrt{r^2 - (r\sin\theta)^2}$$
$$= r\cos\theta \tag{4.41}$$

can be used. The endpoints (limits of integration) run from $x = 0$ to $x = r$, that is $\sin\theta = 0$ and $\sin\theta = 1$, therefore the new limits of integration are $\theta = 0$ and $\theta = \frac{\pi}{2}$ respectively. One can write

$$M_{xQC} = \int_0^r \sqrt{r^2 - x^2}dx = \int_0^{\pi/2} (\sqrt{r^2 - (r\sin\theta)^2})(r\cos\theta d\theta),$$
$$= \int_0^{\pi/2} r^2\cos\theta\sqrt{1 - \sin^2\theta}d\theta,$$
$$= \int_0^{\pi/2} r^2\cos\theta\sqrt{\cos^2\theta}d\theta,$$
$$= \int_0^{\pi/2} r^2\cos^2\theta d\theta$$
$$= r^2 \int_0^{\pi/2} \cos^2\theta d\theta,$$
$$= r^2 \left(\frac{\theta}{2} + \frac{\sin 2\theta}{4}\right) \Big|_0^{\pi/2},$$
$$= \frac{\pi r^2}{4}. \tag{4.42}$$

The quarter of the circle area OBC, was computed in MATLAB with:

```
clear all; clc; close all
syms r b x y

dOB=r; dOC=r;
dAO=b;

quartercircle_Or = sqrt(r^2-x^2);
M_xQC=int(quartercircle_Or,0,r);
fprintf('The area of the quarter of the circle ');
fprintf('with respect to x is\n');
fprintf('M_xQC = %s   \n',M_xQC);
```

In Equation (4.42) the integral $\int_0^{\pi/2} \cos^2\theta d\theta$ was calculated using

$$\int_0^{\pi/2} \cos^2\theta d\theta = \int_0^{\pi/2} \frac{1 + \cos(2\theta)}{2} d\theta$$

$$= \frac{1}{2} \int_0^{\pi/2} (1 + \cos(2\theta))d\theta$$

$$= \frac{1}{2} \int_0^{\pi/2} d\theta + \frac{1}{2} \int_0^{\pi/2} \cos(2\theta)d\theta$$

$$= \left(\frac{\theta}{2} + \frac{\sin 2\theta}{4}\right)\Big|_0^{\pi/2}$$

$$= \frac{\pi}{4}. \tag{4.43}$$

Furthermore, one can compute the integral

$$I_{xQC} = \int_0^r x\sqrt{r^2 - x^2}dx \tag{4.44}$$

using the substitution

$$u = r^2 - x^2 \sin\theta,$$

$$du = -2xdx \Leftrightarrow dx = \frac{du}{-2x}. \tag{4.45}$$

The endpoints (limits of integration) run from $x = 0$ to $x = r$. For $x = 0$ one can calculate $u = r^2$ and for $x = r$ one can obtain $u = 0$, therefore the new limits of integration run from r to 0. One can write

$$I_{xQC} = \int_0^r x\sqrt{r^2 - x^2}dx = \int_{r^2}^0 x\left(\sqrt{u}\frac{du}{-2x}\right)$$

$$= \int_{r^2}^0 \sqrt{u}\frac{du}{-2} = \frac{1}{2} \int_0^{r^2} \sqrt{u}du = \frac{1}{2}\frac{\sqrt{u^3}}{\frac{3}{2}}\Big|_0^{r^2}$$

$$= \frac{r^3}{3}. \tag{4.46}$$

In MATLAB the integral denoted by I_{xQC} is calculated with:

```
I_xQC=int(x*quartercircle_Or,0,r);
fprintf('The integral I_xQC is calculated with \n');
fprintf('I_xQC = %s  \n',I_xQC);
```

The centroidal coordinate x_C of the quarter of the circle can be calculated using

$$x_{CQC} = \frac{I_{xQC}}{M_{xQC}} = \frac{\int_0^r x\sqrt{r^2 - x^2}dx}{\int_0^r \sqrt{r^2 - x^2}dx} = \frac{\frac{r^3}{3}}{\frac{\pi r^2}{4}} = \frac{4r}{3\pi}. \tag{4.47}$$

Due to the symmetry $y_{CQC} = x_{CQC}$ the centroidal coordinates of the quarter of the circle are

$$x_{CQC} = \frac{4r}{3\pi}$$

$$y_{CQC} = \frac{4r}{3\pi}. \tag{4.48}$$

The centroidal coordinates are calculated in MATLAB with:

```
x_CQC=I_xQC/M_xQC;
y_CQC=x_CQC;
fprintf('The x_CQC centroidal coordinate ');
fprintf('of the quarter of the circle is \n');
fprintf('x_CQC = %s  \n',x_CQC);
fprintf('Due to the symmetry y_CQC=x_CQC that is \n');
fprintf('y_CQC = %s  \n\n',x_CQC);
```

The triangle *OAB* is bounded by the segment *AB*, by the segment *OA* and by the segment *OB*. The segment *AB* is located on the line passing through *A* and *B* defined by the equation $y = \sqrt{\dfrac{x}{k}} + r$, the segment *OA* is located on the *Ox* axis defined by the equation $y = 0$ and the segment *OB* is located on the *Oy* axis. The area of the triangle *OAB* can be obtained by integration using

$$
\begin{aligned}
M_{xT} &= \int_{-b}^{0} \frac{r}{b}x + r\,dx = \int_{-b}^{0} \frac{r}{b}x\,dx + \int_{-b}^{0} r\,dx \\
&= \frac{r}{b}\int_{-b}^{0} x\,dx + r\int_{-b}^{0} dx = \frac{r}{b}\frac{x^2}{2}\Big|_{-b}^{0} + rx\Big|_{-b}^{0} \\
&= -\frac{r}{b}\frac{b^2}{2} + rb = \frac{rb}{2}.
\end{aligned}
\tag{4.49}
$$

Furthermore, one can compute the integral

$$
\begin{aligned}
I_{xT} &= \int_{-b}^{0} x\left(\frac{r}{b}x + r\right) dx = \int_{-b}^{0} \left(\frac{r}{b}x^2 + rx\right) dx \\
&= \frac{r}{b}\int_{-b}^{0} x^2\,dx + r\int_{-b}^{0} x\,dx = \frac{r}{b}\frac{x^3}{3}\Big|_{-b}^{0} + r\frac{x^2}{2}\Big|_{-b}^{0} \\
&= \frac{r}{b}\frac{b^3}{3} - r\frac{b^2}{2} = \frac{rb^2}{3} - r\frac{b^2}{2} = -r\frac{b^2}{6}.
\end{aligned}
\tag{4.50}
$$

The centroidal coordinate x_{CT} of the triangle can be calculated using

$$
x_{CT} = \frac{I_{xT}}{M_{xT}} = \frac{-r\dfrac{b^2}{6}}{\dfrac{rb}{2}} = -\frac{b}{3}.
\tag{4.51}
$$

The integral I_{xT}, the area M_xT and the centroidal coordinate x_{CT} of the triangle *OAB* are calculated in MATLAB with:

```
line_ABx = x*r/b + r;
M_xT=int(line_ABx,-b,0);
I_xT=int(x*line_ABx,-b,0);
x_CT=I_xT/M_xT;
fprintf('The area of the triangle ');
fprintf('(with respect to x) is\n');
fprintf('M_xT = %s  \n',M_xT);
fprintf('The I_xT with respect to x is\n');
```

```
fprintf('I_xT = %s  \n',I_xT);
fprintf('The x_CT centroidal coordinate ');
fprintf('of the triangle is\n');
fprintf('x_CT = %s  \n\n',x_CT);
```

The area of the triangle *OAB* can be obtained by integration using

$$M_{yT} = \int_0^r \frac{b}{r}y - b\,dy = \int_0^r \frac{b}{r}y\,dy - \int_0^r \frac{b}{r}b\,dy$$

$$= \frac{b}{r}\int_0^r y\,dy - b\int_0^r dy = \frac{b}{r}\frac{y^2}{2}\Big|_0^r - by\Big|_0^r$$

$$= \frac{b\,r^2}{r\,2} - br = -b\frac{r}{2}. \tag{4.52}$$

Also, one can compute the integral

$$I_{yT} = \int_0^r y\left(\frac{b}{r}y - b\right)dy = \int_0^r \frac{b}{r}y^2 - by\,dy$$

$$= \int_0^r \frac{b}{r}y^2\,dy - \int_0^r by\,dy = \frac{b}{r}\int_0^r y^2\,dy - b\int_0^r y\,dy$$

$$= \frac{b}{r}\frac{y^3}{3}\Big|_0^r - b\frac{y^2}{2}\Big|_0^r = \frac{b\,r^3}{r\,3} - b\frac{r^2}{2} = b\frac{r^2}{3} - b\frac{r^2}{2} = -b\frac{r^2}{6}. \tag{4.53}$$

The centroidal coordinate y_{CT} of the triangle can be calculated using

$$y_{CT} = \frac{I_{yT}}{M_{yT}} = \frac{-b\dfrac{r^2}{6}}{-b\dfrac{r}{2}} = \frac{r}{3}. \tag{4.54}$$

Therefore, the x_{CT} and y_{CT} coordinates of the triangle *OAB* are

$$x_{CT} = -\frac{b}{3}$$

$$y_{CT} = \frac{r}{3}. \tag{4.55}$$

The integral I_{yT}, the area M_yT and the centroidal coordinate y_cT of the triangle *OAB* are calculated in MATLAB with:

```
line_ABy = y*b/r - b;
M_yT=int(line_ABy,0,r);
I_yT=int(y*line_ABy,0,r);
y_CT=I_yT/M_yT;
fprintf('The area of the triangle ');
fprintf('with respect to y is\n');
fprintf('M_yT = %s  \n',M_yT);
fprintf('The I_yT with respect to y is\n');
fprintf('I_yT = %s  \n',I_yT);
fprintf('The y_CT centroidal coordinate ');
fprintf('of the triangle is\n');
fprintf('y_CT = %s  \n\n',y_CT);
```

Table 4.5 Table for x_i centroidal coordinates.

Element number	x_i	A_i	$M_{y_i} = x_i A_i$
Triangle area	$-\dfrac{b}{3}$	$b\dfrac{r}{2}$	$-\dfrac{b^2 r}{6}$
Quarter-circular area	$\dfrac{4r}{3\pi}$	$\dfrac{\pi r^2}{4}$	$\dfrac{r^3}{3}$
Σ	–	ΣA_i	$\Sigma x_i A_i$

Table 4.6 Table for y_i centroidal coordinates.

Element number	y_i	A_i	$M_{x_i} = y_i A_i$
Triangle area	$\dfrac{r}{3}$	$-b\dfrac{r}{2}$	$-\dfrac{b^2 r}{6}$
Quarter-circular area	$\dfrac{4r}{3\pi}$	$\dfrac{\pi r^2}{4}$	$\dfrac{r^3}{3}$
Σ	–	ΣA_i	$\Sigma y_i A_i$

To calculate the centroidal coordinate X_C of the homogeneous planar plate ABC shown in Figure 4.10 Table 4.5 is set. One can calculate

$$\sum A_i = b\frac{r}{2} + \frac{\pi r^2}{4} = \frac{r}{2}\left(b + \frac{\pi r}{2}\right),$$
$$\sum x_i A_i = -\frac{b^2 r}{6} + \frac{r^3}{3} = \frac{r}{3}\left(-\frac{b^2}{2} + r^2\right)$$

(4.56)

and in MATLAB:

```
A1x=M_xQC; M1x=I_xQC;
A2x=M_xT; M2x=I_xT;
sum_Ax = A1x+A2x; sum_Mx = M1x+M2x;
fprintf('The next sums are calculated with \n');
fprintf('sum_Ax = %s  \n',simplify(sum_Ax));
fprintf('sum_Mx = %s  \n\n',simplify(sum_Mx));
```

The X_C centroidal coordinate of the homogeneous planar plate ABC is

$$X_C = \frac{\sum x_i A_i}{\sum A_i} = \frac{-\dfrac{b^2 r}{6} + \dfrac{r^3}{3}}{\dfrac{br}{2} + \dfrac{\pi r^2}{4}}.$$

To calculate the centroidal coordinate Y_C of the homogeneous planar plate ABC shown in Figure 4.10 Table 4.6 is set.
One can calculate

$$\sum A_i = -\frac{br}{2} + \frac{\pi r^2}{4},$$

$$\sum y_i A_i = -\frac{b^2 r}{6} + \frac{r^3}{3} \qquad (4.57)$$

and in MATLAB:

```
A1y=A1x; M1y=M1x;
A2y=M_yT; M2y=I_yT;
sum_Ay = A1y+A2y; sum_My = M1y+M2y;
fprintf('The next sums are calculated with \n');
fprintf('sum_Ay = %s \n',simplify(sum_Ay));
fprintf('sum_My = %s \n\n',simplify(sum_My));
```

The centroidal coordinate Y_C of the homogeneous planar plate ABC is

$$Y_C = \frac{\sum y_i A_i}{\sum A_i} = \frac{-\dfrac{b^2 r}{6} + \dfrac{r^3}{3}}{-\dfrac{br}{2} + \dfrac{\pi r^2}{4}}.$$

The centroidal coordinates X_C and Y_C of the homogeneous planar plate ABC are calculated in MATLAB with:

```
X_C=sum_Mx/sum_Ax; Y_C=sum_My/sum_Ay;
fprintf('The X_C and Y_C centroidal coordinates ');
fprintf('of the plate are \n');
fprintf('X_C = %s \n',X_C);
fprintf('Y_C = %s \n\n',Y_C);
```

The numerical data are given in MATLAB with:

```
% numerical results
lists = {r,b};
listn = {1,4};
```

The numerical results are calculated and printed in MATLAB using:

```
X_C = eval(subs(X_C,lists,listn));
Y_C = eval(subs(Y_C,lists,listn));
fprintf('The numerical values of the centroidal coordinates ');
fprintf('of the plate are\n');
fprintf('X_C = %f [m] \n',X_C);
fprintf('Y_C = %f [m] \n\n',Y_C);
```

The MATLAB numerical results of the centroidal coordinates of the plate ABC are:

```
The numerical values of the
centroidal coordinates of the plate are
X_C = -0.837702 [m]
Y_C = 0.274438 [m]
```

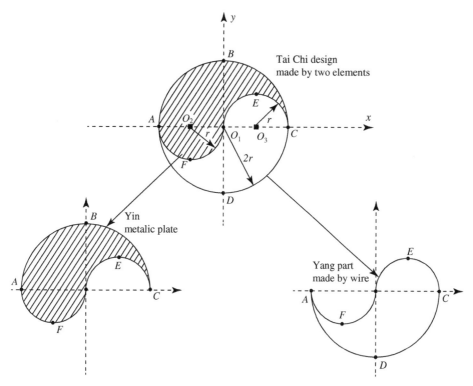

Figure 4.11 Tai Chi design made by a Yin metallic plate and a Yang part made by wire.

Example 4.7

An engineer is 3D printing an Tai Chi design $ABCEOFADC$ shown in Figure 4.11. The printing is made by two elements, a negligible thickness Yin metallic plate $ABCEOFA$ and a Yang part $AFOECDA$ made of wire, which are later attached together. The Tai Chi design has the radius of the main circle $ABCD$ centred at O_1 equal to r (that is $O_1A = O_1B = O_1C = O_1D = r$), the radius of the semicircle AFO_1O_2A centred at O_2 equal to $\frac{r}{2}$ (that is $O_2A = O_2O = O_2F = \frac{r}{2}$), and the radius of the semicircle $O_1ECO_3O_1$ centred at O_1 equal to $\frac{r}{2}$ (that is $O_1O = O_1E = O_1C = \frac{r}{2}$). Calculate the centroid of each element, namely the Yin metallic plate $ABCEOFA$ and the Yang wire $AFOECDA$ with respect to the considered Oxy Cartesian frame.

Numerical application: $r = 0.4$ m.

Solution

Considering the decomposition method one can calculate the centroidal coordinate x_i and y_i for each corresponding area $A_i Yin$, as well as the first moment $M_{ix} Yin$ and $M_{iy} Yin$ with respect to the O_1x and O_1y axes, for all three semicircular sectors shown in Figure 4.12 namely the semicircular area ABC centred at O_1, the semicircular area O_1EC (to be removed) centred at O_3, and the semicircular area AFO_1 centred at O_2. The results of the calculation are given in Table 4.7.

The centroidal coordinate $x_i Yin$ and $y_i Yin$ for the sector 1, 2 and 3 of the Yin metallic plate is calculated in MATLAB using:

Table 4.7 Table for x_i and y_i centroidal coordinates.

Element number	$x_i Yin$	$y_i Yin$	$A_i Yin$	$x_i Yin\, A_i Yin$	$y_i Yin\, A_i Yin$
Sector 1 – semicircle centred at O_1	0	$\dfrac{8r}{3\pi}$	$2\pi r^2$	0	$\dfrac{16r^3}{3}$
Sector 2 – semicircle centred at O_2	$-r$	$-\dfrac{4r}{3\pi}$	$\dfrac{\pi r^2}{2}$	$-\dfrac{\pi r^3}{2}$	$-\dfrac{2r^3}{3}$
Sector 2 – semicircle centred at O_3	r	$\dfrac{4r}{3\pi}$	$-\dfrac{\pi r^2}{2}$	$-\dfrac{\pi r^3}{2}$	$-\dfrac{2r^3}{3}$
Σ	–	–	$\sum A_i Yin = 2\pi r^2$	$\sum x_i Yin\, A_i Yin = -\pi r^3$	$\sum y_i Yin\, A_i Yin = 4r^3$

```
clear all; clc; close all
syms r

dO1A=2*r; dO2A=r; dO2O1=r;
dO1O3=r; dO3C=r;

% Yin metalic plate
x1Yin=0; y1Yin=(4*dO1A)/(3*pi);
x2Yin=-dO2O1; y2Yin=-(4*dO2A)/(3*pi);
x3Yin= dO1O3; y3Yin= (4*dO3C)/(3*pi);
fprintf('The x_i and y_i coordinates of sectors 1 to 3 ');
fprintf('of the Yin metalic plate are \n');
fprintf('x1Yin = %f, y1Yin = %s,   \n',x1Yin,y1Yin);
fprintf('x2Yin = %s, y2Yin = %s,   \n',x2Yin,y2Yin);
fprintf('x3Yin = %s, y3Yin = %s,   \n\n',x3Yin,y3Yin);
```

The corresponding areas $A_i Yin$ of each sector 1, 2 and 3 of the Yin metallic plate is calculated in MATLAB using:

```
A1Yin=(pi*dO1A^2)/2;
A2Yin=(pi*dO2A^2)/2;
A3Yin=-(pi*dO3C^2)/2;
fprintf('The area of the Yin sectors 1 to 3 are\n');
fprintf('A1Yin = %s  \n',A1Yin);
fprintf('A2Yin = %s  \n',A2Yin);
fprintf('A3Yin = %s  \n\n',A3Yin);
```

The moments related to each sector 1, 2 and 3 of the Yin metallic plate are calculated in MATLAB with:

```
M1yYin=x1Yin*A1Yin; M1xYin=y1Yin*A1Yin;
M2yYin=x2Yin*A2Yin; M2xYin=y2Yin*A2Yin;
M3yYin=x3Yin*A3Yin; M3xYin=y3Yin*A3Yin;
fprintf('The moments of the Yin sectors 1 to 3 are \n');
```

```
fprintf('M1xYin = %s   \n',M1xYin);
fprintf('M1yYin = %s   \n',M1yYin);
fprintf('M2xYin = %s   \n',M2xYin);
fprintf('M3yYin = %s   \n',M2yYin);
fprintf('M3xYin = %s   \n',M3xYin);
fprintf('M3yYin = %s   \n',M3yYin);
```

The sums $\sum_{A_i Yin}$, $\sum_{M_{ix} Yin}$ and $\sum_{M_{iy} Yin}$ related to the Yin metallic plate are calculated in MATLAB with:

```
sum_AiYin = A1Yin + A2Yin + A3Yin;
sum_MixYin = M1xYin + M2xYin + M3xYin;
sum_MiyYin = M1yYin + M2yYin + M3yYin;
fprintf('The next sums are calculated \n');
fprintf('sum_AiYin = %s   \n',simplify(sum_AiYin));
fprintf('sum_MixYin = %s   \n',simplify(sum_MixYin));
fprintf('sum_MiyYin = %s   \n\n',simplify(sum_MiyYin));
```

The centroidal coordinates $x_{C_{Yin}}$ and $y_{C_{Yin}}$ of the Yin metallic plate are calculated from Table 4.7 as

$$x_{C_{Yin}} = \frac{\sum x_i Yin\, A_i Yin}{\sum A_i Yin} = \frac{-\pi r^3}{2\pi r^2} = -\frac{r}{2},$$

$$y_{C_{Yin}} = \frac{\sum y_i Yin\, A_i Yin}{\sum A_i Yin} = \frac{4r^3}{2\pi r^2} = \frac{2r}{\pi}, \tag{4.58}$$

and in MATLAB with:

```
X_CYin=sum_MiyYin/sum_AiYin;
Y_CYin=sum_MixYin/sum_AiYin;
fprintf('The Y_C and x_C centroidal coordinates ');
fprintf('of the metalic Yin plate are \n');
fprintf('Y_CYin = %s   \n',simplify(Y_CYin));
fprintf('X_CYin= %s   \n\n',X_CYin);
```

The calculation of the centroidal coordinates x_i and y_i, length $L_i Yang$ and associated moments $x_i Yang\, L_i Yang$ and $y_i Yang\, L_i Yang$ of the Yang part $AFO_1 ECDA$ made of wire, is presented in Table 4.8. The centroidal coordinates $x_i Yang$ and $y_i Yang$ for each sector 1, 2 and 3 of the Yang part made by wire are calculated in MATLAB using:

```
% Yang part made by wire
x1Yang=0;  y1Yang=-(2*d01A)/(pi);
x2Yang=-d02O1;  y2Yang=-(2*d02A)/(pi);
x3Yang= d01O3;  y3Yang= (2*d03C)/(pi);
fprintf('The x_i and y_i coordinates of sectors 1 to 3 ');
fprintf('of the Yang part made by wire are \n');
fprintf('x1Yang = %f, y1Yang = %s,    \n',x1Yang,y1Yang);
fprintf('x2Yang = %s, y2Yang = %s,    \n',x2Yang,y2Yang);
fprintf('x3Yang = %s, y3Yang = %s,    \n\n',x3Yang,y3Yang);
```

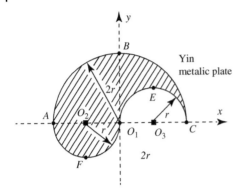

Figure 4.12 The Yin metallic plate of the Tai Chi design.

Table 4.8 Table for x_i and y_i centroidal coordinates.

Element number	x_i Yang	y_i Yang	L_i Yang	x_i Yang L_i Yang	y_i Yang L_i Yang
Sector 1 – semicircle centred at O_1	0	$-\dfrac{4r}{\pi}$	$2\pi r$	0	$-8r^2$
Sector 2 – semicircle centred at O_2	$-r$	$-\dfrac{2r}{\pi}$	πr	$-\pi r^2$	$-2r^2$
Sector 2 – semicircle centred at O_3	r	$\dfrac{2r}{\pi}$	πr	πr^2	$2r^2$
Σ	–	–	$\sum L_i Yang = 4\pi r$	$\sum x_i Yang\, L_i Yang = 0$	$\sum y_i Yang\, L_i Yang = -8r^2$

The length of each sector 1, 2 and 3 of the Yang part is calculated in MATLAB with:

```
L1Yang=(2*pi*dO1A)/2;
L2Yang=(2*pi*dO2A)/2;
L3Yang=(2*pi*dO3C)/2;
fprintf('The length of the sectors 1 to 3 are \n');
fprintf('L1Yang = %s  \n',L1Yang);
fprintf('L2Yang = %s  \n',L2Yang);
fprintf('L3Yang = %s  \n\n',L3Yang);
```

The moments related to each sector 1, 2 and 3 of the Yang part are calculated in MATLAB with:

```
M1yYang=x1Yang*L1Yang; M1xYang=y1Yang*L1Yang;
M2yYang=x2Yang*L2Yang; M2xYang=y2Yang*L2Yang;
M3yYang=x3Yang*L3Yang; M3xYang=y3Yang*L3Yang;
fprintf('The area of the Yang sectors 1 to 3 are \n');
fprintf('M1xYang = %s  \n',M1xYang);
fprintf('M1yYang = %s  \n',M1yYang);
fprintf('M2xYang = %s  \n',M2xYang);
```

```
fprintf('M3yYang = %s  \n',M2yYang);
fprintf('M3xYang = %s  \n',M3xYang);
fprintf('M3yYang = %s  \n',M3yYang);
```

The sums $\sum_{L_i Yang}$, $\sum_{M_{ix} Yang}$ and $\sum_{M_{iy} Yang}$ related to the Yang part made of wire are calculated in MATLAB with:

```
sum_LiYang = L1Yang + L2Yang + L3Yang;
sum_MixYang = M1xYang + M2xYang + M3xYang;
sum_MiyYang = M1yYang + M2yYang + M3yYang;
fprintf('The next sums are calculated with \n');
fprintf('sum_LiYang = %s  \n',simplify(sum_LiYang));
fprintf('sum_MixYang = %s  \n',simplify(sum_MixYang));
fprintf('sum_MiyYang = %s  \n\n',simplify(sum_MiyYang));
```

The centroidal coordinates $x_{C_{Yang}}$ and $y_{C_{Yang}}$ of the Yang part made by wire are calculated from Table 4.8 as

$$x_{C_{Yang}} = \frac{\sum x_i Yang\, L_i Yang}{\sum L_i Yang} = \frac{0}{4\pi r} = 0,$$

$$y_{C_{Yang}} = \frac{\sum y_i Yang\, L_i Yang}{\sum L_i Yang} = \frac{-8r^2}{4\pi r} = -\frac{2r}{\pi}, \tag{4.59}$$

and in MATLAB with:

```
X_CYang=sum_MiyYang/sum_LiYang;
Y_CYang=sum_MixYang/sum_LiYang;
fprintf('The Y_C and X_C centroidal coordinate ');
fprintf('of the metalic Yang plate is \n');
fprintf('X_CYang = %s  \n',simplify(X_CYang));
fprintf('Y_CYang = %s  \n\n',Y_CYang);
```

The numerical data considered in MATLAB is:

```
% numerical results
lists = {r};
listn = {0.4};
```

The numerical results are computed and printed in MATLAB using:

```
X_CYin = eval(subs(X_CYin,lists,listn));
Y_CYin = eval(subs(Y_CYin,lists,listn));
X_CYang = eval(subs(X_CYang,lists,listn));
Y_CYang = eval(subs(Y_CYang,lists,listn));
fprintf('The numerical values of the centroidal coordinates ');
fprintf('of the Yin metalic plate are \n');
fprintf('X_CYin = %f [m] \n',X_CYin);
fprintf('Y_CYin = %f [m]\n\n',Y_CYin);
fprintf('The numerical values of the centroidal coordinates ');
fprintf('of the Yang part made by wire are\n');
fprintf('X_CYang = %f [m] \n',X_CYang);
fprintf('Y_CYang = %f [m]\n',Y_CYang);
```

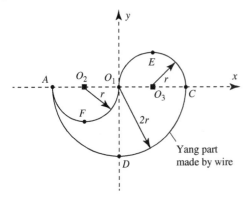

Figure 4.13 The wire Yang part of the Tai Chi design.

The MATLAB numerical results of the centroidal coordinates of the plate Yin metallic plate are:

```
X_CYin = -0.200000 [m]
Y_CYin = 0.254648 [m]
```

The MATLAB numerical results of the centroidal coordinates of the Yang part made of wire are:

```
X_CYang = 0.000000 [m]
Y_CYang = -0.254648 [m]
```

Example 4.8
Determine the mass centre position of the 3D body in Figure 4.14 with respect to the $Oxyz$ Cartesian frame. The body is composed of a homogeneous hemisphere of radius r and constant density $\rho_s = k$ which is connected to a cylinder having radius r, height h and a variable density $\rho_{cyl} = kz$. The homogeneous hemisphere has the origin O_s located on the Oz axis at the distance h from O, that is $d_{OO_s} = h$.
Numerical application: $r = 0.3$ m, and $h = 0.8$ m.

Solution
To calculate the centre of mass for the cylinder shown in Figure 4.15 the reference frame $Oxyz$ is considered. Using cylindrical coordinates the differential volume element dV_c can be calculated with $dV_c = \rho d\rho d\theta dz$ and the mass element is calculated using $dm = \rho_{cyl}dV_c$ where ρ_{cyl} is the variable density of the cylinder. The total mass of the cylinder can be calculated with

$$m_{Cyl} = \int_{V_c} dm = \int_{V_c} \rho_{cyl}dV_c = \int_{V_c} kzdV_c = \int_0^r \int_0^{2\pi} \int_0^h kz\rho d\rho d\theta dz$$

$$= k\int_0^r \rho d\rho \int_0^{2\pi} d\theta \int_0^h zdz = k\frac{r^2}{2}\frac{h^2}{2}2\pi = \frac{1}{2}k\pi r^2 h^2. \tag{4.60}$$

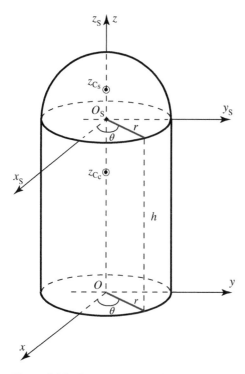

Figure 4.14 Body composed of a homogeneous hemisphere connected to a cylinder.

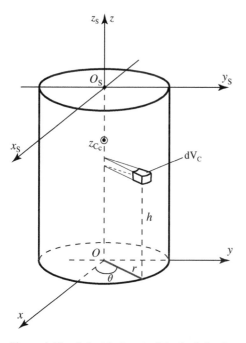

Figure 4.15 Cylindrical part of the body having a variable density.

The mass centre coordinate z_{C_c} of the cylinder is calculated in the reference frame $Oxyz$ using

$$z_{C_c} = \frac{\int_{V_c} z \, dm}{\int_{V_c} dm} = \frac{\int_{V_c} z \, dm}{m_{cyl}}. \tag{4.61}$$

One can calculate

$$\int_{V_c} z \, dm = \int_{V_c} z \rho_{cyl} dV_c = \int_{V_c} k z^2 dV_c = \int_0^r \int_0^{2\pi} \int_0^h k z^2 \rho \, d\rho \, d\theta \, dz$$

$$= k \int_0^r \rho \, d\rho \int_0^{2\pi} d\theta \int_0^h z^2 dz$$

$$= \frac{1}{3} k \pi r^2 h^3. \tag{4.62}$$

From Equations (4.60)–(4.62), one can calculate

$$z_{C_c} = \frac{\int_{V_c} z \, dm}{\int_{V_c} dm} = \frac{\frac{1}{3} k \pi r^2 h^3}{\frac{1}{2} k \pi r^2 h^2} = \frac{2}{3} h. \tag{4.63}$$

The mass centre coordinate z_{C_c} of the cylinder is calculated in MATLAB with:

```
clear all; clc; close all
syms rho theta phi r h k z zC

rho_cyl=k*z;
m_cyl = int(int(int(rho, rho, 0, r),...
    theta, 0, 2*pi)*rho_cyl,z, 0, h);
fprintf('The mass m_cyl of the cylinder calculated as\n');
fprintf('m_cyl = %s \n\n', char(m_cyl));

Mc = int(int(int(rho, rho, 0, r),...
    theta, 0, 2*pi)*rho_cyl*z,z, 0, h);
fprintf('The integral Mc is calculated as\n');
fprintf('Mc = %s \n\n', char(Mc))

zCc = Mc/m_cyl;
fprintf('The center of the mass of the cylinder ');
fprintf('with respect to the Oxyz reference is \n');
fprintf('zCc = Mc/m_cyl = %s \n\n', char(zCc))
```

To calculate the centre of the mass for the hemisphere shown in Figure 4.16 the reference frame $O_s x_s y_s z_s$ is considered. The axis $O_s z_s$, which is a symmetry axis for the hemisphere, is coincident with the Oz axis, which is a symmetry axis for the 3D body. The $O_s x_s y_s z_s$ frame is parallel with and at a distance h from the reference frame $Oxyz$, that is $d_{OO_s} = h$ (Figure 4.16). For the spherical coordinates $z = \rho \sin \varphi$ the differential volume element dV_s

can be calculated using $dV_s = \rho^2 \cos\varphi\, d\rho\, d\theta\, d\varphi$. The z_{C_s} coordinate is calculated in the reference frame $O_s x_s y_s z_s$ using

$$z_{C_s} = \frac{\displaystyle\int_{V_s} z V_s}{\displaystyle\int_{V_s} dV_s} = \frac{\displaystyle\int_0^r \int_0^{2\pi} \int_0^{\pi/2} \rho \sin\varphi\, \rho^2 \cos\varphi\, d\rho\, d\theta\, d\varphi}{\displaystyle\int_0^r \int_0^{2\pi} \int_0^{\pi/2} \rho^2 \cos\varphi\, d\rho\, d\theta\, d\varphi} \tag{4.64}$$

or equivalent

$$z_{C_s} \int_0^r \int_0^{2\pi} \int_0^{\pi/2} \rho^2 \cos\varphi\, d\rho\, d\theta\, d\varphi = \int_0^r \int_0^{2\pi} \int_0^{\pi/2} \rho^3 \sin\varphi \cos\varphi\, d\rho\, d\theta\, d\varphi. \tag{4.65}$$

From Equation (4.65) one can calculate

$$z_{C_s} \int_0^r \rho^2 d\rho \int_0^{2\pi} d\theta \int_0^{\pi/2} \cos\varphi\, d\varphi = \int_0^r \rho^3 d\rho \int_0^{2\pi} d\theta \int_0^{\pi/2} \sin\varphi \cos\varphi\, d\varphi,$$

or

$$z_{C_s} = \frac{\displaystyle\int_0^r \rho^3 d\rho \int_0^{2\pi} d\theta \int_0^{\pi/2} \sin\varphi \cos\varphi\, d\varphi}{\displaystyle\int_0^r \rho^2 d\rho \int_0^{2\pi} d\theta \int_0^{\pi/2} \cos\varphi\, d\varphi}$$

$$= \frac{\left(\dfrac{\rho^4}{4}\right)\Big|_0^r \left(-\dfrac{1}{2}\cos^2\varphi\right)\Big|_0^{\pi/2}}{\dfrac{\rho^3}{3}\Big|_0^r \sin\varphi\Big|_0^{\pi/2}}$$

$$= \frac{\left(\dfrac{r^4}{4} - \dfrac{0^4}{4}\right)\left(-\dfrac{1}{2}\cos^2\dfrac{\pi}{2} + \dfrac{1}{2}\cos^2\dfrac{0}{2}\right)}{\left(\dfrac{r^3}{3} - \dfrac{0^3}{3}\right)\left(\sin\dfrac{\pi}{2} - \sin\dfrac{0}{2}\right)}$$

$$= \frac{\left(\dfrac{r^4}{4}\right)\left(\dfrac{1}{2}\right)}{\dfrac{r^3}{3}} = \frac{3r}{8}. \tag{4.66}$$

The centre of the mass for the hemisphere in the reference frame $O_s x_s y_s z_s$ is calculated in MATLAB with:

```
% dVs = rho^2 cos(phi) drho dtheta dphi
% 0<rho<R 0<theta<2pi 0<phi<pi/2
Vs=int(int(int(rho^2, rho, 0, r),...
    theta, 0, 2*pi)*cos(phi), phi, 0, pi/2);
fprintf('The volume of the hemisphere is\n');
fprintf('Vs = %s \n', char(Vs));

% dMs = rho rho^2 cos(phi) drho dtheta dphi
% 0<rho<R 0<theta<2pi 0<phi<pi/2
Ms = int(int(int(rho^3, rho, 0, r),...
```

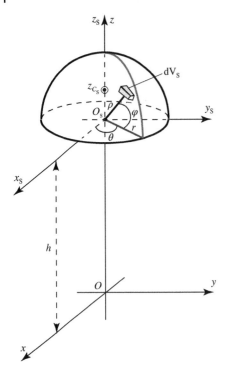

Figure 4.16 Homogeneous hemispherical part of the body.

```
       theta, 0, 2*pi)*sin(phi)*cos(phi),phi, 0, pi/2);
fprintf('The integral Ms is calculated as\n');
fprintf('Ms = %s \n\n', char(Ms))
```

```
zCsOs = Ms/Vs;
fprintf('The center of the mass of the hemisphere ');
fprintf('with respect to the O_sx_sy_sz_s reference is \n');
fprintf('zCsOs = MsOs/VsOs = %s \n\n', char(zCsOs))
```

The centre of the mass for the hemisphere shown in Figure 4.16 can be calculated in the reference frame *Oxyz* as

$$z_{C_s} = h + \frac{3r}{8}.$$

where the mass of the hemisphere is $m_s = \rho_s V_s = k\dfrac{2\pi r^3}{3}$.

The mass and the centre of the mass for the hemisphere in the reference frame *Oxyz* is calculated in MATLAB with:

```
rho_s=k;
m_s = rho_s*Vs;
zCs = h+ Ms/Vs;
fprintf('The mass of the hemisphere is\n');
fprintf('m_s = %s \n\n', char(m_s))
```

```
fprintf('The center of the mass of the hemisphere ');
fprintf('with respect to the Oxyz reference is \n');
fprintf('zCs = Ms/Vs = %s \n\n', char(zCs))
```

The centre of the mass for the whole body can be calculated using

$$z_{C_s} m_s + z_{C_c} m_{cyl} = z_C * (m_s + m_{cyl}) \tag{4.67}$$

that is

$$z_C = \frac{z_{C_s} m_s + z_{C_c} m_{cyl}}{m_s + m_{cyl}} = \frac{\left(h + \dfrac{3r}{8}\right)\dfrac{2\pi k r^3}{3} + \dfrac{\pi k r^2 h^3}{2}}{\dfrac{2\pi k r^3}{3} + \dfrac{\pi k r^2 h^2}{2}}$$

$$= \frac{\dfrac{2hr}{3} + \dfrac{r^2}{4} + \dfrac{h^3}{2}}{\dfrac{2r}{3} + \dfrac{h^2}{2}} = \frac{h\left(\dfrac{2r}{3} + \dfrac{h^2}{2}\right) + \dfrac{r^2}{4}}{\dfrac{2r}{3} + \dfrac{h^2}{2}}$$

$$= h + \frac{\dfrac{r^2}{4}}{\dfrac{2r}{3} + \dfrac{h^2}{2}} \tag{4.68}$$

and in MATLAB with:

```
eq_zC=zCs*m_s+zCc*m_cyl-zC*(m_s+m_cyl);
zC=solve(eq_zC,zC);
fprintf('The center of the mass of the body is \n');
fprintf('zC = %s \n\n', char(zC))
```

The input numerical data are introduced in MATLAB with:

```
% numerical results
lists = {h,r};
listn = {0.8,0.3};
```

The numerical results are calculated and printed in MATLAB using:

```
zCc=eval(subs(zCc,lists,listn));
zCs=eval(subs(zCs,lists,listn));
zC=eval(subs(zC,lists,listn));

fprintf('The numerical value of the center of the cylinder is \n');
fprintf('zCc = %f \n\n', zCc)
fprintf('The numerical value of the center of the hemisphere is \n');
fprintf('zCs = %f \n\n', zCs)
fprintf('The numerical value of the center of the mass of the body is \n');
fprintf('zC = %f \n\n', zC)
```

The MATLAB numerical results for centre of the mass coordinates for the cylinder, hemisphere and whole body are:

```
The numerical value of the center of the cylinder is
zCc = 0.533333
```

The numerical value of the center of the hemisphere is

zCs = 0.912500

The numerical value of the center of the mass of the body is

zC = 0.679167

Example 4.9

Figure 4.17 shows a mechanism with five moving links. The links 1 and 4 are homogeneous triangular plates. The following dimensions are given:

```
OA  = 0.100;   % m
OC  = 0.085;   % m
AC  = 0.070;   % m
AB  = 0.230;   % m
O1D = 0.140;   % m
CD  = 0.115;   % m
DP  = 0.160;   % m
CP  = 0.250;   % m
```

The origin of the system is at O and the coordinates of the joint O_1 are

```
xO1 = -0.095;   % m
yO1 =  0.120;   % m
```

Find the centroid of the triangular plates 1 and 4 for $\phi = \pi/3$.

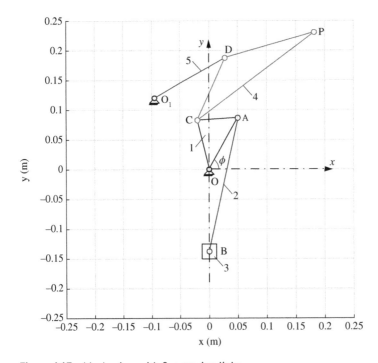

Figure 4.17 Mechanism with five moving links.

Solution

The joint A coordinates are calculated in MATLAB with:

```
xA = OA*cos(phi);
yA = OA*sin(phi);
rA_ = [xA yA 0]; % position vector of A
% rA_ = [ 0.050,   0.087, 0] (m)
```

For the position of B the following equation is used:

$$(x_A - x_B)^2 + (y_A - y_B)^2 = AB^2,$$

or with MATLAB:

```
xB = 0;
syms yBs
eqB = (xA - xB)^2 + (yA - yBs)^2 - AB^2;
solB = vpasolve(eqB, yBs);
yB1 = solB(1);
yB2 = solB(2);
```

There are two solutions for y_B and the correct one if obtained with the constraint relation:

```
if yB1 < xA yB=yB1;  else
xB=xB2; end
rB_ = [xB yB 0]; % position vector of B
% rB_ = [ 0, -0.138, 0] (m)
```

The joint C position is calculated from two quadratic equations:

$$x_C^2 + y_C^2 = OC^2 \quad \text{and} \quad (x_A - x_C)^2 + (y_A - y_C)^2 = AC^2.$$

The solutions for the two equations are obtained using the MATLAB commands:

```
syms xCs yCs
eqOC = xCs^2 + yCs^2 - OC^2;
eqAC = (xA - xCs)^2 + (yA - yCs)^2 - AC^2;
solC = vpasolve(eqOC, eqAC, xCs, yCs);
xC1 = solC.xCs(1);
xC2 = solC.xCs(2);
yC1 = solC.yCs(1);
yC2 = solC.yCs(2);
```

and the correct solution for the mechanism is:

```
if xC1 < 0 xC=xC1;yC=yC1; else
xC=xC2; yC=yC2; end
rC_ = [xC yC 0]; % position vector of C
% rC_ = [-0.020, 0.083, 0] (m)
```

The joint D position is calculated from two quadratic equations:

$$(x_C - x_D)^2 + (y_C - y_D)^2 = CD^2 \quad \text{and} \quad (x_{O_1} - x_D)^2 + (y_{O_1} - y_D)^2 = O_1D^2.$$

The MATLAB solution is:

```
syms xDs yDs
eqCD  = (xC - xDs)^2 + (yC - yDs)^2 - CD^2;
eqO1D = (xO1 - xDs)^2 + (yO1 - yDs)^2 - O1D^2;

solD = vpasolve(eqCD, eqO1D, xDs, yDs);
xD1 = solD.xDs(1);
xD2 = solD.xDs(2);
yD1 = solD.yDs(1);
yD2 = solD.yDs(2);

if xD1 > xO1 xD=xD1;yD=yD1; else
xD=xD2; yD=yD2; end

rD_ = [xD yD 0]; % position vector of D
% rD_ = [0.028, 0.187, 0] (m)
```

The position of point P is calculated from the following quadratic equations:

$$(x_C - x_D)^2 + (y_C - y_D)^2 = CD^2 \text{ and } (x_{O_1} - x_D)^2 + (y_{O_1} - y_D)^2 = O_1D^2.$$

The MATLAB solution is:

```
syms xPs yPs
eqCP = (xC - xPs)^2 + (yC - yPs)^2 - CP^2;
eqDP = (xD - xPs)^2 + (yD - yPs)^2 - DP^2;
solP = vpasolve(eqCP, eqDP, xPs, yPs);
xP1 = solP.xPs(1);
xP2 = solP.xPs(2);
yP1 = solP.yPs(1);
yP2 = solP.yPs(2);
if xP1 > 0 xP=xP1;yP=yP1; else
xP=xP2; yP=yP2; end
rP_ = [xP yP 0]; % position vector of P
% rP_ = [ 0.182, 0.230, 0] (m)
```

The mass centre for links 1 and 4 is calculated using the coordinates of the vertices of the triangles:

$$x_{C_1} = \frac{0 + x_A + x_B}{3} \text{ and } y_{C_1} = \frac{0 + y_A + y_B}{3},$$

$$x_{C_4} = \frac{x_C + x_D + x_P}{3} \text{ and } y_{C_4} = \frac{y_C + y_D + y_P}{3},$$

or in MATLAB:

```
xC1 = (0+xA+xC)/3; yC1 = (0+yA+yC)/3;
rC1_ = [xC1 yC1 0]; % position vector of C1
% rC1_ = [ 0.010, 0.056, 0] (m)
```

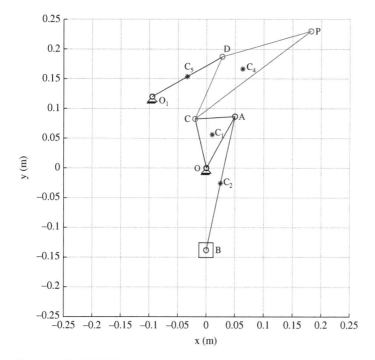

Figure 4.18 MATLAB plot of the mechanism with the mass centres.

```
xC4 = (xC+xD+xP)/3; yC4 = (yC+yD+yP)/3;
rC4_ = [xC4 yC4 0]; % position vector of C4
% rC4_ = [ 0.063, 0.167, 0] (m)
```

The MATLAB plot of the mechanism with the mass centres is shown in Figure 4.18 and is obtained with:

```
axis manual
axis equal
hold on
grid on
sx = 0.25;
sy = 0.25;
axis([-sx sx -sy sy])
xlabel('x (m)'),ylabel('y (m)')
pM = plot([0,xA],[0,yA],'k-o',...
       [xA,xC],[yA,yC],'k-o',...
       [0,xC],[0,yC],'k-o',...
       [xA,xB],[yA,yB],'b-o',...
       [xC,xD],[yC,yD],'r-o',...
       [xO1,xD],[yO1,yD],'b-o',...
       [xC,xP],[yC,yP],'r-o',...
       [xD,xP],[yD,yP],'r-o');
```

```
plot (xC1,yC1,'b*')
plot (xC2,yC2,'b*')
plot (xC4,yC4,'b*')
plot (xC5,yC5,'b*')

text (0,0,' O')
text (xA,yA,' A')
text (xB,yB,' B')
text (xO1,yO1,' O_1')
text (xC,yC,' C')
text (xD,yD,' D')
text (xP,yP,' P')

text (xC1,yC1,' C_1')
text (xC2,yC2,' C_2')
text (xC4,yC4,' C_4')
text (xC5,yC5,' C_5')
```

Example 4.10

For the triangular area in Figure 4.19 with $b = 6$ in and $h = 9$ in calculate:

(a) The moment of inertia of the triangular area about the x axis and about the y axis
(b) The product of inertia of each of triangular area about about the x–y axes
(c) The maximum and minimum moments of inertia with respect to the x–y axes through the origin O
(d) The angles α_1 and α_2 measured from the x-axis to the axes of maximum and minimum moments of inertia.

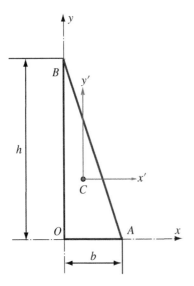

Figure 4.19 Triangular area.

Solution

(a) The equation of the AB line is

$$\frac{x}{b} + \frac{y}{h} - 1 = 0.$$

The moment of inertia of the triangular area about x axis is

$$I_{xx} = \int_A y^2 \, dA = \int_0^b \int_0^{h\,(1-x/b)} y^2 \, dx \, dy = \int_0^b dx \int_0^{h\,(1-x/b)} y^2 \, dy,$$

or in MATLAB:

```
syms  b h x y real
Ixx1 = int(y^2, y, 0, h*(1-x/b));
Ixx  = int(Ixx1, x, 0, b);
% Ixx = (b*h^3)/12
```

The moment of inertia of the triangular area about y axis is

$$I_{yy} = \int_A x^2 \, dA = \int_0^b \int_0^{h\,(1-x/b)} x^2 \, dx \, dy = \int_0^b x^2 \, dx \int_0^{h\,(1-x/b)} dy,$$

or in MATLAB:

```
Iyy1 = int(1, y, 0, h*(1-x/b));
Iyy  = int(x^2*Iyy1, x, 0, b);
% Iyy = (b^3*h)/12
```

(b) The product of inertia of the triangular area about x–y axes is

$$I_{xy} = \int_A x y \, dA = \int_0^h \int_0^{h\,(1-x/b)} x y \, dx \, dy = \int_0^b x \, dx \int_0^{h\,(1-x/b)} y \, dy,$$

or in MATLAB:

```
Ixy1 = int(y, y, 0, h*(1-x/b));
Ixy  = int(x*Ixy1, x, 0, b);
% Ixy = (b^2*h^2)/24
```

The moment of inertia of about the z axis (the polar moment about O) is

$$I_O = I_{zz} = I_{xx} + I_{yy}.$$

The inertia matrix of the plane figure (triangular area) with respect to the x–y axes through   the origin O is

$$[I] = \begin{bmatrix} I_{xx} & I_{xy} & I_{xz} \\ I_{yx} & I_{yy} & I_{yz} \\ I_{zx} & I_{zy} & I_{zz} \end{bmatrix}.$$

The numerical values are:

```
lists ={b, h};
listn ={6, 9}; % (in)
```

```
Ixxn=subs(Ixx, lists, listn);
Iyyn=subs(Iyy, lists, listn);
Izzn=subs(Izz, lists, listn);
Ixyn=subs(Ixy, lists, listn);

% Ixx = (b*h^3)/12 = 364.500 (in^4)
% Iyy = (b^3*h)/12 = 162.000 (in^4)
% Izz = (b*h^3)/12 + (b^3*h)/12 = 526.500 (in^4)
% Ixy = (b^2*h^2)/24 = 121.500 (in^4)
```

(c) The maximum or minimum moments of inertia for the area are

$$I_{1,\,2} = I_{max,\,min} = \frac{I_{xx} + I_{yy}}{2} \pm \sqrt{\left(\frac{I_{xx} - I_{yy}}{2}\right)^2 + I_{xy}^2},$$

$$I_1 = I_{max} = \frac{I_{xx} + I_{yy}}{2} + \sqrt{\left(\frac{I_{xx} - I_{yy}}{2}\right)^2 + I_{xy}^2},$$

$$I_2 = I_{min} = \frac{I_{xx} + I_{yy}}{2} - \sqrt{\left(\frac{I_{xx} - I_{yy}}{2}\right)^2 + I_{xy}^2}.$$

The polar moment of inertia of the triangle about O is $I_1 + I_2$. The maximum and minimum moments of inertia in MATLAB are

```
I1=(Ixx+Iyy)/2+sqrt((Ixx-Iyy)^2/4+Ixy^2);
I1n=subs(I1, lists, listn);
I2=(Ixx+Iyy)/2-sqrt((Ixx-Iyy)^2/4+Ixy^2);
I1n=subs(I1, lists, listn);
I2n=subs(I2, lists, listn);
% I1 = Imax = 421.408 (in^4)
% I2 = Imin = 105.092 (in^4)
% I1+I2 = 526.5 (in^4)
```

(d) The principal directions are calculated with

$$\tan 2\alpha_0 = \frac{2I_{xy}}{I_{yy} - I_{xx}},$$

and in MATLAB:

```
tanalpha0=simplify(2*Ixy/(Iyy-Ixx));
alpha1n=atan(2*Ixyn/(Iyyn-Ixxn))/2;
alpha2n=alpha1n+pi/2;
% tan (2 alpha0) = (b*h)/(b^2 - h^2)
% alpha1 = -25.0972 (degrees)
% alpha2 = 64.9028 (degrees)
```

The principal directions are shown in Figure 4.20 and in MATLAB are obtained with:

```
% graph
xA = b; yA = 0;
```

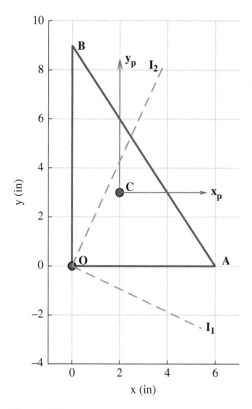

Figure 4.20 MATLAB plot showing the principal directions of the mechanism.

```
xB = 0; yB = h;
xC = b/3;
yC = h/3;

xAn = subs(xA, lists, listn);
yAn = subs(yA, lists, listn);
xBn = subs(xB, lists, listn);
yBn = subs(yB, lists, listn);
xCn = subs(xC, lists, listn);
yCn = subs(yC, lists, listn);

figure(1)
axis manual
axis equal
hold on
grid on
sa = 10;
axis([-1 7 -4 sa])
plot(...
```

```
[xAn,xBn],[yAn,yBn],'b-','LineWidth',2)
plot(...
[0,xBn],[0,yBn],'b-','LineWidth',2)
plot(...
[0, xAn],[0,yAn],'b-','LineWidth',2)

plot(0,0,...
'o','MarkerSize',8,...
'MarkerEdgeColor','k',...
'MarkerFaceColor','b')

plot(xCn,yCn,...
'o','MarkerSize',8,...
'MarkerEdgeColor','k',...
'MarkerFaceColor','r')

xlabel('x (in)'), ylabel('y (in)')
text(0,0+0.25,' O','fontsize',12,'fontweight','b')
text(xAn,yAn+0.25,' A','fontsize',12,'fontweight','b')
text(xBn,yBn,' B','fontsize',12,'fontweight','b')
text(xCn,yCn+0.25,' C','fontsize',12,'fontweight','b')

% quiver(x,y,u,v) plots vectors as arrows with
% components (u,v) at the points (x,y)
quiver(xCn,yCn,sa-6,0,...
'Color','r','LineWidth',1.0)
quiver(xCn,yCn,0,sa-4,...
'Color','r','LineWidth',1.0)

text(xCn+sa-6.25,yCn,' x_p','fontsize',12,'fontweight','b')
text(xCn,yCn+sa-4.2,' y_p','fontsize',12,'fontweight','b')

x1 = b*cos(alpha1n);
y1 = b*sin(alpha1n);
x2 = h*cos(alpha2n);
y2 = h*sin(alpha2n);

x1n = subs(x1, lists, listn);
y1n = subs(y1, lists, listn);
x2n = subs(x2, lists, listn);
y2n = subs(y2, lists, listn);

line([0 x1n],[0 y1n],...
    'LineStyle','-','LineWidth',1.5)
line([0 x2n],[0 y2n],...
```

```
'LineStyle','-','LineWidth',1.5)
```

```
text(x1n,y1n,'  I_1','fontsize',12,'fontweight','b')
text(x2n-0.7,y2n,'I_2','fontsize',12,'fontweight','b')
```

The moments and the product of inertia I_{uu}, I_{vv}, and I_{uv} are calculated about inclined axes with an angle θ. The variation of the moments of inertia and the product of inertia in terms of the angle θ are:

$$I_{uu}(\theta) = \frac{I_{xx} + I_{yy}}{2} + \frac{I_{xx} - I_{yy}}{2} \cos 2\theta - I_{xy} \sin 2\theta,$$

$$I_{vv}(\theta) = \frac{I_{xx} + I_{yy}}{2} - \frac{I_{xx} - I_{yy}}{2} \cos 2\theta + I_{xy} \sin 2\theta,$$

$$I_{uv}(\theta) = \frac{I_{xx} - I_{yy}}{2} \sin 2\theta + I_{xy} \cos 2\theta,$$

or in MATLAB:

```
syms theta
Iuu=0.5*(Ixx+Iyy)+0.5*(Ixx-Iyy)*cos(2*theta)-Ixy*sin(2*theta);
Ivv=0.5*(Ixx+Iyy)-0.5*(Ixx-Iyy)*cos(2*theta)+Ixy*sin(2*theta);
Iuv=0.5*(Ixx-Iyy)*sin(2*theta)+Ixy*cos(2*theta);
```

The variations of I_{uu} and I_{uv} for $-90 \le \theta \le 90$ are shown in Figure 4.21. The maximum and minimum moments of inertia, $I_{max} = I_1$ and $I_{min} = I_2$ are obtained for $\alpha_1 = -25.097°$ and $\alpha_2 = 64.903°$. The moments of inertia and the product of inertia of the triangle about x' axis and y' axis through the centroid C of coordinates x_C and y_C are obtained using the parallel axis theorem

$$I_{Cx'x'} = I_{yy} - y_C^2 A \text{ and } I_{Cy'y'} = I_{yy} - x_C^2 A,$$

and with MATLAB

```
A  = b*h/2;
xC = b/3;
yC = h/3;


ICxx = Ixx - yC^2*A;
ICyy = Iyy - xC^2*A;
ICxy = Ixy -xC*yC*A;


ICxxn=subs(ICxx, lists, listn);
ICyyn=subs(ICyy, lists, listn);
ICxyn=subs(ICxy, lists, listn);


% ICxx = (b*h^3)/36 = 121.500 (in^4)
% ICyy = (b^3*h)/36 = 54.000 (in^4)
% ICxy = -(b^2*h^2)/72 = -40.500 (in^4)
```

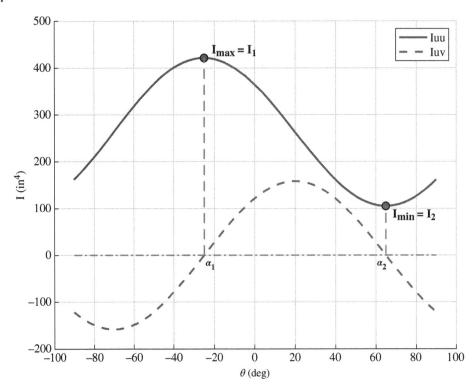

Figure 4.21 MATLAB plot showing the maximum and minimum moments of inertia.

5

Stress, Strain and Deflection

5.1 Stress

A normal force F is applied to the ends of a bar. The bar, which is homogeneous and straight, has its cross section area denoted by A. The normal force F is in tension or in compression. The line of action of the force includes the centroid of the section. The bar is cut at a distance from one of its end and that part is eliminated. The eliminated part is then replaced by a uniformly distributed force (UDF) σA, where σ is the *normal stress*. The stress σ is calculated with

$$\sigma = \frac{F}{A}. \tag{5.1}$$

For a bar in shear the *shear stress* is

$$\tau = \frac{F}{A}, \tag{5.2}$$

where F is the shear force.

A stress element is depicted in Figure 5.1(a). On the figure one can observe three normal stresses, denoted by σ_x, σ_y, and σ_z, and six shear stresses denoted by $\tau_{zy}, \tau_{zx}, \tau_{yz}, \tau_{yx}, \tau_{xy}$, and τ_{xz}. Using equilibrium it results that

$$\tau_{zx} = \tau_{xz}, \quad \tau_{zy} = \tau_{yz}, \quad \tau_{yx} = \tau_{xy}. \tag{5.3}$$

The tensile stresses are the normal stresses σ_x, σ_y, and σ_z. Figure 5.1(b) represents a plane stress element. The normal stresses are positive. The shear stresses are positive if they operate clockwise ($\tau_{yx} > 0$ and $\tau_{xy} < 0$).

5.2 Elastic Strain

The total strain is the amount of elongation, δ, of a straight bar, with the length l, when a tensile force is applied. The strain, ε, defined as the elongation per length, is calculated using

$$\varepsilon = \frac{\delta}{l}. \tag{5.4}$$

Engineering Applications: Analytical and Numerical Calculation with MATLAB, First Edition.
Mihai Dupac and Dan B. Marghitu.
© 2021 John Wiley & Sons Ltd. Published 2021 by John Wiley & Sons Ltd.

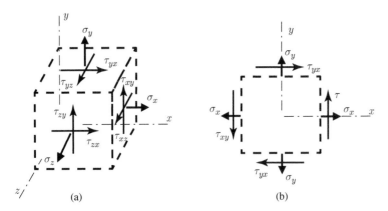

Figure 5.1 Centroid representation of a set of points.

The change of a right angle of an element exposed to pure shear stresses is the shear strain, γ. The elasticity is an attribute of materials and can be denoted with

$$\sigma = E\,\epsilon, \qquad \tau = G\,\gamma, \tag{5.5}$$

where G is the modulus of rigidity (or shear modulus of elasticity), and E is the modulus of elasticity. With $\sigma = F/A$, $\epsilon = \delta/l$, and Equation (5.5) one can calculate the total deformation

$$\delta = \frac{F\,l}{A\,E}. \tag{5.6}$$

with the bar in axial compression or tension.

The ratio $\nu = \dfrac{\text{lateral strain}}{\text{axial strain}}$ of the lateral strain to the axial strain, is named Poisson's ratio. The elasticity modulus, the shear modulus of elasticity, and Poisson's ratio are related by

$$E = 2\,(1+\nu)\,G. \tag{5.7}$$

The principal strains are in the direction of the principal stresses.

5.3 Shear and Moment

Figure 5.2(a) shows a beam, $AB = l$, supported by two positive reaction joints R_1 and R_2. A negative concentrated transversal force F acts on the beam at a distance a. To solve for the joint reactions two equations are needed, sum of the moments of the forces with respect to A and sum of the forces for the beam, or in MATLAB:

```
syms l a F R1 R2
% joint reaction force R1 R2
% sum of moments of the beam about A
SMA = R2*l - F*a;
% sum of forces of the beam
SF = R1 + R2 - F;
```

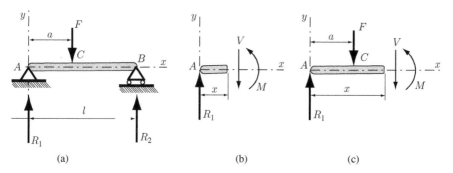

Figure 5.2 Beam (a) supported by two positive reaction joints, (b) free-body diagram of left-hand part at distance x, (c) free-body diagram of left-hand part at a different distance x.

and the symbolical solution is obtained with:

```
solR = solve (SMA==0, SF==0, R1, R2);
R1 = solR.R1;
R2 = solR.R2;
% R1 = -(F*(a - 1))/l
% R2 = (F*a)/l
```

The free-body diagram of the left-hand part at a distance x, section 1 (AC), is represented in Figure 5.2(b). For static equilibrium the internal shear force $V = V_1$ and the internal bending moment $M = M_1$ act at $x, 0 \leq x < a \leq a$. The summing of the forces of the left-hand segment is

$$\Sigma F = R_1 - F - V = 0,$$

and the the sum of the moments of all the forces of part with respect to an axis through the section is

$$\Sigma M = R_1\, x + M = 0,$$

and with MATLAB:

```
syms  x
% section 1 - AC
% SF1 = R1 - V1 == 0;
V1 = R1;
% SM1 = -R1*x + M1 == 0;
M1 = R1*x;
% V1 = -(F*(a - 1))/l
% M1 = -(F*x*(a - 1))/l
```

Figure 5.2(c) shows the free-body diagram of the left-hand part at a distance x, section 2 (CB). The shear force $V = V_2$ and the bending moment $M = M_2$ at x, $a < x < a \leq l$, are

```
% section 2 - CB
% SF2 = R1 - F - V2 == 0;
```

```
V2 = R1 - F;
% SM1 = -R1*x + F*(x-a) + M2 == 0;
M2 = R1*x - F*(x-a);
% V2 = -(F*a)/l
% M2 = (F*a*(l - x))/l
```

For the numerical application

```
lists = {l, a, F};
listn = {10, 4, 200};
% l = 10   (m)
% a =  4   (m)
% F = 200 (N)
R1n = subs(R1, lists, listn);
R2n = subs(R2, lists, listn);
V1n = subs(V1, lists, listn);
M1n = subs(M1, lists, listn);
V2n = subs(V2, lists, listn);
M2n = subs(M2, lists, listn);
% R1 = 120 (N)
% R2 = 80 (N)
% V1 = 120 (N)
% M1 = 120*x (N m)
% V2 = -80 (N)
% M2 = 800 - 80*x (N m)
```

Figure 5.3 shown the shear and moment diagrams. The MATLAB commands for the diagrams are obtained with:

```
l = 10;
a =  4;
figure
subplot(2,1,1);
V = piecewise(...
    0<=x & x<=a, V1n, ...
     a<x & x<=l, V2n);
fplot(V,[0, l],'-k','LineWidth',2);
hold on
fplot(0,[0, l],'-.k','LineWidth',1);
grid on;
xlabel('x (m)');
ylabel('V (N)');
subplot(2,1,2);
M = piecewise(...
    0<=x & x<=a, M1n,...
     a<x & x<=l, M2n);
fplot(M,[0, l],'-k','LineWidth',2);
```

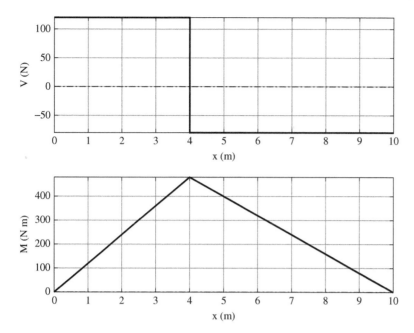

Figure 5.3 Beam shear and moment diagrams.

```
hold on;
fplot(0,[0, 1],'-.k','LineWidth',1);
grid on;
xlabel('x (m)');
ylabel('M (N m)');
```

5.4 Deflections of Beams

Consider an infinitesimal beam element between two points A and B of a deflected beam (Figure 5.4(a)) having the curvature radius denoted by γ. The distance along the curve between the point A and the point B is ds and amid A and B measured on the x axis direction is dx. The increment angle of rotation is $d\theta$ and the increment in deflection between point A and point B is dy. The following relations can be written

$$\gamma d\theta = ds,$$

or

$$\frac{1}{\gamma} = \frac{d\theta}{ds}. \tag{5.8}$$

The deflection curve slope is in fact the first derivative $\dfrac{dy}{dx}$. The slope is calculated using the tangent of the rotation angle θ, because dx is infinitesimally small

$$\frac{dy}{dx} = \tan \theta.$$

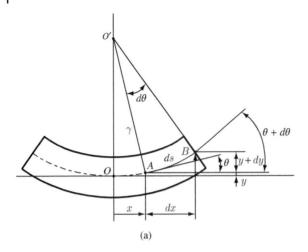

(a)

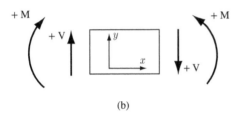

(b)

Figure 5.4 (a) Infinitesimal beam element between two points A and B of a deflected beam, (b) signs for bending moment and shear force.

For small curvatures, the angle θ is a very small quantity, and some approximations are made to simplify the analysis.

$$ds = \frac{dx}{\cos \theta}.$$

Since for a very small angle θ the cosine is approximated by $\cos \theta \approx 1$, one can calculate

$$ds \approx dx.$$

Therefore, Equation (5.8) becomes

$$\frac{1}{\gamma} = \frac{d\theta}{dx}.$$ (5.9)

Also, for a very small angle θ the $\tan \theta \approx \theta$, thus

$$\theta = \tan \theta = \frac{dy}{dx}.$$ (5.10)

Differentiating Equation (5.10) with respect to x,

$$\frac{d\theta}{dx} = \frac{d^2y}{dx^2}.$$

Combining Equations (5.9) and (5.10), yields

$$\frac{1}{\gamma} = \frac{d\theta}{dx} = \frac{d^2y}{dx^2}. \tag{5.11}$$

This equation relates the curvature to the deflection y when the rotations are infinitesimal (small). If the beam material is linearly elastic the curvature is

$$\frac{1}{\gamma} = \frac{M}{EI}, \tag{5.12}$$

where M is the bending moment and $E\,I$ is the flexural rigidity of the beam. The signs for bending moment and shear force are portrayed in Figure 5.4(b). Equations (5.11) and (5.12) yield

$$\frac{d^2y}{dx^2} = \frac{M}{E\,I},$$

or

$$M = E\,I\frac{d^2y}{dx^2}. \tag{5.13}$$

Consider an element of the beam cut out between the cross-sections that are dx apart as seen in Figure 5.5. On the left-hand section of the element are shown the shear force V and the bending moment M, acting in their positive directions. In general, V and M are functions of the distance x measured along the axis of the beam. If the increments in V and M are denoted by dV and dM respectively, then the corresponding resultants on the right-hand section are $V + dV$ and $M + dM$. The load acting on the top of the element is considered distributed dF. The resultant force due to the external loading has the magnitude $dF\ dx$.

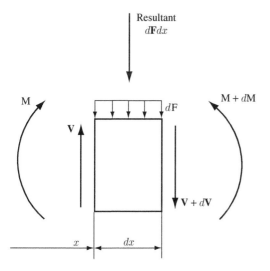

Figure 5.5 Element of the beam cut out between the cross-sections that are dx apart.

The momentum equilibrium equation about an axis through the left-hand section of the element and perpendicular to the plane of the figure is

$$-M - dF\ dx\left(\frac{dx}{2}\right) - (V + dV)dx + M + dM = 0.$$

The differential products are negligible and the shear force can be calculated with

$$V = \frac{dM}{dx}.$$

(5.14)

The force equilibrium equation in the vertical direction is

$$V - (V + dV) - dF dx = 0,$$

or

$$dF = -\frac{dV}{dx}.$$

(5.15)

Taking into account Equation (5.13), yields

$$dF = -\frac{d}{dx}\left(EI\frac{d^2y}{dx^2}\right).$$

(5.16)

If a load w is uniformly distributed is acting on the beam then

$$\frac{dV}{dx} = \frac{d^2M}{dx^2} = w.$$

(5.17)

Integrating (5.17) between any two locations on the beam having the coordinates x_1 and x_2 yields

$$\int_{V_1}^{V_2} dV = \int_{x_1}^{x_2} w\ dx = V_2 - V_1.$$

(5.18)

The difference between moment 1 and moment 2 is the same as the area of the shear force diagram between x_1 and x_2

$$\int_{M_1}^{M_2} dM = \int_{x_1}^{x_2} V\ dx = M_2 - M_1.$$

(5.19)

The stress is calculated with

$$\sigma = -\frac{E\ y}{\gamma}.$$

(5.20)

The deformation of the beam from the neutral axis is distance y, and γ is the radius of curvature, $1/\gamma = d\theta/ds$. The bending moment generated by the stress σ is

$$M = \int \sigma y\ dA = \frac{1}{\gamma}E\int y^2\ dA,$$

(5.21)

where the second integral is the second moment of area I, about the z axis

$$I = \int y^2\ dA.$$

(5.22)

The stress, σ, is a function of the bending moment M and the distance to the neutral axis

$$\sigma = -\frac{My}{I}.$$

(5.23)

The maximum stress is obtained when where $y = y_{max} = h$

$$\sigma_{max} = \frac{M h}{I}. \tag{5.24}$$

5.5 Examples

Example 5.1

Figure 5.6(a) shows three links 1, 2, and 3. Link i has the Young's modulus E_i the length l_i, and the diameter d_i, $i = 1, 2$. A force F acts at the middle of the rigid link 3. The length of the link 3 is $AD = 2\,d$. The rods 1 and 2 will stretch under the action of force F. Calculate the angular displacement of link 3.

For the numerical application use: $E_1 = 30\,(10^6)$ psi (steel), $E_2 = 15\,(10^6)$ psi (brass), $l_1 = 55$ in,
$l_2 = 55$ in, $d_1 = 1/4$ in, $d_2 = 3/8$ in, $a = 5$ in, $b = 20$ in, $c = 15$ in, $F = 4\,(10^3)$ lb.

Solution

To calculate the forces on links 1 and 3 the free-body diagram of link 3 is depicted in Figure 5.6(b). The force F_1 is the reaction force of the link 3 on link 1 and the force F_2 is the reaction force of the link 3 on link 2. The sum of moments for the link 3 with respect to application point C of force F_2 gives

$$\sum M_C = F_1\,b - F\,[(a + b + c)/2 - c] = 0. \tag{5.25}$$

The sum of moments for the link 3 with respect to application point B of force F_1 gives

$$\sum M_B = F_2\,b - F\,[(a + b + c)/2 - a] = 0. \tag{5.26}$$

The reaction forces F_1 and F_2 are

$$F_1 = F\,[(a + b + c)/2 - c]/b,$$
$$F_2 = F\,[(a + b + c)/2 - a]/b. \tag{5.27}$$

The cross section areas of the links 1 and 2 are

$$A_1 = \pi\,d_1^2/4,$$
$$A_2 = \pi\,d_2^2/4. \tag{5.28}$$

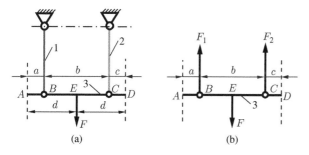

(a)   (b)

Figure 5.6 System of connected links under the action of a force F.

The elastic displacement of link 1 at B is

$$\delta_1 = F_1\, l_1/(A_1\, E_1), \tag{5.29}$$

and the elastic displacement of link 2 at C is

$$\delta_2 = F_2\, l_1/(A_2\, E_2). \tag{5.30}$$

The angular displacement of link 3 is calculated as

$$\phi = (\delta_2 - \delta_1)/b. \tag{5.31}$$

The MATLAB program for the angular displacement ϕ is:

```
clear all; clc; close all

% Young's modulus
E1 = 30*10^6; % (psi) steel
E2 = 15*10^6; % (psi) brass

% bar lengths
l1 = 55; % (in)
l2 = 55; % (in)

% bar diameter
d1 = 1/4; % (in)
d2 = 3/8; % (in)

% distance to loads
a =  5; % (in)
b = 20; % (in)
c = 15; % (in)

% load
F = 4*10^3; % (lb)

% sum moments about C
% F1*b-F*((a+b+c)/2-c) = 0;
% sum moments about B
% F2*b-F*((a+b+c)/2-a) = 0;

F1 = F*((a+b+c)/2-c)/b;
F2 = F*((a+b+c)/2-a)/b;

fprintf('F1 = %6.3f (lb) \n',F1)
fprintf('F2 = %6.3f (lb) \n',F2)

A1 = pi*d1^2/4;
A2 = pi*d2^2/4;
```

```
% rod deflections
delta1 = F1*l1/(A1*E1);
delta2 = F2*l2/(A2*E2);

fprintf('delta1 = %6.3f (in) \n',delta1)
fprintf('delta2 = %6.3f (in) \n',delta2)

% angular deflection
phi = (delta2-delta1)/b;

fprintf...
('phi=%6.6f(rad)=%6.6f(deg)\n',phi,phi*180/pi)
```

The final results are:

```
% F1 = 1000.000 (lb)
% F2 = 3000.000 (lb)
% delta1 =   0.037 (in)
% delta2 =   0.100 (in)
% phi=0.003112(rad)=0.178325(deg)
```

Example 5.2

Figure 5.7(a) shows a simply supported beam with the length $l = 30$ in, the diameter $d = 1$ in, and Young's modulus $E = 30 \ (10^6)$ psi, The distance to the extreme fibre is $c = d/2$. A load $F = 20$ lb acts at a distance $a = l/3$, as seen in the Figure 5.7(a). Determine the maximum deflection and the maximum stress in the beam.

Solution

The beam shown in Figure 5.7(a) is loaded with a concentrated force F. The input data for the beam in MATLAB are:

```
l = 30; %(in)
d = 1; %(in)
a = 1/3; %(in) % a<=l/2
E = 30*10^6; % (psi)
F = 20; %(lb)
c = d/2;
```

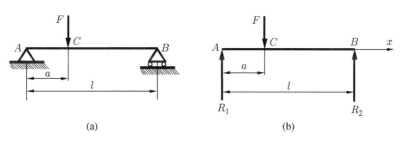

(a)   (b)

Figure 5.7   Simple supported beam under the action of a force F.

The unknown reactions at the supports, R_1 and R_2, are concentrated loads, as shown in Figure 5.7(b). The equation for the load function has the following expression

$$q(x) = R_1 \, \delta(x) - F \, \delta(x - a) + R_2 \, \delta(x - l), \tag{5.32}$$

where $\delta(x)$ is the Dirac delta function

$$\delta(x) = \begin{cases} \infty & x = 0, \\ 0 & x \neq 0. \end{cases} \tag{5.33}$$

The shear function is obtained integrating Equation (5.32)

$$V(x) = \int q(x) \, dx. \tag{5.34}$$

A second integration gives the moment function

$$M(x) = \int V(x) \, dx. \tag{5.35}$$

The MATLAB commands for the load function, the shear function, and the moment function are:

```
syms x R1 R2
q = R1*dirac(x)-F*dirac(x-a)+R2*dirac(x-l);
V = int(q, x);
M = int(V, x);
```

where `dirac(x)` is the Dirac delta function. The functions $V(x)$ and $M(x)$ at x slightly larger than l are zero. The reactions R_1 and R_2 are determined from $V(l+) = 0$ and $M(l+) = 0$:

```
Vlp = subs(V, x, l+10^(-6));
Mlp = subs(M, x, l+10^(-6));
solR = vpasolve(Vlp, Mlp, R1, R2);
R1n = solR.R1;
R2n = solR.R2;


% R1 = 13.333 (lb)
% R2 =  6.667 (lb)
```

and the functions $V(x)$ and $M(x)$ will be:

```
Vn = subs(V, {R1, R2}, {R1n, R2n});
Mn = subs(M, {R1, R2}, {R1n, R2n});
```

The moment equation is integrated to obtain the term $E\,I\,\theta$, where θ is the slope and I is the mass moment of inertia. The term $E\,I\,y$, where y is the deflection is obtained integrating the expression $E\,I\,\theta$:

```
syms C1 C2
EItheta = int(Mn, x) + C1;
EIy = int(theta, x) + C2;
```

The constants of integration C_1 and C_2 are evaluated by setting the deflection equal to zero at $x = 0$ and $x = l$:

```
y0 = subs(EIy, x, 0);
yl = subs(EIy, x, l);

solC = vpasolve(y0, yl, C1, C2);
C1n = solC.C1;
C2n = solC.C2;

% C1 = -1111.111
% C2 =   0.000
```

The slope, θ, and the deflection y are obtained with:

```
% moment of inertia
I = pi*d^4/64; %(in^4)
% slope
thetan = ...
  subs(EItheta,{C1,C2},{C1n,C2n})/(E*I);
% deflection
yn = ...
  subs(EIy,{C1,C2},{C1n,C2n})/(E*I);
```

The maximum deflection, y_M is obtain when the slope θ is zero:

```
% maximum deflection
xM = vpasolve(thetan, x, [0 l]);
yM = subs(yn, x, xM);
% xM = 13.670 (in)
% yM = -0.00657 (in)
```

For this type of beam the maximum deflection has the analytical expression:

```
yMAX = F*a*(l^2-a^2)^(3/2)/(9*sqrt(3)*l*E*I);
% |yMAX| = 0.00657 (in)
```

To plot $V(x)$ and $M(x)$ the following MATLAB commands are used:

```
x = linspace(0,l,1000); % 0 < x < l
% q = R1*dirac(x)-F*dirac(x-a)+R2*dirac(x-l);
Vm = R1n*heaviside(x)...
     -F*heaviside(x-a)...
     +R2n*heaviside(x-l);
Mm = R1n*x.*heaviside(x)...
     -F*(x-a).*heaviside(x-a)...
     +R2n*(x-l).*heaviside(x-l);
```

The MATLAB function heaviside(x) is the Heaviside step function or unit step function

$$H(x) = \begin{cases} 0 & x < 0, \\ 1 & x \geq 0. \end{cases} \tag{5.36}$$

The differential of the Heaviside step function is

$$\frac{d}{dt}H(x-a) = \delta(x-a). \tag{5.37}$$

The following relations exit

$$\int_{-\infty}^{x} \delta(x-a) = H(x-a) \text{ and } \int_{-\infty}^{x} H(x) = x\,H(x). \tag{5.38}$$

The maximum shear force Vmax and the maximum moment Mmax are:

```
Vmax = max(abs(Vm));
Mmax = max(abs(Mm));
```

```
% |Vmax| = 13.333 (lb)
% |Mmax| = 133.333 (lb in)
```

The plots of the bending moment and shear force are shown in Figure 5.8 The MATLAB commands for the shear force and the bending moment plots are:

```
figure(1)
subplot(2,1,1);
fplot(Vn, [0, 1],'-k','LineWidth',2)
hold on
fplot(0,[0, 1],'-.k','LineWidth',1)
xlabel('x (in)')
ylabel('V (lb)')
title('shear force')
grid
subplot(2,1,2);
fplot(Mn, [0, 1],'-k','LineWidth',2)
hold on
fplot(0,[0, 1],'-.k','LineWidth',1)
xlabel('x (in)')
ylabel('M (lb in)')
title('bending moment')
grid
```

The slope and the deflection plots are shown in Figure 5.9. The MATLAB commands for the slope and the deflection plots are:

```
figure(2)
subplot(2,1,1);
fplot(thetan, [0, 1],'-k','LineWidth',2)
```

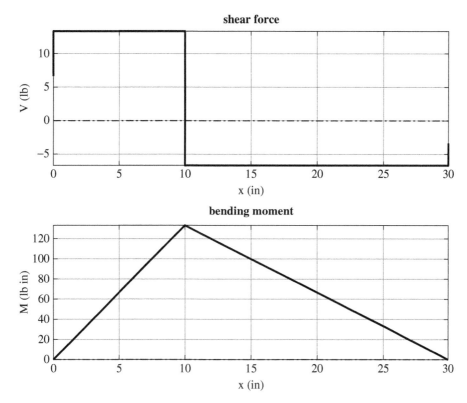

Figure 5.8 MATLAB plots of the bending moment and shear force.

```
hold on
fplot(0,[0, 1],'-.k','LineWidth',1)
xlabel('x (m)')
ylabel('slope \theta (rad)')
title('\theta')
grid
subplot(2,1,2);
fplot(yn,[0, 1],'-k','LineWidth',2)
hold on
fplot(0,[0, 1],'-.k','LineWidth',1)
xlabel('x (m)')
ylabel('y (m)')
title('deflection y')
grid
```

The maximum bending stress occurs where the moment is maximum at $x = a$:

```
sigmaM = Mmax*c/I;
% sigmaM = 1358.122 (psi)
```

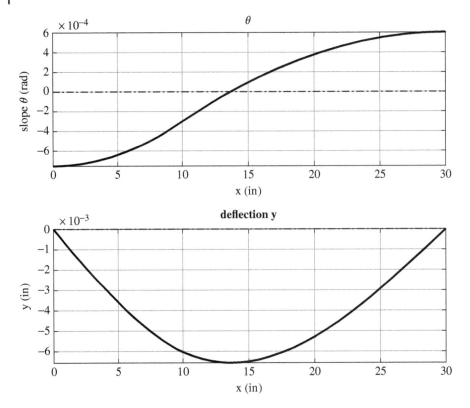

Figure 5.9 MATLAB slope and deflection plots of the beam.

Example 5.3

Figure 5.10(a) shows a beam supported and loaded with a force F. Find the reaction forces, maximum shear force, and maximum moment. The length of the beam is l and the distance between the supports is a. The numerical data are:

```
l = 30;    % (in)
a = 1/3;  % (in)
F = 20;    % (lb)
```

Solution

Figure 5.10(b) shows the beam with the concentrated forces F, and the support reactions R_1 and R_2. The load function equation is

$$q(x) = R_1\, \delta(x) + R_2\, \delta(x - a) - F\, \delta(x - l). \tag{5.39}$$

The shear function and moment function are obtained as

$$V(x) = \int q(x)\, dx \text{ and } M(x) = \int V(x)\, dx. \tag{5.40}$$

The moment function, load function, and the shear function are calculated with MATLAB:

```
syms x R1 R2
q = R1*dirac(x-0)+R2*dirac(x-a)-F*dirac(x-l);
```

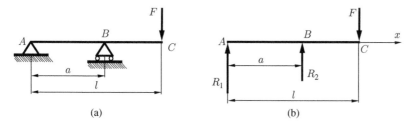

Figure 5.10 Beam supported and loaded with a force F.

```
V = int(q, x);
M = int(V, x);
```

The reactions R_1 and R_2 are determined using the conditions $V(l+) = 0$ and $M(l+) = 0$:

```
Vlp = subs(V, x, 1+10^(-6));
Mlp = subs(M, x, 1+10^(-6));
solR = vpasolve(Vlp, Mlp, R1, R2);
R1n = solR.R1;
R2n = solR.R2;

% R1 = -40.000 (lb)
% R2 = 60.000 (lb)
```

The functions $V(x)$ and $M(x)$ are:

```
Vn = subs(V, {R1, R2}, {R1n, R2n});
Mn = subs(M, {R1, R2}, {R1n, R2n});
```

To plot $V(x)$ and $M(x)$ a range for x is defined:

```
x = linspace(0, 1, 1000); % 0 < x < 1
Vm = R1n*heaviside(x)...
     +R2n*heaviside(x-a)...
     -F*heaviside(x-1);
Mm = R1n*x.*heaviside(x)...
     +R2n*(x-a).*heaviside(x-a)...
     -F*(x-1).*heaviside(x-1);
```

The maximum shear force Vmax and the maximum moment Mmax are:

```
Vmax = max(abs(Vm));
Mmax = max(abs(Mm));

% |Vmax| = 40.000 (lb)
% |Mmax| = 400.000 (lb in)
```

The shear force and the bending moment plots are shown in Figure 5.11

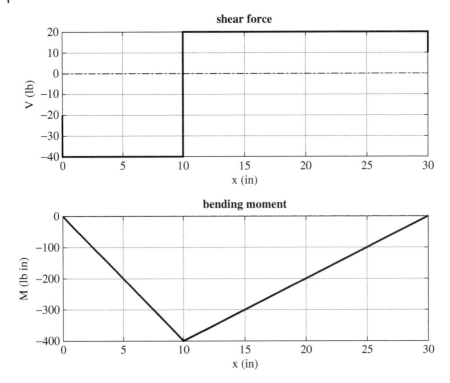

Figure 5.11 MATLAB plots of the bending moment and shear force.

Example 5.4

Figure 5.12(a) shows a supported beam and a distributed load of magnitude w. The beam has a length l and a diameter d. The distance to the distributed load is l_1. The modulus of elasticity of the beam is E. Find the reaction forces, maximum shear force, maximum bending moment, and maximum bending stress. The numerical data are:

```
l   = 1;      % (m) beam length
l1  = 0.3;    % (m) distance to distributed load
d   = 0.02;   % (m) beam diameter
w   = 800;    % (N/m) distributed load magnitude
E   = 207*10^9; % (Pa) modulus of elasticity
```

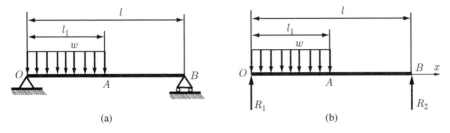

(a)   (b)

Figure 5.12 Supported beam with a distributed load.

Solution

Figure 5.12 shows the beam with the distributed load w and the support reactions R_1 and R_2. The load function equation is

$$q(x) = R_1 \, \delta(x) - w \, H(x) + w \, H(x - l_1) + R_2 \, \delta(x - l). \tag{5.41}$$

The moment function, load function, and the shear function are calculated in MATLAB with:

```
syms x R1 R2
% load function
q = R1*dirac(x)...
    -w*heaviside(x)+w*heaviside(x-l1)...
    +R2*dirac(x-l);
% shear function
V = int(q, x);
% moment function
M = int(V, x);
```

The reactions R_1 and R_2 are determined using the conditions $V(l+) = 0$ and $M(l+) = 0$:

```
Vlp = subs(V, x, l+10^(-6));
Mlp = subs(M, x, l+10^(-6));
solR = vpasolve(Vlp, Mlp, R1, R2);
R1n = solR.R1;
R2n = solR.R2;

% R1 = 204.000 (N)
% R2 = 36.000 (N)
```

The functions $V(x)$ and $M(x)$ are:

```
Vn = subs(V, {R1, R2},{R1n, R2n});
Mn = subs(M, {R1, R2},{R1n, R2n});
```

The moment equation is integrated to obtain $E\,I\,\theta$, and $E\,I\,\theta$ is integrated to obtain $E\,I\,y$:

```
syms C1 C2 % constants of integration
EItheta = int(Mn, x) + C1;
EIy = int(EItheta, x) + C2;

% determine C1 and C2
% deflection y=0 at x=0
y0 = subs(EIy, x, 0);
% deflection y=0 at x=1
yl = subs(EIy, x, l);
solC = vpasolve(y0, yl, C1, C2);
C1n = solC.C1;
C2n = solC.C2;
```

```
% C1 =  -8.670
% C2 =   0.000
```

With the constants of integration determined, the slope, θ, and the deflection, y, are obtained with:

```
% moment of inertia
I = pi*d^4/64; % (m^4)
% slope
thetan = ...
 subs(EItheta,{C1,C2},{C1n,C2n})/(E*I);
% deflection
yn = ...
 subs(EIy,{C1,C2},{C1n,C2n})/(E*I);
```

The maximum deflection, y_M is obtain when the slope θ is zero:

```
% maximum deflection where theta = 0
xM = vpasolve(thetan, x, [0 1]);
yM = subs(yn, x, xM);
```

```
% xM =   0.436 (m)
% yM = -0.00133 (m)
```

The maximum shear force Vmax and the maximum moment Mmax are obtained with:

```
% Vmax and Mmax
x = linspace(0,1,1000); % 0 < x < 1
% q = R1*dirac(x)...
%      -w*heaviside(x)+w*heaviside(x-a)...
%      +R2*dirac(x-1);
Vm = R1n*heaviside(x)...
       -w*x.*heaviside(x)...
       +w*(x-l1).*heaviside(x-l1)...
       +R2n*heaviside(x-1);
Mm = R1n*x.*heaviside(x)...
       -w*x.^2.*heaviside(x)/2 ...
       +w*(x-l1).^2.*heaviside(x-l1)/2 ...
       +R2n*(x-1).*heaviside(x-1);
Vmax = max(abs(Vm)); % maximum shear
Mmax = max(abs(Mm)); % maximum moment
```

```
% |Vmax| = 203.199 (N)
% |Mmax| =  26.010 (N m)
```

The shear force and the bending moment plots are shown in Figure 5.13. The slope and the deflection plots are shown in Figure 5.14. The maximum bending stress is calculated where the moment is a maximum:

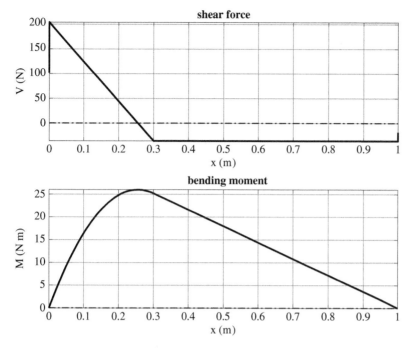

Figure 5.13 MATLAB plots of the bending moment and shear force.

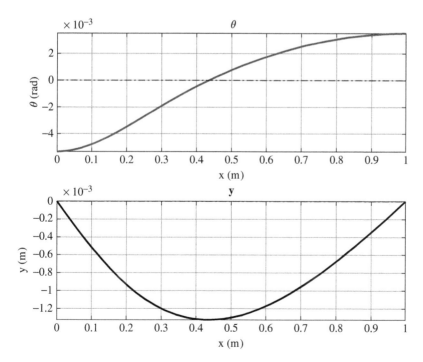

Figure 5.14 MATLAB slope and deflection plots of the beam.

```
% distance to extreme fiber
c = d/2;
% maximum bending stress where the moment is a maximum
sigmaM = Mmax*c/I;

% sigmaM = 33.117 (MPa)
```

Example 5.5

A cantilever beam of length l has a uniformly distributed load of magnitude w at a distance l_1, as shown in Figure 5.15(a). The beam with the modulus of elasticity E has a rectangular cross section with the dimensions $b \times h$. Find the maximum shear force, the maximum bending moment, and the maximum bending stress.

The numerical data are:

```
% bar length
l = 1.2; % (m)
% distance to distributed load
l1 = 0.4; % (m)
% distributed load magnitude
w = 500; % (N/m)
% cross section b x h
b = 4*10^-2; % (m)
h = 4*10^-2; % (m)
% modulus of elasticity
E = 207*10^9; % (Pa)
```

Solution

The distance to the extreme fibre and the moment of inertia are:

```
c = h/2; % distance to extreme fiber
I = b*h^3/12; % mass moment of inertia
```

Figure 5.15(b) shows the beam with the distributed load w, the reaction force R_1 and the reaction moment M_1. The load function equation is

$$q(x) = M_1 \, d\delta(x)/dt + R_1 \, \delta(x) - w \, H(x - l_1). \tag{5.42}$$

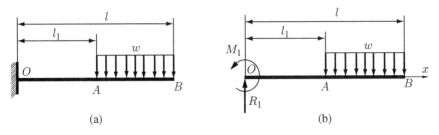

(a)   (b)

Figure 5.15 MATLAB cantilever beam of length l with a uniformly distributed load.

The load function, the shear function, and the moment function with MATLAB are:

```
syms x R1 M1
% load function equation
q = M1*diff(dirac(x),x)...
    +R1*dirac(x)...
    -w*heaviside(x-l1);
% shear force V(x): int from -Inf to x
V = int(q, x);
% moment M(x): int from -Inf to x
M = int(V, x);
```

The reactions R_1 and M_1 are evaluated considering the shear and moment equations are zero at a point just to the right of $x = l$.

```
% reactions R1 and M1
Vlp = subs(V,x,l+10^(-6));
Mlp = subs(M,x,l+10^(-6));
solR = vpasolve(Vlp, Mlp, R1, M1);
R1n = solR.R1;
M1n = solR.M1;

% R1 =  400.000  (N)
% M1 = -320.001  (N m)
```

The functions $V(x)$ and $M(x)$ are:

```
Vn = subs(V,{R1,M1},{R1n,M1n});
Mn = subs(M,{R1,M1},{R1n,M1n});
```

The moment equation is integrated to obtain $EI\theta$, and $EI\theta$ is integrated to obtain EIy. At the point $x = 0$ the slope and the deflection are zero:

```
syms C1 C2 % constants of integration
EItheta = int(Mn, x) + C1;
EIy = int(EItheta, x) + C2;

% determine C1 and C2
% slope theta=0 at x=0
theta0 = subs(EItheta, x, 0);
C1n = vpasolve(theta0, C1);
% deflection y=0 at x=0
y0 = subs(EIy, x, 0);
C2n = vpasolve(y0, C2);

% C1 =   0.000
% C2 =   0.000
```

The slope, θ, and the deflection, y, are:

```
% slope
thetan = ...
  subs(EItheta,{C1,C2},{C1n,C2n})/(E*I);
% deflection
yn = ...
  subs(EIy,{C1,C2},{C1n,C2n})/(E*I);
```

The maximum slope, thetaM, and the maximum deflection, yM, are obtained when $x = l$:

```
% maximum slope when x = 1
thetaM = subs(thetan, x, 1);
% maximum deflection when x = 1
yM = subs(yn, x, 1);
```

```
% thetaM=-0.003140(rad)=-0.179915(deg)
% yM = -0.0028 (m)
```

The maximum shear force Vmax and the maximum moment Mmax are calculated with:

```
%   Vmax and Mmax
x = linspace(0,1,10000); % 0 < x < 1
% q = M1*diff(dirac(x),x)...
%       +R1*dirac(x)...
%       -w*heaviside(x-a);

Vm = R1n*heaviside(x)  ...
     -w*(x-l1).*heaviside(x-l1);

Mm = M1n*heaviside(x)  ...
     +R1n*x.*heaviside(x)  ...
     -w*(x-l1).^2.*heaviside(x-l1)/2;

Vmax = max(abs(Vm));
Mmax = max(abs(Mm));
```

```
% |Vmax| = 400.000 (N)
% |Mmax| = 319.953 (N m)
```

The shear force and the bending moment plots are shown in Figure 5.16. The slope and the deflection plots are shown in Figure 5.17. The maximum bending stress is calculated where the moment is a maximum:

```
sigmaM = Mmax*c/I;
```

```
% sigmaM = 29.996 (MPa)
```

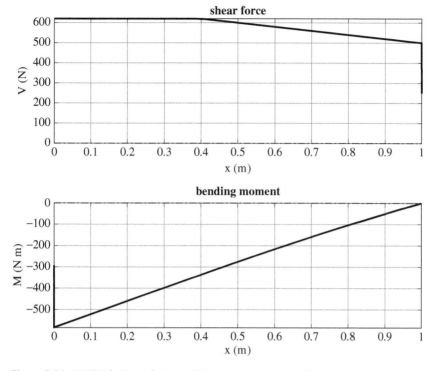

Figure 5.16 MATLAB plots of the bending moment and shear force.

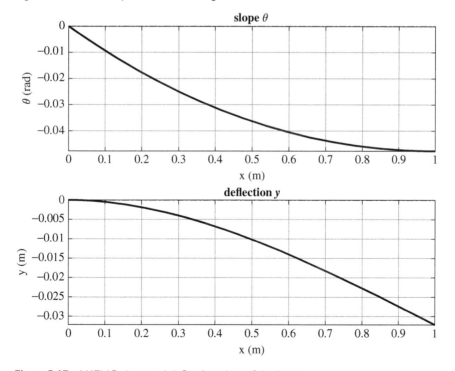

Figure 5.17 MATLAB slope and deflection plots of the beam.

6

Friction

Friction, coming from the Latin word *fricare*, is the resistive force caused by the motion between two surfaces (objects) that are sliding or trying to slide across one another. This resistive force, i.e. friction force, which acts always opposite to the direction the object is moving or is trying to move is due to the irregularities and roughness of the surfaces in contact. Therefore, the magnitude of the resistive force depends on the materials from which the two surfaces are made.

The standard friction laws were discovered but not published by Leonardo da Vinci (1452–1519) a very famous Italian astronomer, engineer, mathematician, architect and sculptor. Considered one of the greatest painters of all time he also studied diverse engineering mechanisms such as screws, gears and bearings, and is considered a pioneer in tribology. The three laws of friction were reformulated by Guillaume Amontons in 1699 and verified and further developed by the French physicist Charles-Augustin de Coulomb in 1791. In an important work, namely "Théorie des machines simples", considering the laws of friction, namely: (i) Amontons first law – the friction force is directly proportional to the normal force, (ii) Amontons second law – the friction force is independent of the apparent area of contact, and (iii) Coulomb's Law – friction is independent of the relative velocity of sliding, Charles-Augustin de Coulomb investigated static and dynamic friction of sliding surfaces, rolling friction and the bending friction of cords.

John Leslie (1766–1832) and Arthur Jules Morin (1795–1880) further developed and extended Coulomb's work on rolling versus sliding friction under the supervision of Jean-Victor Poncelet.

A very simple way to attempt to discuss friction is the motion of a body on a horizontal surface (Figure 6.1). If the body is pushed by a *sufficiently large* horizontal force $\mathbf{F}$, the body starts to move, but if the force is not *big enough* the body remains at rest, that is, a resistive force opposes the desired motion.

The free-body diagram in Figure 6.1 shows the rigid body weight $\mathbf{W}$, the normal reaction force $\mathbf{F}_N$ and the friction force $\mathbf{F}_f$ exerted by the horizontal surface on the rigid body. If the body is in equilibrium the friction force named static friction $\mathbf{F}_s$ force balances the pushing force, that is, the magnitudes are equal $F_s = F$. When the force $\mathbf{F}$ increases but the body does not move (is in equilibrium) the friction force $\mathbf{F}_f$ should proportionally increase in order to balance the force $\mathbf{F}$.

Once the pushing horizontal force F reaches a *sufficiently large* value, the body is no longer in equilibrium, the friction force named kinetic friction can no longer balance the

Engineering Applications: Analytical and Numerical Calculation with MATLAB, First Edition.
Mihai Dupac and Dan B. Marghitu.
© 2021 John Wiley & Sons Ltd. Published 2021 by John Wiley & Sons Ltd.

pushing force and the body starts moving, i.e. slips on the surface. When the rigid body is moving the force required to keep it moving is smaller than the force needed to start the rigid body to slide on the horizontal surface. This can be explained partly by the requirement to break the contacting surface asperities before sliding can start.

Overall, Coulomb's theory of dry friction foresees:

- The maximum (static) friction forces exerted on stationary and dry contacting surfaces
- The (kinetic) friction forces exerted by moving or sliding surfaces.

6.1 Coefficient of Static Friction

When the external applied force F balances the static friction force, F_s, and the body is in equilibrium but on the verge of slipping, the coefficient of friction denoted by μ_s is named the coefficient of static friction. The magnitude of the static friction can be expressed by

$$F_s = \mu_s \|\mathbf{F_N}\| \tag{6.1}$$

where $\mathbf{F_N}$ is the normal (contact) force between the body and the horizontal surface, and $\|\mathbf{F_N}\|$ is the magnitude of the normal force.

When the body is in static equilibrium (rest), i.e. not sliding relative to the horizontal surface, and not on the verge of slipping, the friction force can be expressed as in [28] by

$$\|\mathbf{F_f}\| \leq F_s. \tag{6.2}$$

When the body is in static equilibrium (rest) but on the verge of slipping, i.e. state of impending motion to the horizontal surface, the friction force can be expressed as in [28] by

$$\|\mathbf{F_f}\| = F_s \tag{6.3}$$

where $\|\mathbf{F_f}\|$ is the magnitude of the friction force.

The value μ_s of the static coefficient of friction, is related to

- The irregularities/degree of contamination and roughness/smoothness of the surfaces in contact
- The materials of the two contacting surfaces.

Table 6.1 Typical μ_s values of classical metallic and non-metallic materials.

Materials	μ_s
Metallic material on metallic material	0.15–0.20
Metallic material on wood	0.20–0.60
Metallic material on masonry	0.30–0.70
Wood on wood	0.25–0.50
Masonry on masonry	0.60–0.70
Rubber on concrete	0.50–0.90

Some representative values of the stator coefficient of friction μ_s for different metallic and non-metallic materials are shown in Table 6.1.

Equation (6.1) gives the magnitude of the static friction force the body can exert on the horizontal surface without slipping, but not its direction. If the coefficient of static friction μ_s between two friction surfaces is established, the maximum external force F magnitude that can be applied without producing slipping is $F = \mu_s F_N$.

6.2 Coefficient of Kinetic Friction

When the external applied force F balances the force of kinetic friction, F_c, and the body is sliding relative to the horizontal surface, the coefficient of friction denoted by μ_k is named the coefficient of kinetic friction. The kinetic friction force magnitude between dry surfaces can be expressed by

$$F_c = \mu_k \|\mathbf{F_N}\| \tag{6.4}$$

where $\mathbf{F_N}$ is the normal force between the surfaces in contact, $\|\mathbf{F_N}\|$ is the magnitude, and F_c denotes the magnitude of Coulomb friction.

When the body is sliding relative to the horizontal surface i.e. $v_r \neq 0$, the friction force can be expressed as in [28] by

$$\mathbf{F_f} = -F_c \frac{\mathbf{v_r}}{\|\mathbf{v_r}\|}, \tag{6.5}$$

where $\mathbf{v_r}$ is the relative (tangential) velocity of the body against the horizontal surface.

The coefficient of kinetic friction μ_k is generally smaller than the coefficient of static friction μ_s. To preserve the sliding of the body in Figure 6.1 on the horizontal surface the applied external force must be at least $F = \mu_k N$. When the body shown in Figure 6.2(a) is moving along the positive direction of the x axis, the friction force acts in the opposite direction to its motion as shown in Figure 6.2(b).

6.3 Friction Models

Different friction models ([31, 46]) have been considered in the literature to account for the influence of parameters (contact geometry, materials properties, dry vs. lubricated surface) on friction.

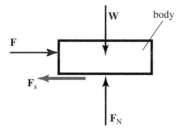

Figure 6.1 Motion of a body on a horizontal surface.

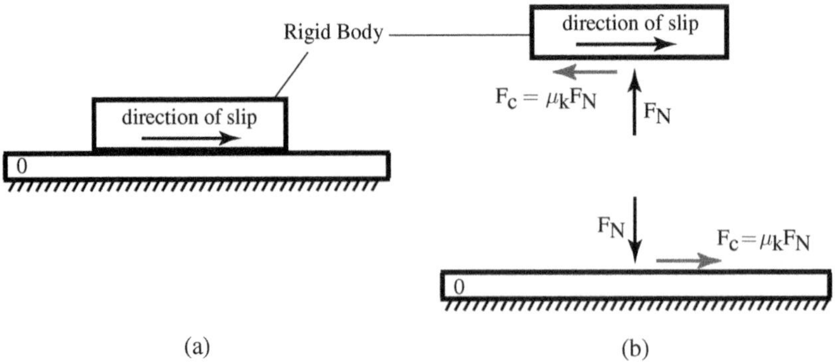

Figure 6.2 Friction force and coefficient of kinetic friction.

6.3.1 Coulomb Friction Model

When the kinetic friction force is $F_c = \mu_k \|\mathbf{F_N}\|$ and the relative motion between the body and the horizontal surface is described through the relative velocity $\mathbf{v_r} \neq 0$, the friction model (Figure 6.3.(a)) with a constant dynamic coefficient of friction [28] can be expressed by

$$
\mathbf{F_f} = \begin{cases} -F_c \dfrac{\mathbf{v_r}}{\|\mathbf{v_r}\|} & \text{if } \|\mathbf{v_r}\| \neq 0 \\ -\mathbf{F_{app}} & \text{if } \|\mathbf{v_r}\| = 0 \end{cases} \tag{6.6}
$$

where $\mathbf{v_r}$ is the relative (tangential) velocity of the body against the horizontal surface, $\|\mathbf{v_r}\|$ is the magnitude of the relative velocity, $\mathbf{F_{app}} = \min(\|\mathbf{F}\|, F_c)\, sgn(\mathbf{F})$, $\|\mathbf{F}\|$ is the magnitude of the resultant of all the forces acting on the tangential direction of the contact, and the *signum* function is expressed as

$$
sgn(\mathbf{F}) = \begin{cases} \dfrac{\mathbf{F}}{\|\mathbf{F}\|} & \text{if } \|\mathbf{F}\| \neq 0 \\ 0 & \text{if } \|\mathbf{F}\| = 0 \end{cases}. \tag{6.7}
$$

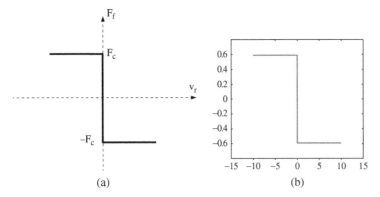

Figure 6.3 Coulomb model (a) Friction force vs. velocity and (b) MATLAB implementation.

The friction model (Figure 6.3) with a constant dynamic coefficient of friction is implemented in MATLAB using:

```
clear all; close all; clc;

Vmax=10;              % maximum velocity
M=3;                  % mass of the body
mu_s=0.03;            % coefficient of static friction
mu_k=0.02;            % coefficient of kinetic friction
kv=0.025;             % viscous coefficient
Fn=M*9.81;            % normal force
t(1)=-Vmax;
step=0.001;           % step used in the simulation
k=1;q=0;              % k and q are step counters
n=0;                  % n=number of steps at zero velocity
vZero=0.01;
for V=-Vmax:step:Vmax
    if(abs(V)<vZero)
    q=q+1;
    n=2*vZero/step;
    end
    %Run Coulomb model
    %Ff(k+1)=Coulomb(V,Fn,mu_s,mu_k,step,vZero,q,n);
    %Run Coulomb viscous model
    %Ff(k+1)=Coulomb_w_v(V,kv,Fn,mu_s,mu_k,step,vZero,q,n);
    %Run Coulomb with stiction model
    Ff(k+1)=Coulomb_w_S(V,Fn,mu_s,mu_k,step,vZero,q,n);
    t(k+1)=t(k)+step;
    k=k+1;
end
plot(t,Ff)
```

Next a MATLAB function that calculates the friction force using the Coulomb friction model in Equation (6.6) is presented:

```
function Ff=Coulomb(V,Fn,mu_s,mu_k,step,vZero,q,n)
if(abs(V)>vZero)
    Fc=-mu_k*Fn*sign(V);
    Ff=Fc;
end
if(abs(V)<=vZero)
    Fc=-mu_k*Fn*sign(V);
  if V<0
    Fapp=Fc-2*(q-1)*Fc/n
    Ff=min(Fc,Fapp);
  else
```

```
    Fapp=-Fc+2*(q-1)*Fc/n
    Ff=-abs(max(Fc,Fapp));
  end
end
```

The challenge in the implementation of the friction model is due to the unknown value of the friction force at zero velocity. The function `Coulomb(V,Fn,mu_s,mu_k,step,vZero,q,n)` approximates the applied force F_{app} for the zero relative velocity inside the variation of Coulomb friction force given by Equation (6.4). The graphical representation of the friction force obtained using the Coulomb friction model is shown in Figure 6.3(b).

6.3.2 Coulomb Model with Viscous Friction

In many engineering applications it might be necessary to reduce friction. A good example is given by the engine of a car, where the moving parts inside are lubricated with oil in order to decrease the inherent wear between parts as well as the amount of generated heat. Such a friction model (Figure 6.4(a)) based on Coulomb's friction law in Equation (6.6) and a viscous friction component ([28]) can be expressed by

$$\mathbf{F}_f = \begin{cases} -F_c \dfrac{\mathbf{v}_r}{\|\mathbf{v}_r\|} - k_v \mathbf{v}_r & \text{if } \|\mathbf{v}_r\| \neq 0 \\ -\mathbf{F}_{app} & \text{if } \|\mathbf{v}_r\| = 0 \end{cases} \tag{6.8}$$

where the viscous coefficient k_v is related to the viscosity of the lubricated oil/fluid, and $\mathbf{F}_{app}$ is calculated using $\mathbf{F}_{app} = \min(\|\mathbf{F}\|, F_c)sgn(\mathbf{F})$.

The MATLAB function `Coulomb_w_v(V,kv,Fn,mu_s,mu_k,step,vZero,q,n)` presented below calculates the friction force in Equation (6.8) using the Coulomb friction model with stiction:

```
function Ff=Coulomb_w_v(V,kv,Fn,mu_s,mu_k,step,vZero,q,n)
if(abs(V)>vZero);
    Fc=mu_k*Fn*sign(V)+kv*V;
    Ff=-Fc;
end
if(abs(V)<=vZero);
    Fc=-mu_k*Fn*sign(V);
    if V<0
    Fapp=Fc-2*(q-1)*Fc/n
    Ff=min(Fc,Fapp);
    else
    Fapp=-Fc+2*(q-1)*Fc/n
    Ff=-abs(max(Fc,Fapp));
    end
end
```

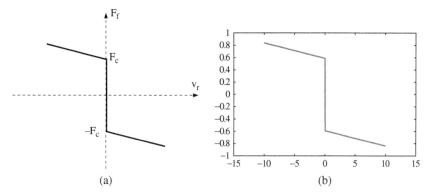

(a)   (b)

Figure 6.4 Coulomb model with viscous friction. (a) Friction force vs. velocity and (b) MATLAB implementation.

The graphical representation of the friction force obtained using the Coulomb model with viscous friction is shown in Figure 6.4(b).

6.3.3 Coulomb Model with Stiction

A friction model [Figure 6.5(a)] based on Coulomb's friction law and the combination of the static and kinetic friction coefficients ([28]) can be expressed using

$$
\mathbf{F}_f = \begin{cases} -F_c \dfrac{\mathbf{v}_r}{\|\mathbf{v}_r\|} & \text{if } \|\mathbf{v}_r\| \neq 0 \\[2ex] -\mathbf{F}_{app} & \text{if } \|\mathbf{v}_r\| = 0 \end{cases} \tag{6.9}
$$

where $\mathbf{F}_{app} = \min\,(\|\mathbf{F}\|, F_s)sgn(\mathbf{F})$.

It can be seen that the only difference between the classical Coulomb friction model and the stiction model (Figure 6.5) is the replacement of F_c with F_s in the calculation of friction force when $\|\mathbf{v}_r\| = 0$.

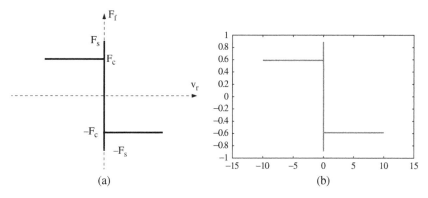

(a)   (b)

Figure 6.5 Coulomb friction model with stiction. (a) Friction force vs. velocity and (b) MATLAB implementation.

The MATLAB function `Coulomb_w_S(V,Fn,mu_s,mu_k,step,vZero,q,n)` presented below calculates the friction force in Equation (6.9) using the Coulomb friction model with stiction:

```
function Ff=Coulomb_w_S(V,Fn,mu_s,mu_k,step,vZero,q,n)
if (abs(V)>vZero);
    Fc=mu_k*Fn*sign(V);
    Ff=-Fc;
end
if (abs(V)<=vZero);
    Fc=-mu_s*Fn*sign(V);
  if V<0
    Fapp=Fc-2*(q-1)*Fc/n
    Ff=min(Fc,Fapp);
  else
    Fapp=-Fc+2*(q-1)*Fc/n
    Ff=-abs(max(Fc,Fapp));
  end
end
```

The graphical representation of the friction force obtained using the Coulomb friction model with stiction is shown in Figure 6.5(b).

6.4 Angle of Friction

For any two surfaces (Figure 6.6) one can define the angle of friction, θ, as the angle between the surface normal $\mathbf{F_N}$ and the friction force, $\mathbf{F_f}$. The magnitudes of the normal force and friction force can be expressed by

$$F_f = F_R \sin\theta, \tag{6.10}$$

$$F_N = F_R \cos\theta \tag{6.11}$$

where the magnitude of the resultant F_R is $\|\mathbf{F_R}\| = |\mathbf{F_N} + \mathbf{F_f}|$.

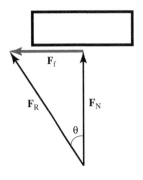

Figure 6.6 Angle of friction.

When slip is impending, the angle of friction denoted θ_s and called *the static angle of friction* can be expressed as

$$\tan \theta_s = \mu_s. \tag{6.12}$$

In the case of sliding friction, the angle of friction denoted by θ_k and called *the kinetic angle of friction*, can be expressed as

$$\tan \theta_k = \mu_k. \tag{6.13}$$

6.5 Examples

Example 6.1
The worker shown in Figure 6.7 is using a rope and an inclined plane to lift a weight A to a desired height. The rope, which makes an angle β with the vertical direction, turns around a fixed drum D of radius r before connecting with the weight A placed on the inclined plane S which makes an angle α with the horizontal direction. The friction coefficient between the rope and the fixed drum D is μ_D, and between the block A and the inclined plane S is μ_S respectively. Knowing that weight of block A determine:

(a) The minimal force the worker must apply to the end of the rope for the block A not to slide down the inclined plane.
(b) The minimal force the worker must apply to the end of the rope for the block A to start moving up the inclined plane.

Numerical application: $W_A = 9$ N, $\mu_S = 0.6$, $\mu_D = 0.45$ m, $r = 0.1$ m, $\alpha = \dfrac{\pi}{4}$, and $\beta = \dfrac{\pi}{5}$.

Solution
As shown in Figure 6.8 the rope that makes an angle β with the vertical direction is making contact with the drum at D_F. After a partial turn around the drum, the rope exit the drum at D_A from where is directed to the weight A. One can calculate the angular displacement of the rope in contact with the drum by

$$\theta = \theta_{D_F O_D D_A} = \left(\frac{\pi}{2} - \beta + \alpha \right), \tag{6.14}$$

```
clear all; clc; close all
```

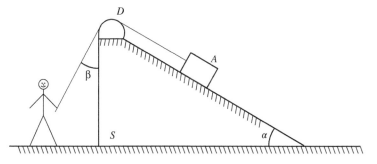

Figure 6.7 Inclined plane with a sliding block.

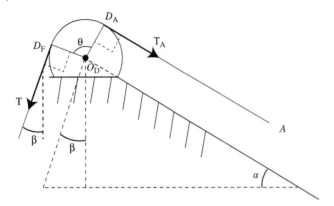

Figure 6.8 Rope and drum on the inclined plane.

```
syms N_A N_B T T_A W_A T F_bdown F_bup e_muD_theta
syms alpha beta R mu_S mu_D r

theta=(pi/2-beta+alpha);
fprintf('The angular displacemet at the drum is \n')
fprintf('Theta = %s \n\n',theta)
```

The relation between forces in the drum rope can be expressed (depending on the direction of the motion) either as

$$T_A = T e^{\mu_D \theta} = T e^{\mu_D \left(\frac{\pi}{2} - \beta + \alpha\right)}$$ (6.15)

or

$$T = T_A e^{\mu_D \theta} = T_A e^{\mu_D \left(\frac{\pi}{2} - \beta + \alpha\right)}.$$ (6.16)

To calculate the minimal force the worker must apply to the end of the rope for the block A not to slide down the inclined plane one can consider that the block A is moving down the inclined plane. Considering the equilibrium of block A, as shown in the free body diagram in Figure 6.9, one can write

$$\begin{cases} N_A - W_A \cos \alpha = 0 \\ T_A - W_A \sin \alpha + F_A = 0 \end{cases} \Leftrightarrow \begin{cases} N_A - W_A \cos \alpha = 0 \\ T_A - W_A \sin \alpha + \mu_S N_A = 0 \end{cases}$$ (6.17)

where $F_A = \mu_S N_A$ is the friction force acting on block A, calculated in MATLAB with:

```
F_A=mu_S*N_A;
fprintf('The friction force acting ')
fprintf('on block A is \n')
fprintf('F_A = %s \n\n',F_A)
```

Considering the (minimal) force the worker must apply to the end of the rope one can write

$$T - F = 0.$$ (6.18)

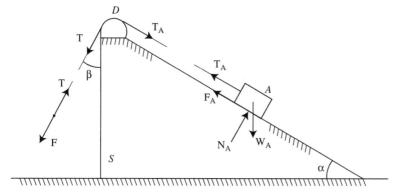

Figure 6.9 Free-body diagram of sliding block, drum and rope on the inclined plane – block not to slide down the inclined plane.

From Equations (6.15)–(6.18) one can write

$$\begin{cases} T_A = Te^{\mu_D\left(\frac{\pi}{2} - \beta + \alpha\right)} \\ N_A - W_A \cos\alpha = 0 \\ T_A - W_A \sin\alpha + \mu_S N_A = 0 \\ T - F = 0 \end{cases},$$

$$\Leftrightarrow \begin{cases} T_A = Fe^{\mu_D\left(\frac{\pi}{2} - \beta + \alpha\right)} \\ T_A - W_A \sin\alpha + \mu_S W_A \cos\alpha = 0 \end{cases},$$

$$\Leftrightarrow Fe^{\mu_D\left(\frac{\pi}{2} - \beta + \alpha\right)} - W_A \sin\alpha + \mu_S W_A \cos\alpha = 0,$$

$$\Leftrightarrow Fe^{\mu_D\left(\frac{\pi}{2} - \beta + \alpha\right)} = W_A(\sin\alpha - \mu_S \cos\alpha). \tag{6.19}$$

In MATLAB the equilibrium equations can be written by:

```
%e_muD_theta=exp(mu_D*theta);
eq1r=N_A-W_A*cos(alpha);
eq2r=T_A-W_A*sin(alpha)+F_A;
eq3r=T_A-T*e_muD_theta;
eq4r=T-F_bdown;
fprintf('Case 1: The equilibrium equations are \n')
fprintf('%s = 0 \n',eq1r);
fprintf('%s = 0 \n',eq2r);
fprintf('%s = 0 \n',eq3r);
fprintf('%s = 0 \n\n',eq4r);
fprintf('where e_muD_Theta is \n\n')
fprintf('e_muD_Theta = %s \n\n',exp(mu_D*theta))
```

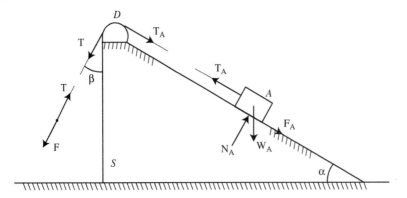

Figure 6.10 Free-body diagram of sliding block, drum and rope on the inclined plane – block moving up the inclined plane.

It results (minimal force the worker must apply to the end of the rope for the block A not to slide down the inclined plane) that

$$F = W_A(\sin \alpha - \mu_S \cos \alpha)e^{-\mu_D\left(\frac{\pi}{2} - \beta + \alpha\right)}. \tag{6.20}$$

The minimal force the worker must apply at the end of the rope for the block A not to slide down the inclined plane is calculated in MATLAB with:

```
sol_r=solve(eq1r,eq2r,eq3r,eq4r,...
    'N_A', 'T', 'F_bdown', 'T_A');
fprintf('The minimal force the worker must apply ')
fprintf('at the end of the rope \n')
fprintf('for the block not to slide ')
fprintf('down the inclined plane \n\n')
fprintf('F_bdown = %s \n\n',sol_r.F_bdown)
```

To calculate the minimal force the worker must apply to the end of the rope for the block A to start moving up the inclined plane one can consider the equilibrium of block A, as shown in the free-body diagram in Figure 6.10, that is

$$\begin{cases} N_A - W_A \cos \alpha = 0 \\ T_A - W_A \sin \alpha - F_A = 0 \end{cases} \Leftrightarrow \begin{cases} N_A - W_A \cos \alpha = 0 \\ T_A - W_A \sin \alpha - \mu_S N_A = 0 \end{cases} \tag{6.21}$$

where $F_A = \mu_S N_A$ is the friction force acting on block A. Considering the (minimal) force the worker must apply to the end of the rope one can write

$$T - F = 0. \tag{6.22}$$

From Equations (6.16), (6.21) and (6.22) one can write

$$\begin{cases} T = T_A e^{\mu_D\left(\frac{\pi}{2} - \beta + \alpha\right)} \\ N_A - W_A \cos \alpha = 0 \\ T_A - W_A \sin \alpha - \mu_S N_A = 0 \\ T - F = 0 \end{cases},$$

$$\Leftrightarrow \begin{cases} F = T_A e^{\mu_D\left(\frac{\pi}{2} - \beta + \alpha\right)} \\ T_A - W_A \sin\alpha - \mu_S W_A \cos\alpha = 0 \end{cases},$$

$$\Leftrightarrow F e^{-\mu_D\left(\frac{\pi}{2} - \beta + \alpha\right)} - W_A \sin\alpha - \mu_S W_A \cos\alpha = 0,$$

$$\Leftrightarrow F e^{-\mu_D\left(\frac{\pi}{2} - \beta + \alpha\right)} = W_A \sin\alpha + \mu_S W_A \cos\alpha. \tag{6.23}$$

In MATLAB the equilibrium equations can be written by:

```
%e_muD_theta=exp(mu_D*theta);
eq11=N_A-W_A*cos(alpha);
eq21=T_A-W_A*sin(alpha)-F_A;
eq31=T-T_A*e_muD_theta;
eq41=T-F_bup;
fprintf('Case 2: The equilibrium equations are \n')
fprintf('%s = 0 \n',eq11);
fprintf('%s = 0 \n',eq21);
fprintf('%s = 0 \n',eq31);
fprintf('%s = 0 \n\n',eq41);
```

It results that

$$F = W_A(\sin\alpha + \mu_S \cos\alpha)e^{\mu_D\left(\frac{\pi}{2} - \beta + \alpha\right)}. \tag{6.24}$$

The minimal force the worker must apply at the end of the rope for the block A to start moving up the inclined plane is calculated in MATLAB with:

```
sol_1=solve(eq11,eq21,eq31,eq41, ...
    'N_A', 'T', 'F_bup', 'T_A');
fprintf('The minimal force the worker must apply ')
fprintf('at the end of the rope \n');
fprintf('for the block to start moving ');
fprintf('up the inclined plane is \n\n');
fprintf('F_bup = %s \n\n',sol_1.F_bup);
```

The input numerical data are introduced in MATLAB with:

```
% numerical results
lists = {W_A,mu_S,mu_D,r,beta,alpha};
listn = {150,0.6,0.45,0.1,pi/5,pi/4};
```

The numerical results are calculated and printed in MATLAB using:

```
e_muD_theta=eval(subs(exp(mu_D*theta),lists,listn));
Theta = eval(subs(theta,lists,listn));
F_bdown = eval(subs(sol_r.F_bdown,lists,listn));
F_bup = eval(subs(sol_1.F_bup,lists,listn));
fprintf('theta = [%f] (deg)\n',Theta*180/pi);
```

```
fprintf('F_bdown = [%f] (N)\n',F_bdown);
fprintf('F_bup = [%f] (N)\n',F_bup);
```

The MATLAB numerical results for the forces and the angle θ are:

```
theta = [99.000000] (deg)
F_bdown = [19.496341] (N)
F_bup = [369.300071] (N)
```

Example 6.2

The worker shown in Figure 6.11 is using the wedge A and the block C to lift a load B, guided by frictionless rollers, to a desired height h. A rope attached to the load B turns around the fixed drum D_{CB} of radius r before connecting to the load C. The wedge A sits on frictionless rollers, the friction coefficient between the rope and the fixed drum D_{CB} is μ_D, and the friction coefficient between the load B and the wedge A is μ_A. Knowing the mass m_C of block C and mass m_A of wedge A determine the minimal horizontal force the worker must apply to the wedge A to move the load B upwards. Numerical application: $F_B = 9$ N, $\mu_{AB} = 0.6$, $\mu_D = 0.45$ m, $g = 9.81$ m s^{-2}, $m_B = 4$ kg, $m_C = 6$ kg, $\alpha = \frac{\pi}{6}$, and $\theta_{Dbc} = \pi$.

Solution

To calculate the maximal minimal force F the worker should apply one can consider that the load B should be moving up. Considering the equilibrium of the system, as shown in the free body diagram in Figure 6.12, the sum of all forces acting on B can be expressed as

$$\sum \mathbf{F}_B = 0 \Leftrightarrow \mathbf{T}_B + \mathbf{W}_B + \mathbf{R}_B + \mathbf{F}_B + \mathbf{N}_B = 0 \tag{6.25}$$

where $\mathbf{T}_B$ is the rope tension, $\mathbf{W}_B$ is the weight of the body B, $\mathbf{F}_B$ is the friction force, $\mathbf{N}_B$ is the reaction force with the wall, and $\mathbf{R}_B$ is the force of wedge A on body B. The rope tension magnitude acting at drum D_{CB} can be expressed by

$$T_B = T_C e^{\mu_D \theta_{DBC}} = T_C e^{\mu_D \pi}. \tag{6.26}$$

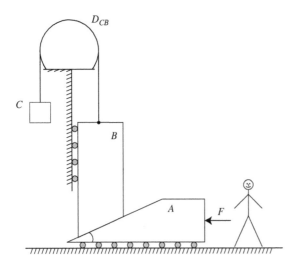

Figure 6.11 System composed of a wedge A, a rope and drum, a block C, and a load B to be lifted.

The sum of all forces acting on wedge A can be expressed as

$$\sum \mathbf{F}_{\text{wedge}A} = 0 \quad \Leftrightarrow \quad \mathbf{F} + \mathbf{W}_A + \mathbf{R}_A + \mathbf{F}_A + \mathbf{N}_A = 0 \tag{6.27}$$

where $\mathbf{F}$ is the applied force by the worker, $\mathbf{W}_A = m_A\mathbf{g}$ is the weight of the wedge A, $\mathbf{F}_A$ is the friction force between the wedge A and body B, $\mathbf{N}_A$ is the reaction force with the ground, and $\mathbf{R}_A$ is the force of body B on the wedge A. The sum of all forces acting on the body C can be expressed as

$$\sum \mathbf{F}_{\text{body}C} = 0 \quad \Leftrightarrow \quad \mathbf{W}_C + \mathbf{T}_C = 0 \tag{6.28}$$

where $\mathbf{W}_C = m_C\mathbf{g}$ is the weight of the body C.

All the forces, weights and tensions acting on the system are computed in MATLAB with:

```
clear all; clc; close all
syms alpha mu_A N_A W_B R_B F_B N_B F W_A R_A
syms T_B T_bodyB T_C e_muD_ThetaDbc theta_Dbc mu_D mu_AB
syms m_A m_B m_C g

W_C = m_C*g;
W_B = m_B*g;
W_A = m_A*g;
F_A = F_B;
N_B_  = [N_B 0 0];
T_B_  = [0 T_B 0];
R_B_  = [-R_B*sin(alpha) R_B*cos(alpha) 0];
F_Bf_ = [mu_AB*F_B*cos(alpha) mu_AB*F_B*sin(alpha) 0];
W_B_  = [0 -W_B 0];
T_C_  = [0 T_C 0];
W_C_  = [0 -W_C 0];
F_    = [-F 0 0];
W_A_  = [0 -W_A 0];
R_A_  = - R_B_;
F_Af_ = -F_Bf_;
N_A_  = [0 -N_A 0];
```

From Equations (6.25), (6.27) and (6.28) the equilibrium equations can be written as

$$\mathbf{T}_B + \mathbf{W}_B + \mathbf{R}_B + \mathbf{F}_B + \mathbf{N}_B = 0$$
$$\mathbf{F} + \mathbf{W}_A + \mathbf{R}_A + \mathbf{F}_A + \mathbf{N}_A = 0$$
$$\mathbf{W}_C + \mathbf{T}_C = 0 \tag{6.29}$$

or equivalent

$$N_B - R_B \sin \alpha + F_B \mu_{AB} \cos \alpha = 0$$
$$T_B - W_B + R_B \cos \alpha + F_B \mu_{AB} \sin \alpha = 0$$
$$T_B - T_C e^{\pi \mu_D} = 0$$
$$R_B \sin \alpha - F - F_B \mu_{AB} \cos \alpha = 0$$
$$-N_A - W_A - R_B \cos \alpha - F_B \mu_{AB} \sin \alpha = 0$$
$$T_C - W_C = 0. \tag{6.30}$$

In MATLAB the equilibrium equations can be written by:

```
eq1_sumFB = T_B_ + W_B_ + R_B_ + F_Bf_ + N_B_;
eq2_ropeT = T_B_ - T_C_*exp(mu_D*theta_Dbc);
eq3_sumFA = F_ + W_A_ + R_A_ + F_Af_ + N_A_;
eq4_sumFC = W_C_ + T_C_;
fprintf('The equilibrium equations are \n\n')
fprintf('%s = 0 \n',eq1_sumFB(1));
fprintf('%s = 0 \n',eq1_sumFB(2));
fprintf('%s = 0 \n',eq2_ropeT(2));
fprintf('%s = 0 \n',eq3_sumFA(1));
fprintf('%s = 0 \n',eq3_sumFA(2));
fprintf('%s = 0 \n',eq4_sumFC(2));
```

Solving Equation (6.30) one can obtain the minimal horizontal force the worker must apply to the wedge to move the load B upwards as

$$F = \frac{W_B \sin \alpha - F_B \mu_{AB} - W_C e^{\pi \mu_D}}{\cos \alpha}. \tag{6.31}$$

The minimal horizontal force the worker must apply to the wedge to move the load B upwards is calculated and printed in MATLAB with:

```
sol=solve(eq1_sumFB(1),eq1_sumFB(2),...
    eq2_ropeT(2),eq3_sumFA(1),eq3_sumFA(2),...
    eq4_sumFC(2),'F', 'T_C', 'T_B', 'R_B', 'N_B', 'N_A');
fprintf('Solving the equilibrium equations ')
fprintf('the maximal horizontal force F the worker ')
fprintf('must apply to the wedge is \n\n')
fprintf('F = %s \n\n',simplify(sol.F))
pretty(simplify(sol.F))
```

The input numerical data are introduced in MATLAB with:

```
% numerical results
lists = {F_B,mu_AB,mu_D,g,m_B,m_C,alpha,theta_Dbc};
listn = {9,0.6,0.45,9.81,4,6,pi/6,pi};
```

The numerical results are calculated and printed in MATLAB using:

```
e_muD_ThetaDbc=eval(subs(exp(mu_D*theta_Dbc),lists,listn));
F = eval(subs(sol.F,lists,listn));
fprintf('F = [%f] (N)\n',F);
```

The MATLAB numerical result for the force F is:

```
F = [-123.290640] (N)
```

Example 6.3
The belt in Figure 6.13 connects the shaft B having the radius r_B with the shaft of an electric motor having the radius r_A and acting torque T_A. A polishing pad C having the radius r_C is

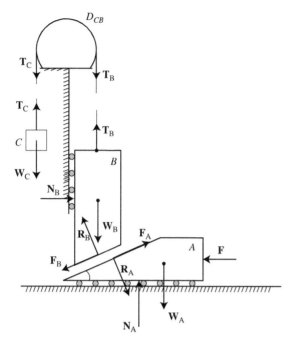

Figure 6.12 Free-body diagram of the system composed of the wedge *A*, the rope and drum, the block *C*, and the load *B* to be lifted.

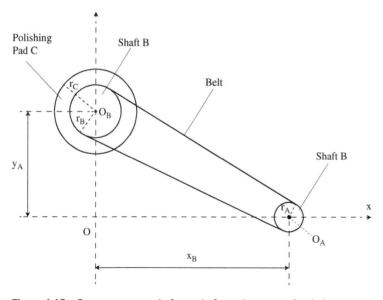

Figure 6.13 System composed of two shafts and a connecting belt.

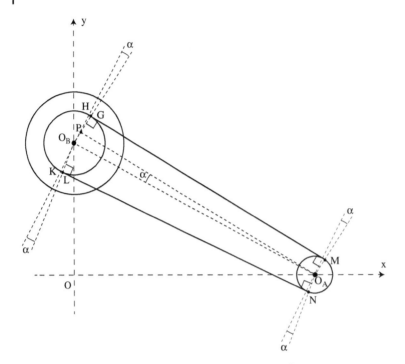

Figure 6.14 Representation of the angle α for the system composed of two shafts and a connecting belt.

attached to the rotational shaft B at one of its end. The pad is pressing against a dry polishing surface D with a force F. Knowing that the friction coefficients between the belt and the shaft A, belt and the shaft B, pad C and surface D, are μ_A, μ_B, and μ_{CD} respectively (where $\mu = \mu_A = \mu_B$), determine the resisting torque T at the shaft B. Numerical application: $F = 15$ N, $\mu = 0.45$, $\mu_{CD} = 0.55$, $r_A = 0.1$ m, $r_B = 0.25$ m, $r_C = 0.35$ m, $x_A = 1$ m, $y_A = 0$ m, $x_B = 0$ m, $y_B = 0.7$ m, and $T_A = 25$ N.

Solution
One can calculate the angle $\alpha = \angle(O_B O_A P)$ from the right angle triangle $O_B P O_A$ having $PO_A \parallel GM$ (Figure 6.14) using the sin rule, that is

$$\sin \alpha = \frac{l_{O_B P}}{l_{O_A O_B}} = \frac{l_{O_B G} - l_{GP}}{\sqrt{(x_A - x_B)^2 + (y_A - y_B)^2}} = \frac{r_B - r_A}{\sqrt{x_B^2 + y_A^2}}$$

$$\Leftrightarrow \alpha = \sin^{-1}\left(\frac{r_B - r_A}{\sqrt{x_B^2 + y_A^2}}\right). \tag{6.32}$$

The input data is declared in MATLAB with:

```
clear all; clc; close all
syms r_A r_B r_C x_B x_A y_B y_A alpha
syms mu mu_CD T_1 T_2 T_A T_B F r
```

```
l_OaM = r_A;
l_ObG = r_B;
l_GP = r_A;
l_ObP = l_ObG - l_GP;
l_OaOb = sqrt((x_A-x_B)^2 + (y_A-y_B)^2);
fprintf('The radius of the shaft A is \n');
fprintf('R_shaft = %s \n\n',l_OaM);
fprintf('The radius of the shaft B is \n');
fprintf('R_shaft = %s \n\n',l_ObG);
fprintf('The distance between the two shafts is \n');
fprintf('d_ShaftPulley = %s \n\n',l_OaOb);
```

and the angle α is calculated with:

```
eqAlpha = sin(alpha) - l_ObP/l_OaOb;
alpha = solve(eqAlpha,alpha);
alpha = alpha(2);
fprintf('The angle alpha is \n');
fprintf('alpha = %s \n\n',alpha);
```

Since a clockwise torque is acting on the shaft A, the shaft would slip in a clockwise sense relative to the rope. The equilibrium equation for the shaft, as shown in the free body diagram in Figure 6.15, can be expressed as

$$\sum M_A = 0 \Leftrightarrow T_A + T_2 r_A - T_1 r_A = 0 \tag{6.33}$$

where the minimum allowable tension in each part of the rope with respect to the shaft (if the rope is not to slip) can be calculated as

$$T_1 = T_2 e^{\mu\theta} = T_2 e^{\mu(\pi+2\alpha)} \tag{6.34}$$

From Equations (6.33) and (6.34) one can write

$$T_1 = \frac{e^{\mu(\pi+2\alpha)}}{e^{\mu(\pi+2\alpha)} - 1} \frac{T_A}{r_A}$$

$$T_2 = \frac{1}{e^{\mu(\pi+2\alpha)} - 1} \frac{T_A}{r_A}. \tag{6.35}$$

The angle θ and the tensions T_1 and T_2 are calculated in MATLAB with:

```
Theta=pi+2*alpha;
fprintf('The theta theta is \n');
fprintf('Theta = %s \n\n',Theta);
eqM_A = T_A - T_1*r_A + T_2*r_A;
eqTension = T_1 - T_2*exp(mu*Theta);
solM_A = solve(eqM_A,eqTension,'T_1','T_2');
fprintf('The allowable tension T1 is \n\n')
pretty(simplify(solM_A.T_1))
fprintf('The allowable tension T2 is \n\n')
pretty(simplify(solM_A.T_2))
```

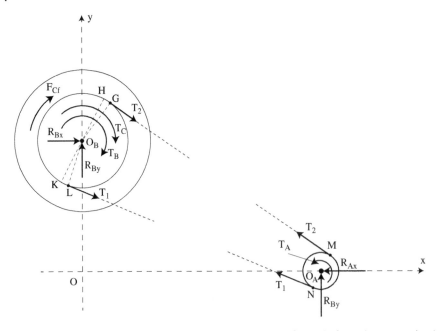

Figure 6.15 Free-body diagram of the system composed of two shafts and a connecting belt.

The equilibrium equation for the shaft, as shown in the free-body diagram in Figure 6.15, can be expressed as

$$\sum M_B = 0 \ \Leftrightarrow \ T_B + T_2 r_B - T_1 r_B = 0. \tag{6.36}$$

From Equations (6.36) and (6.35) one can calculate the resisting torque T_B due to the flat belt as

$$T_B = r_B(T_1 - T_2)$$

$$= r_B \left(\frac{e^{\mu(\pi + 2\alpha)}}{e^{\mu(\pi + 2\alpha)} - 1} - \frac{1}{e^{\mu(\pi + 2\alpha)} - 1} \right) \frac{T_A}{r_A} +$$

$$= \frac{r_B}{r_A} T_A. \tag{6.37}$$

The resisting torque T_B is calculated in MATLAB with:

```
eqM_B = T_B - T_1*r_B + T_2*r_B;
T_B = solve(eqM_B,T_B);
T_B=subs(T_B,{T_1,T_2},{solM_A.T_1,solM_A.T_2});
fprintf('The resisting torque T_B is \n');
fprintf('T_B = %s \n\n',simplify(T_B));
```

The resisting torque T_C due to the friction of the pad with the polished surface can be expressed as

$$T_C = \mu_{CD} \frac{F}{\pi r_C^2} \int_0^r 2\pi r^2 dr$$

$$= \frac{2}{3} \mu_{CD} F r_C. \tag{6.38}$$

The resisting torque T_C is calculated in MATLAB with:

```
T_C = mu_CD*(F/(pi*r_C^2))*int(2*pi*r^2,0,r_C);
fprintf('The resisting torque T_C is \n');
fprintf('T_C = %s \n\n',simplify(T_C));
```

From Equations (6.37) and (6.38) one can calculate the resisting torque T as the sum of the resisting torque due to the friction of the pad and the resisting torque due to the flat belt as

$$T = T_B + T_C$$
$$= \frac{r_B}{r_A}T_A + \frac{2}{3}\mu_{CD}Fr_C. \qquad (6.39)$$

The total resisting torque T is calculated in MATLAB with:

```
T = T_B + T_C;
fprintf('The total resisting torque T is \n');
fprintf('T = %s \n\n',simplify(T));
```

The input numerical data are introduced in MATLAB with:

```
% numerical results
lists = {F,mu,mu_CD,r_A,r_B,r_C,x_A,y_A,x_B,y_B,T_A};
listn = {15,0.45,0.55,0.1,0.25,0.35,1,0,0,0.7,25};
```

The numerical results are calculated and printed in MATLAB using:

```
r_A = eval(subs(r_A,lists,listn));
r_B = eval(subs(r_B,lists,listn));
d_ShaftPulley = eval(subs(l_OaOb,lists,listn));
alpha = eval(subs(alpha,lists,listn));
theta = 180+2*alpha;
T_1 = eval(subs(solM_A.T_1,lists,listn));
T_2 = eval(subs(solM_A.T_2,lists,listn));
T_B = eval(subs(T_B,lists,listn));
T_C = eval(subs(T_C,lists,listn));
T = eval(subs(T,lists,listn));

fprintf('r_A = [%f] (m)\n',r_A);
fprintf('r_B = [%f] (m)\n',r_B);
fprintf('d_ShaftPulley = [%f] (m)\n',d_ShaftPulley);
fprintf('alpha = [%f] (deg)\n',alpha);
fprintf('theta = [%f] (deg)\n',theta);
fprintf('T_1 = [%f] (N)\n',T_1);
fprintf('T_2 = [%f] (N)\n',T_2);
fprintf('T_B = [%f] (Nm)\n',T_B);
fprintf('T_C = [%f] (Nm)\n',T_C);
fprintf('T = [%f] (Nm)\n',T);
```

The numerical results obtained in MATLAB are:

```
r_A = [0.100000] (m)
r_B = [0.250000] (m)
d_ShaftPulley = [1.220656] (m)
alpha = [3.018396] (deg)
theta = [186.036793] (deg)
T_1 = [254.085306] (N)
T_2 = [4.085306] (N)
T_B = [62.500000] (Nm)
T_C = [1.925000] (Nm)
T = [64.425000] (Nm)
```

Example 6.4

The worker shown in Figure 6.16 is using an inextensible rope and a massless pulley to lift a weight A to a desired height. The rope, which makes an angle α with the vertical direction, turns around a fixed pulley (suspended from the ceiling) before connecting to the slider A placed on a vertical rod. The friction coefficient between the slider A and the rod is μ. Knowing that weight of sliding block A determine:

(a) The minimal force the worker must apply to the end of the rope for the block A not to slide down the rod.
(b) The minimal force the worker must apply to the end of the rope for the block A to start moving up the sliding rod.

Numerical application: $g = 9.81$ m s^{-2}, $\mu = 0.45$, $m_A = 7$ kg, $\alpha = \frac{\pi}{4}$ rad, and $T_A = 25$ N.

Solution

The equilibrium of the weight A when the direction of motion is downwards as shown in Figure 6.17 can be written as

$$\sum \mathbf{F} = 0 \Leftrightarrow \mathbf{W}_A + \mathbf{T}_A + \mathbf{F}_f + \mathbf{N}_A = 0 \tag{6.40}$$

or equivalent

$$\sum F_x = 0 \Leftrightarrow N_A - T_A \sin \alpha = 0$$

$$\sum F_y = 0 \Leftrightarrow F_f + T_A \cos \alpha - W_A = 0. \tag{6.41}$$

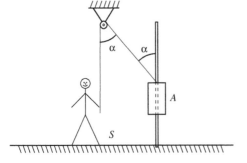

Figure 6.16 System composed of a pulley, rope and slider placed on a vertical rod.

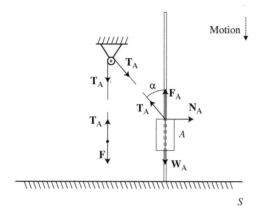

Figure 6.17 Free-body diagram of the system composed of a pulley, rope and slider placed on a vertical rod – direction of motion of the slider is downward.

The input data is declared in MATLAB with:

```
clear all; clc; close all
syms m_A g T_A alpha F_f N_A mu

W_A = m_A*g;
T_A_ = [-T_A*sin(alpha) T_A*cos(alpha) 0];
F_f_ = [0 F_f 0];
N_A_ = [N_A 0 0];
W_A_ = [0 -W_A 0];
```

Since $T_A = F$ one can rewrite Equation (6.41) as

$$\begin{cases} N_A - F \sin \alpha = 0 \\ F_f + F \cos \alpha - W_A = 0 \end{cases} \Leftrightarrow \begin{cases} N_A = F \sin \alpha \\ F_f = W_A - F \cos \alpha \end{cases}. \tag{6.42}$$

The block equilibrium equations are written in MATLAB with

```
% Case 1: Direction of motion is Downward
eq_blockD = W_A_ + T_A_ + F_f_ + N_A_;
fprintf('The block equilibrium equations ');
fprintf('when the direction of motion is downward are \n');
fprintf('%s = 0 \n',eq_blockD(1));
fprintf('%s = 0 \n\n',eq_blockD(2));
```

Solving the cart equilibrium equations in MATLAB one can obtain:

```
solD=solve(eq_blockD(1),eq_blockD(2),'N_A', 'F_f');
fprintf('Solving the cart equilibrium equations ');
fprintf('when the direction of motion is downward: \n');
fprintf('The reaction N_A is \n');
fprintf('N_A = %s \n',solD.N_A);
```

```
fprintf ('The friction force is \n');
fprintf ('F_f = %s \n\n', solD.F_f);
```

Considering the condition for the block A not to slide down the rod, i.e. $F_f \leq \mu N_A$, and equilibrium Equation (6.42) one can deduce

$$\mu N_A \geq F_f \Leftrightarrow \mu F \sin \alpha \geq W_A - F \cos \alpha$$

$$\Leftrightarrow \mu F \sin \alpha + F \cos \alpha \geq W_A$$

$$\Leftrightarrow F(\mu \sin \alpha + \cos \alpha) \geq W_A \tag{6.43}$$

or equivalent

$$F \geq \frac{W_A}{\cos \alpha + \mu \sin \alpha}. \tag{6.44}$$

The magnitude of the minimal force the worker must apply for the block not to slide down the rod is calculated in MATLAB with:

```
eqD=mu*solD.N_A-solD.F_f;
minFD=solve (eqD,'T_A');
fprintf ('The minimal force the worker must apply ');
fprintf ('for the block not to slide down the rod is \n');
fprintf ('F >= %s \n\n', minFD);
```

The equilibrium of the weight A when the direction of motion is upward, as shown in Figure 6.18, can be written as

$$\sum F_x = 0 \Leftrightarrow N_A - T_A \sin \alpha = 0$$

$$\sum F_y = 0 \Leftrightarrow T_A \cos \alpha - F_f - W_A = 0. \tag{6.45}$$

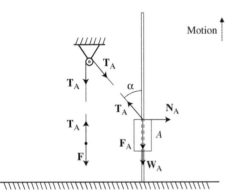

Figure 6.18 Free-body diagram of the system composed of a pulley, rope and slider placed on a vertical rod – direction of motion of the slider is upward.

The equilibrium equations when the direction of motion is upward are written in MATLAB with:

```
% Case 2: Direction of motion is Upward
F_f   = [0 -F_f 0];
eq_blockU = W_A_  + T_A_  + F_f_  + N_A_;
fprintf('The block equilibrium equations ');
fprintf('when the direction of motion is upward are \n');
fprintf('%s = 0 \n',eq_blockU(1));
fprintf('%s = 0 \n\n',eq_blockU(2));
```

Since $T_A = F$ one can rewrite Equation (6.45) as

$$N_A = F \sin \alpha$$
$$F_f = F \cos \alpha - W_A. \tag{6.46}$$

The equilibrium equations are solved in MATLAB with:

```
solU=solve(eq_blockU(1),eq_blockU(2),'N_A', 'F_f');
fprintf('Solving the cart equilibrium equations ');
fprintf('when the direction of motion is upward: \n');
fprintf('The reaction N_A is \n');
fprintf('N_A = %s \n',solU.N_A);
fprintf('The friction force is \n');
fprintf('F_f = %s \n\n',solU.F_f);
```

Considering the condition for the block A to start moving up the sliding rod, i.e. $F_f \le \mu N_A$, and equilibrium Equation (6.42) one can deduce

$$\mu N_A \ge F_f \iff \mu F \sin \alpha \ge F \cos \alpha - W_A$$
$$\iff \mu F \sin \alpha - F \cos \alpha \ge -W_A$$
$$\iff F \cos \alpha - \mu F \sin \alpha \ge W_A$$
$$\iff F(\cos \alpha - \mu \sin \alpha) \ge W_A \tag{6.47}$$

that is

$$F \ge \frac{W_A}{\cos \alpha - \mu \sin \alpha}. \tag{6.48}$$

The magnitude of the minimal force the worker must apply for the block to start moving up the rod is calculated in MATLAB with:

```
eqU=mu*solU.N_A-solU.F_f;
minFU=solve(eqU,'T_A');
fprintf('The minimal force the worker must apply ');
fprintf('for the block to start moving up the sliding rod is \n');
fprintf('F >= %s \n\n',minFU);
```

The input numerical data are introduced in MATLAB with:

```
% numerical results
lists = {g,mu,m_A,alpha,T_A};
listn = {9.81,0.45,7,pi/4,25};
```

The numerical results are calculated and printed in MATLAB using:

```
% Case 1: Direction of motion is Downward
minFD = eval(subs(minFD,lists,listn));
fprintf('The minimal force the worker must apply ');
fprintf('for the block not to slide down the rod \n');
fprintf('F >= [%f] (N)\n\n',minFD);
% Case 2: Direction of motion is Upward
minFU = eval(subs(minFU,lists,listn));
fprintf('The minimal force the worker must apply ');
fprintf('for the block to start moving up the sliding rod is \n');
fprintf('F >= [%f] (N)\n',minFU);
```

The MATLAB numerical results for the block not to slide down or for moving up the rod are:

```
The minimal force the worker must apply
for the block not to slide down the rod
F >= [66.975204] (N)

The minimal force the worker must apply
for the block to start moving up the sliding rod is
F >= [176.570992] (N)
```

Example 6.5

The worker shown in Figure 6.19 is using an inextensible rope and a massless double pulley system to lift a block A to a desired height. The worker rope is acting vertically on the outer radius r_B of the pulley and the weight A acts at the inner radius r_A of the pulley. The friction coefficient between the pulley and its shaft is μ_O. Knowing the mass m_A of the block A determine the magnitude of the vertical force $\mathbf{F}_B$ the worker must apply to the end of the rope for the block A to be in equilibrium. Numerical application: $g = 9.81$ m s^{-2}, $\mu = 0.35$, $m_A = 4$ kg, $r_A = 0.15$ m, $r_B = 0.25$ m and $r_O = 0.025$ m.

Solution

To calculate the magnitude of the minimal force the worker must apply to the end of the rope for the block to start moving up one can consider the equilibrium equations. The equilibrium of the pulley and the block A under the applied worker force $\mathbf{F}_B$ 6.19 can be written using the free body diagram in Figure 6.20 by

$$\sum \mathbf{F}_x = 0 \Leftrightarrow F_H = 0$$

$$\sum \mathbf{F}_y = 0 \Leftrightarrow F_V - F_B - W_A = 0$$

$$\sum \mathbf{M}_O = 0 \Leftrightarrow -W_A r_A + F_B r_B - M_{F_f} = 0. \tag{6.49}$$

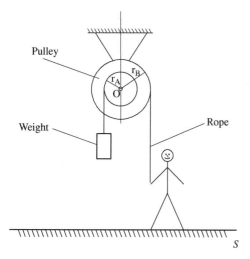

Figure 6.19 System composed of a double pulley, rope and a block.

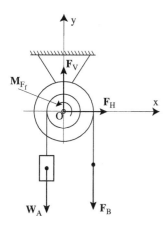

Figure 6.20 Free-body diagram of the system composed of the double pulley, rope and block – block to start moving up.

In MATLAB the equilibrium equations are written using

```
clear all; clc; close all
syms F_H F_V F_B m_A W_A r_A r_B M_Ff
syms mu_O r_O m_A g

sumFx = F_H;
sumFy = F_V-F_B-W_A;
sumMO = -W_A*r_A + F_B*r_B -M_Ff;
fprintf ('The equilibrium equations (block to start moving up) are \n');
fprintf ('%s = 0 \n',sumFx);
fprintf ('%s = 0 \n',sumFy);
fprintf ('%s = 0 \n\n',sumMO);
```

Solving Equation (6.49) one can obtain

$$F_H = 0, \quad F_V = F_B + W_A \text{ and } M_{F_t} = F_B r_B - W_A r_A. \tag{6.50}$$

The equilibrium equations are solved and printed in MATLAB with:

```
solUP=solve(sumFx,sumFy,sumMO,F_H, F_V,M_Ff);
fprintf('Solving the equilibrium equations ');
fprintf('one can obtain: \n');
fprintf('F_H = %s \n',solUP.F_H);
fprintf('F_V = %s \n',solUP.F_V);
fprintf('M_Ff = %s \n\n',solUP.M_Ff);
F_V = solUP.F_V;
M_Ff = solUP.M_Ff;
```

To calculate the minimal force the worker must apply to the end of the rope for the block to start moving up one can consider the additional equation

$$M_{F_t} \leq \mu_0 F_V r_O. \tag{6.51}$$

From Equations (6.50) and (6.51) one can obtain

$$\begin{aligned}
& M_{F_t} \leq \mu F_V r_O \\
\Leftrightarrow\ & F_B r_B - W_A r_A \leq \mu_0 (F_B + W_A) r_O \\
\Leftrightarrow\ & F_B r_B - W_A r_A \leq \mu_0 F_B r_O + \mu_0 W_A r_O \\
\Leftrightarrow\ & F_B r_B - \mu_0 F_B r_O \leq W_A r_A + \mu_0 W_A r_O \\
\Leftrightarrow\ & F_B (r_B - \mu_0 r_O) \leq W_A (r_A + \mu_0 r_O) \\
\Leftrightarrow\ & F_B \leq W_A \frac{r_A + \mu_0 r_O}{r_B - \mu_0 r_O}.
\end{aligned} \tag{6.52}$$

The magnitude of the minimal force the worker should apply to the end of the rope for the block to start moving up is calculated and printed in MATLAB with:

```
eqUP = M_Ff-mu_O*F_V*r_O;
soleqUP=solve(eqUP,F_B);
fprintf('The minimal force F_B the worker must apply ');
fprintf('for the block to start moving up should verify \n');
fprintf('F_B <= %s \n\n',soleqUP);
```

The input numerical data are introduced in MATLAB with:

```
% numerical results
lists = {g,mu_O,m_A,r_A,r_B,r_O};
listn = {9.81,0.35,4,0.15,0.25,0.025};
```

The numerical values of the magnitude of the minimal force the worker must apply to the rope for the block to start moving up is calculated in MATLAB using:

```
W_A=m_A*g;
W_A=eval(subs(W_A,lists,listn));
F_B_UP = eval(subs(soleqUP,lists,listn));
fprintf('The minimal NUMERICAL force the worker ')
```

```
fprintf('should apply is less than: \n')
fprintf('F_B <= %f (N)\n\n',F_B_UP);
```

The MATLAB numerical value representing the minimal force the worker should apply to the rope (for the block to be on the verge to start moving up) should verify:

```
F_B <= 25.821140 (N)
```

To calculate the magnitude of the minimal force the worker must apply to the rope for the block not to go down one can consider the equilibrium equations written using the free body diagram in Figure 6.21 by

$$\sum \mathbf{F}_x = 0 \Leftrightarrow F_H = 0$$
$$\sum \mathbf{F}_y = 0 \Leftrightarrow F_V - F_B - W_A = 0$$
$$\sum \mathbf{M}_O = 0 \Leftrightarrow -W_A r_A + F_B r_B + M_{F_f} = 0 \qquad (6.53)$$

where the only difference with respect to Equation (6.49) is the change in the direction of the moment M_{F_f}. In MATLAB the equilibrium equations are written using

```
clear all; close all
syms F_H F_V F_B m_A W_A r_A r_B M_Ff
syms mu_O r_O m_A g

sumFx = F_H;
sumFy = F_V-F_B-W_A;
sumMO = -W_A*r_A + F_B*r_B + M_Ff;
fprintf('The equilibrium equations (block not to move down) \n');
fprintf('%s = 0 \n',sumFx);
fprintf('%s = 0 \n',sumFy);
fprintf('%s = 0 \n\n',sumMO);
```

Solving Equation (6.53) one can obtain

$$F_H = 0, \ F_V = F_B + W_A \text{ and } M_{F_f} = W_A r_A - F_B r_B. \qquad (6.54)$$

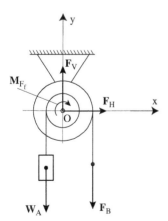

Figure 6.21 Free-body diagram of the system composed of the double pulley, rope and block – block to start moving down.

The equilibrium equations are solved and printed in MATLAB with:

```
solDOWN=solve(sumFx,sumFy,sumMO,F_H,F_V,M_Ff);
fprintf('Solving the equilibrium equations ');
fprintf('one can obtain: \n');
fprintf('F_H = %s \n',solDOWN.F_H);
fprintf('F_V = %s \n',solDOWN.F_V);
fprintf('M_Ff = %s \n\n',solDOWN.M_Ff);
F_V = solDOWN.F_V;
M_Ff = solDOWN.M_Ff;
```

To calculate the magnitude of the minimal force the worker must apply to the rope to keep the block in equilibrium (block not to go down) one can consider the additional equation

$$M_{F_f} \leq \mu_O F_V r_O. \tag{6.55}$$

From Equations (6.54) and (6.55) one can obtain

$$M_{F_f} \leq \mu F_V r_O$$
$$\Leftrightarrow W_A r_A - F_B r_B \leq \mu_O (F_B + W_A) r_O$$
$$\Leftrightarrow W_A r_A - F_B r_B \leq \mu_O F_B r_O + \mu_O W_A r_O$$
$$\Leftrightarrow W_A r_A - \mu_O W_A r_O \leq F_B r_B + \mu_O F_B r_O$$
$$\Leftrightarrow W_A (r_A - \mu_O r_O) \leq F_B (r_B + \mu_O r_O)$$
$$\Leftrightarrow F_B \geq W_A \frac{r_A - \mu_O r_O}{r_B + \mu_O r_O}. \tag{6.56}$$

From Equations (6.52) and (6.56) one can obtain

$$W_A \frac{r_A - \mu_O r_O}{r_B + \mu_O r_O} \leq F_B \leq W_A \frac{r_A + \mu_O r_O}{r_B - \mu_O r_O}. \tag{6.57}$$

The magnitude of the minimal force the worker should apply to the rope for the block not to go down is calculated and printed in MATLAB with:

```
eqDOWN = M_Ff-mu_O*F_V*r_O;
soleqDOWN=solve(eqDOWN,F_B);
fprintf('The minimal force F_B the worker must apply ');
fprintf('for the block not to start going down   should ver-
ify \n');
fprintf('F_B >= %s \n\n',soleqDOWN);
```

The input numerical data are introduced in MATLAB with:

```
% numerical results
lists = {g,mu_O,m_A,r_A,r_B,r_O};
listn = {9.81,0.35,4,0.15,0.25,0.025};
```

The numerical values of the magnitude of the minimal force the worker must apply to the rope for the block to be in equilibrium (not to go down) is calculated using

```
W_A=m_A*g;
```

```
W_A=eval(subs(W_A,lists,listn));
F_B_DOWN = eval(subs(soleqDOWN,lists,listn));
fprintf('The minimal force the worker ')
fprintf('should apply to the piston for equilibrium is big-
ger than: \n')
fprintf('F_B >= %f (N)\n\n',F_B_DOWN);
```

The MATLAB numerical value representing the minimal force the worker should apply to the rope for equilibrium (the block not to go down) should verify:

```
F_B >= 21.420870 (N)
```

Combining the obtained numerical results, the MATLAB values representing the force the worker should apply to the rope for equilibrium should verify:

```
21.420870 (N)   <= F_B   <=   25.821140 (N)
```

Example 6.6

The spool shown in Figure 6.22 is hanging by the wall S by an inextensible string. The string, which makes an angle α with the vertical direction, turns around the spool A of radius r_S and flange radius r_B. The friction coefficient between the spool and the wall S is μ. Knowing that weight of spool (denoted by A), determine the values of angle α for which the spool will not slide down the wall. Numerical application: $\mu = 0.52$, $r_S = 0.015$ m and $r_B = 0.05$ m.

Solution

The equilibrium of the spool (Figure 6.23) can be written as

$$\sum F_x = 0 \Leftrightarrow N - T_{Ax} = 0 \Leftrightarrow N - T_A \sin \alpha = 0$$

$$\sum F_y = 0 \Leftrightarrow F_A - W_A + T_{Ay} = 0 \Leftrightarrow \mu N - m_A g + T_A \cos \alpha = 0 \qquad (6.58)$$

where $T_{Ax} = T_A \sin \alpha$ and $T_{Ay} = T_A \cos \alpha$ are the magnitudes of the Ox and Oy components of the string tension, $\mathbf{N}$ is the normal reaction at the wall, $F_A = \mu N$ is the magnitude of

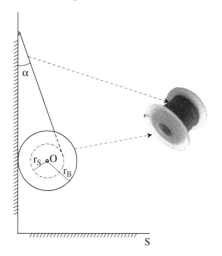

Figure 6.22 Spool hanging by a wall.

the friction force, m_A is the mass of the spool and g is the magnitude of the gravitational acceleration. The equilibrium of the spool (sum of moments about the point O) can be written as

$$T_A r_S - F_A r_B = 0. \tag{6.59}$$

From Equations (6.58) and (6.59) one can write

$$N - T_A \sin \alpha = 0$$
$$\mu N - m_A g + T_A \cos \alpha = 0$$
$$T_A r_S - \mu N r_B = 0. \tag{6.60}$$

The input data is declared in MATLAB with:

```
clear all; clc; close all
syms r_S r_B mu T_A alpha N m_A g

fprintf('The radius of the spool A and spool flange B is \n')
fprintf('R_spool = %s \n',r_S)
fprintf('R_flange = %s \n\n',r_B)


T_Ax = T_A * sin(alpha);
T_Ay = T_A * cos(alpha);
F_A = mu * N;
W_A = m_A * g;
```

The equilibrium equations for the spool are introduced and printed in MATLAB with:

```
SumFOx = N - T_Ax;
SumFOy = F_A - W_A + T_Ay;
SumMO = T_A*r_S - F_A*r_B;
fprintf('The equilibrium equations for the sum of forces ')
fprintf('and for the sum of moments about O are \n')
fprintf('%s = 0 \n',SumFOx)
fprintf('%s = 0 \n',SumFOy)
fprintf('%s = 0 \n\n',SumMO)
```

Solving Equation (6.60) one can calculate

$$\sin \alpha = \frac{N}{T_A} = \frac{N}{\dfrac{\mu N r_B}{r_S}} = \frac{r_S}{\mu r_B}. \tag{6.61}$$

The equilibrium equations are solved in MATLAB with

```
sol = solve(SumFOx,SumMO,'alpha','N');
```

The values of the angle α for which the spool will not slide down the wall can now be expressed as

$$\alpha \le \sin^{-1} \frac{r_S}{\mu r_B}. \tag{6.62}$$

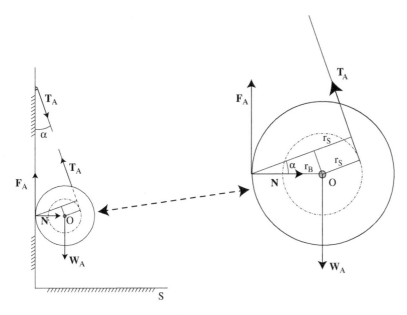

Figure 6.23 Free body diagram of the spool hanging by the wall.

The values of the angle α for which the spool does not slide down are printed in MATLAB using:

```
fprintf('The values of the angle alpha for which ')
fprintf('the spool will not slide down are: \n')
fprintf('alpha <= %s \n\n',sol.alpha(2))
```

The input numerical data are introduced in MATLAB with:

```
% numerical results
lists = {mu,r_S,r_B};
listn = {0.52,0.015,0.05};
```

The numerical results are calculated and printed in MATLAB using:

```
angle = eval(subs(sol.alpha(2),lists,listn));
fprintf('The numerical values of the angle alpha for which ')
fprintf('the spool will not slide down are: \n')
fprintf('alpha< = [%f] (deg)\n',angle*180/pi);
```

The MATLAB numerical values for which the spool does not slide down the wall are:

```
The numerical values of the angle alpha
for which the spool will not slide down are:
alpha< = [35.234418] (deg)
```

Example 6.7
The disk 2, shown in Figure 6.24(a), has a mass of m_2 and a radius r. The coefficient of static friction between the disk and the surface is μ_s. The mass of the homogenous link

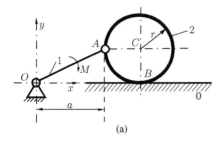

(a)

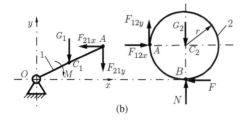

(b)

Figure 6.24 (a) System composed of a link and a disk, (b) free body diagram of the system.

1 is m_1 and the horizontal distance between Oy axis and the disk 2 is a (Figure 6.24(a)). Find the largest moment M on link 1 that can be applied without causing the motion of the system. Numerical application: $m_1 = 20$ kg, $m_2 = 50$ kg, $a = 0.5$ m, $r = 0.2$ m, $\mu_s = 0.3$, and the gravitational acceleration is $g = 9.81$ m s^{-2}.

Solution

The position vectors of the pin joint A and the contact point B are

$$\mathbf{r}_A = a\,\mathbf{\iota} + r\,\mathbf{j} \text{ and } \mathbf{r}_B = (a + r)\,\mathbf{\iota}.$$

The position vectors of the mass centers C_1 and C_2 of the link 1 and the disk 2, as shown in Figure 6.24(b), are

$$\mathbf{r}_{C_1} = \mathbf{r}_A/2 \text{ and } \mathbf{r}_{C_2} = (a + r)\,\mathbf{\iota} + r\,\mathbf{j}.$$

The MATLAB command for the position vectors are:

```
clear all; clc; close all
syms a r rA m1 m2 mus g

rA_  = [a r 0];
rB_  = [a+r 0 0];
rC1_ = rA_/2;
rC2_ = [a+r r 0];
```

The free-body diagram of the of the disk is represented in Figure 6.24(b) where N is the reaction force of the surface on the disk at the contact point B and F is the friction force. The maximum friction force is

$$\mathbf{F} = -\mu_s N\,\mathbf{\iota}.$$

The joint reaction force of link 1 on disk 2 at the pin joint A is

$$\mathbf{F}_{12} = F_{12x}\,\mathbf{1} + F_{12y}\,\mathbf{j}.$$

The vectorial equilibrium equations for the disk are

$$\mathbf{G}_2 + \mathbf{F}_{12} + \mathbf{N} + \mathbf{F} = \mathbf{0},$$
$$\mathbf{r}_{C_2A} \times \mathbf{F}_{12} + \mathbf{r}_{C_2B} \times (\mathbf{N} + \mathbf{F}) = \mathbf{0}.$$

The three scalar equilibrium equations for the disk in MATLAB are:

```
% disk 2
syms   N F12x F12y
G2_    = [0 -m2*g 0];      % @C2 weight
FB_    = [-mus*N N 0];     % @B contact point
F12_   = [F12x F12y 0];    % @A joint reaction

% sum forces on 2
eqF_ = G2_+FB_+F12_;
eqFx = eqF_(1);
eqFy = eqF_(2);

rC2A_ = rA_-rC2_;
rC2B_ = rB_-rC2_;

% sum mom. about C2 for 2
eqM_ = cross(rC2A_, F12_)+cross(rC2B_, FB_);
eqMz = eqM_(3);
```

The joint reaction force at A and the normal force at B are determined and printed in MAT-LAB with:

```
N   = sol.N;
F12x = sol.F12x;
F12y = sol.F12y;
fprintf('The reaction N and joint reactions F1x and F1y are \n');
fprintf('N = %s (N)\n', N);
fprintf('F12x = %s (N)\n', F12x);
fprintf('F12y = %s (N)\n\n', F12x);
```

For the link 1 the moment equilibrium equation with respect to the pin joint at the origin is

$$\mathbf{r}_A \times (-\mathbf{F}_{12}) + \mathbf{r}_{C_1} \times \mathbf{G}_1 + \mathbf{M} = \mathbf{0},$$

and with MATLAB the maximum moment is calculated and printed with:

```
M_ = -cross(rC1_, G1_)-cross(rA_,-F12_);
M = M_(3);
fprintf('The largest moment M that can be applied ');
fprintf('without causing the motion of the system \n');
fprintf('M = %s (N m)\n\n', M);
```

The input numerical data are introduced in MATLAB with:

```
% numerical results
lists = {m1,m2,mus,r,a,g};
listn = {20,50,0.3,0.2,0.5,9.81};
```

The numerical results are calculated and printed in MATLAB using:

```
N = eval(subs(N,lists,listn));
F12x = eval(subs(F12x,lists,listn));
F12y = eval(subs(F12y,lists,listn));
M = eval(subs(M,lists,listn));

fprintf('The obtained numerical values are \n');
fprintf('N = %6.3f (N)\n', N);
fprintf('F12x = %6.3f (N)\n', F12x);
fprintf('F12y = %6.3f (N)\n', F12y);
fprintf('M = %6.3f (N m)\n', M);
```

The MATLAB numerical values for the reaction forces and largest moment that can be applied without causing the motion of the system are:

```
The obtained numerical values are
N = 700.714 (N)
F12x = 210.214 (N)
F12y = -210.214 (N)
M = -98.100 (N m)
```

Example 6.8

The worker shown in Figure 6.25 is using an inextensible rope and a massless pulley to move a block A to a desired location. The rope turns around the pulley – attached to one corner of the moving block A – before connecting to the wall. The friction coefficient between the block A and the floor S is μ. Knowing that weight of block A, determine the minimal force the worker must apply to the end of the rope so the block A will slide without turning. Numerical application: $g = 9.81$ m s^{-2}, $\mu = 0.39$, $m_A = 20$ kg, $d_W = 3.5$ m and $d_h = 2$ m.

Solution

The condition for the block A to start sliding is that the sum of forces acting on the bloc on the horizontal direction (free-body diagram show in Figure 6.26) is greater than the friction force, that is

$$F > F_f$$
$$\Leftrightarrow F > \mu N$$
$$\Leftrightarrow F > \mu(F + m_A g)$$
$$\Leftrightarrow F > \mu F + \mu m_A g$$
$$\Leftrightarrow F - \mu F > \mu m_A g$$
$$\Leftrightarrow F > \frac{\mu m_A g}{1 - \mu} \qquad (6.63)$$

where μ is the coefficient of friction defined as $\mu \in (0, 1)$.

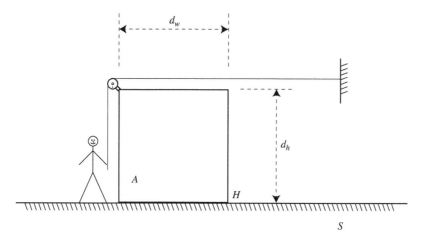

Figure 6.25 System composed of a massless pulley, inextensible rope and a block.

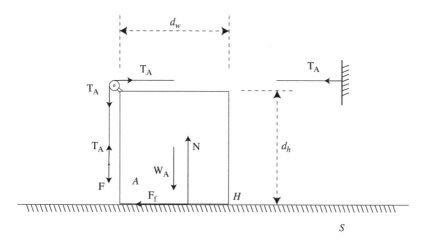

Figure 6.26 Free body diagram of the system composed of a massless pulley, inextensible rope and a block.

The condition for the block A not to turn over the point H (lower right corner denoted by H) is that the sum of moments about H is greater than zero, that is

$$Fd_w + W_A\frac{d_w}{2} \geq T_A d_h$$

$$\Leftrightarrow Fd_w + m_A g\frac{d_w}{2} \geq Fd_h$$

$$\Leftrightarrow m_A g\frac{d_w}{2} \geq Fd_h - Fd_w$$

$$\Leftrightarrow m_A g\frac{d_w}{2} \geq F(d_h - d_w)$$

$$\Leftrightarrow F \leq \frac{m_A g d_w}{2d_h - 2d_w}. \tag{6.64}$$

The conditions for the block A to slide (sum of the horizontal forces) without turning over (sum of the moments about H) are written in MATLAB using:

```
clear all; clc; close all
syms F d_H d_W alpha m_A g mu

fprintf('The dimensions: height and length of the block are \n')
fprintf(' height = %s \n',d_H);
fprintf(' length = %s \n\n',d_W);

T_A = F;
W_A = m_A * g;
N = W_A + F;
F_f= mu * N;
SumFOy = T_A - F_f;
SumMO = -F*d_W - W_A*(d_W/2) + T_A*d_H;
fprintf('The sum of forces about Oy ');
fprintf('and the sum of moments about O ');
fprintf('can be written as: \n');
fprintf('SumFOy = %s \n',SumFOy);
fprintf('SumMO = %s \n\n',SumMO);
```

When $d_w > d_h$ Equation (6.64) always holds, so only Equation (6.63) should be solved, that is, the block A will slide without turning over when

$$\frac{\mu m_A g}{1 - \mu} < F.$$

(6.65)

If $d_w < d_h$ both Equations (6.64) and (6.63) should be solved, that is, the block A will slide without turning when

$$\frac{\mu m_A g}{1 - \mu} < F \text{ and } \frac{\mu m_A g}{1 - \mu} < \frac{m_A g d_w}{2d_h - 2d_w}$$

(6.66)

or equivalent

$$\frac{\mu m_A g}{1 - \mu} < F \text{ and } \mu < \frac{d_w}{2d_h - d_w}.$$

(6.67)

The values of the force F obtained from the force and moment equation are calculated and printed in MATLAB using:

```
ForceS = solve(SumFOy,'F');
ForceM = solve(SumMO,'F');
fprintf('The value of the force F obtained ');
fprintf('from the Force equation is \n');
fprintf('F = %s \n\n',ForceS);
fprintf('The value of the force F obtained ');
fprintf('from the Moment equation is \n');
fprintf('F = %s \n\n',ForceM);
```

The input numerical data are introduced in MATLAB with:

```
% numerical results
lists ={g,mu,m_A,d_W,d_H};
listn = {9.81,0.39,20,2.5,2};
```

The numerical results are calculated and printed in MATLAB using:

```
ForceS_N = eval(subs(ForceS,lists,listn));
ForceM_N = eval(subs(ForceM,lists,listn));
d_W=eval(subs(d_W,lists,listn));
d_H=eval(subs(d_H,lists,listn));
if d_W>d_H
fprintf('The numerical values of the force F for which ');
fprintf('the block will slide without turning over are \n');
fprintf(' F > %f [N] \n\n',ForceS_N);
else
fprintf('The numerical values of the force F for which ');
fprintf('the block will slide without turning over are \n');
fprintf('F > %f [N] \n',ForceS_N);
fprintf('if \n');
fprintf('mu < %f [N] \n\n',d_W/(2*d_H-d_W));
end
```

The MATLAB numerical values for which the block is sliding without turning are:

```
The numerical values of the force F for which
the block will slide without turning over are
 F > 125.439344 [N]
```

Example 6.9

The system in Figure 6.27(a) has two-link mechanisms 1, 2, and a block 3. The mass of link 1 is m_1, the mass of link 2 is m_2, and the mass of the block 3 is m_3. The end of link 2, B, rests on the square block 3, as shown in Figure 6.27(a). The coefficient of friction between link 2 and block 3 is μ_{sB}. The horizontal distance between the origin of the link 1 and point E is a, the horizontal distance between the points E and F is $\frac{1}{2}a$, and the side of the block 3 is $2b$. The block 3 rests on the flat ground 0 and the coefficient of friction between block 3 and the ground is μ_{sD}.

Determine the largest moment M that can be applied to link 1 without causing motion. Numerical application: $m_1 = 8$ kg, $m_2 = 6$ kg, $m_3 = 6$ kg, $g = 9.81$ m s^{-2}, $a = 0.5$ m, $b = 0.1$ m, $\mu_{sB} = 0.7$, $\mu_{sD} = 0.6$, and the angle AOC is 45°.

Solution

The position vectors of the points A, B, C, D, and E, as shown in Figure 6.27(a), are calculated with the following MATLAB commands:

```
b = 0.1;   % (m)
xA = 0.5;  % (m)
yA = 0.5;  % (m)
rA_  = [xA yA 0];

k = 1.5;
xB = k*xA;
yB = 2*b;
rB_  = [xB yB 0];
```

```
xC = k*xA-b;
yC = 0;
rC_ = [xC yC 0];

xD = k*xA+b;
yD = 0;
rD_ = [xD yD 0];

xE = k*xA+b;
yE = yC+2*b;
rE_ = [xE yE 0];
```

The position vectors of the centers of mass of the links are C_i, $i = 1,\ 2,\ 3$ are

```
xC1 = xA/2;
yC1 = yA/2;
rC1_ = [xC1 yC1 0];

xC2 = (xA+xB)/2;
yC2 = (yA+yB)/2;
rC2_ = [xC2 yC2 0];

xC3 = xB;
yC3 = b;
rC3_ = [xC3 yC3 0];
```

The MATLAB figure for the system is obtained with:

```
plot(...
  [0,xA],[0,yA],'k',...
  [xA,xB],[yA,yB],'r',...
  [xB,xC3],[yB,yC3],'b',...
  [xC,xC+2*b],[yC,yC],'k',...
  [xC,xC],[yC,yC+2*b],'k',...
  [xC,xC+2*b],[yC+2*b,yC+2*b],'k',...
  [xC+2*b,xD],[yC+2*b,yD],'k')

text(xA,yA,'  A')
text(xB,yB,'  B')
text(xC,yC,'  C')
text(xD,yD,'  D')
text(xE,yE,'  E')
text(xC1,yC1,'  C1')
text(xC2,yC2,'  C2')
text(xC3,yC3,'  C3')
grid
```

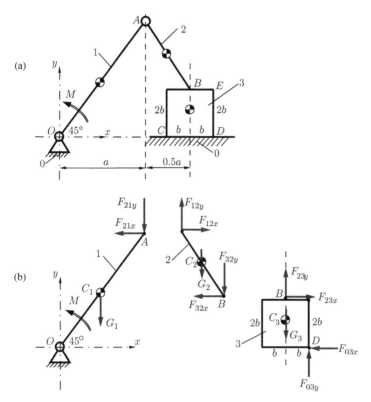

Figure 6.27 (a) Two-link mechanism 1, 2, and a block 3, and (b) free body diagram of the two-link mechanism and block.

The force analysis starts with block 3, Figure 6.27(b). The forces that act on block 3 are: the weight of the block at C_3, $\mathbf{G}_3$, the force of the ground 0 on 3 at D, $\mathbf{F}_{03}$, and the force of the link 2 on 3 at B, $\mathbf{F}_{23}$, or in MATLAB:

```
m3 = 6; % (kg)
g = 9.81; % (m/s^2)
syms F03x F03y F23x F23y
% link 3
G3_  = [0 -m3*g 0];   % @C3
F03_ = [F03x F03y 0]; % @D
F23_ = [F23x F23y 0]; % @B
```

The force equilibrium for block 3 is written as:

```
% sum of forces for link 3
% G3_+F03_+F23_ = 0_
SF3_ = G3_+F03_+F23_;
SF3x = SF3_(1); % Eq. (1)
SF3y = SF3_(2); % Eq. (2)
```

The sum of the moments for block 3 about its mass center C_3 is:

```
% sum of moments for link 3 about C3
SM3_ = cross(rD_-rC3_, F03_)+cross(rB_-rC3_, F23_);
SM3z = SM3_(3); % Eq. (3)
```

There are three scalar equations and four unknowns F_{03x}, F_{03y}, F_{23x}, and F_{23y}.

For link 2 the force of link 1 on link 2 is added and the force equilibrium equations are:

```
syms  F12x F12y
m2 = 6; % (kg)
% link 2
G2_  = [0 -m2*g 0];    % @C2
F12_ = [F12x F12y 0]; % @A
F32_ = -F23_; % @B

% sum of forces for link 2
% G2_+F12_+F32_ = 0
SF2_ = G2_+F12_+F32_;
SF2x = SF2_(1); % Eq. (4)
SF2y = SF2_(2); % Eq. (5)
```

The sum of the moments for link 2 about its mass center C_2 is:

```
% sum of moments for link 2 about C2
SM2_ = cross(rA_-rC2_, F12_)+cross(rB_-rC2_, F32_);
SM2z = SM2_(3); % Eq. (6)
```

For links 2 and 3 there are six scalar equations and six unknowns F_{03x}, F_{03y}, F_{23x}, F_{12x}, and F_{12y}. The system can be solved in MATLAB with:

```
sol32 = vpasolve(SF3x,SF3y,SM3z,SF2x,SF2y,SM2z);

F03x = sol32.F03x;
F03y = sol32.F03y;
F23x = sol32.F23x;
F23y = sol32.F23y;
F12x = sol32.F12x;
F12y = sol32.F12y;
```

The normal and the frictions forces at D and B are:

```
% F03y = ND = 220.725 (N)
% F03x = FfD = -110.362 (N)
% F32y = NB = 161.865 (N)
% F32x = FfB = -110.362 (N)
```

The reaction force $\mathbf{F}_{12}$ is:

```
% F12x = 110.362 (N)
% F12y = -103.005 (N)
```

The maximum friction forces at forces at D and B are calculated with:

```
musB = 0.7;
musD = 0.6;
F03fmax = musD*abs(F03y);
F23fmax = musB*abs(F23y);
```

and the results are:

```
% FfmaxD = 132.435 (N)
% FfmaxB = 113.305 (N)
```

and our assumptions are good:

```
if F03fmax>abs(F03x)
fprintf('FfmaxD > FfD => point D will not slip \n');
else
fprintf('FfmaxD < FfD => point D will slip \n');
end

if F23fmax>abs(F23x)
fprintf('FfmaxB > FfB => point B will not slip \n');
else
fprintf('FfmaxB < FfB => point B will slip \n');
end

% FfmaxD > FfD => point D will not slip
% FfmaxB > FfB => point B will not slip
```

The moment M is calculated from the moment equations for link 1 with respect to point O:

```
m1 = 8; %(kg)
% link 1
G1_  = [0 -m1*g 0];      % @C1
F21_ = [-F12x -F12y 0]; % @A
% sum of moments for link 1 about O
M_  = -cross(rA_, F21_)-cross(rC1_, G1_);
M = M_(3);
% M = -87.064 (N m)
```

7

Work, Energy and Power

7.1   Work

If a particle (or body) moves due to the action of an applied force, the applied force is said to do work on the particle (or body). Considering a particle of mass m and Newton's second law one can write

$$\mathbf{F} = m\mathbf{a}$$

$$\Leftrightarrow \mathbf{F} = m\dot{\mathbf{v}}$$

$$\Leftrightarrow \mathbf{F} \cdot \mathbf{v} = m\dot{\mathbf{v}} \cdot \mathbf{v}$$

$$\Leftrightarrow \mathbf{F} \cdot \frac{d\mathbf{r}}{dt} = m\dot{\mathbf{v}} \cdot \mathbf{v}$$

$$\Leftrightarrow \mathbf{F} \cdot \frac{d\mathbf{r}}{dt} = m\frac{1}{2}\frac{d}{dt}(\mathbf{v} \cdot \mathbf{v})$$

$$\Leftrightarrow \mathbf{F} \cdot d\mathbf{r} = \frac{1}{2}md(\mathbf{v} \cdot \mathbf{v})$$

$$\Leftrightarrow \mathbf{F} \cdot d\mathbf{r} = \frac{1}{2}md(v^2) \tag{7.1}$$

where $\cdot$ is the dot product, $d\mathbf{r}$ it an infinitesimal displacement, $\frac{d}{dt}(\mathbf{v} \cdot \mathbf{v}) = 2\dot{\mathbf{v}} \cdot \mathbf{v}$, and $\mathbf{v} \cdot \mathbf{v} = v^2$.

The quantity $\mathbf{F} \cdot d\mathbf{r}$ in Equation (7.1) denoted by

$$dU = \mathbf{F} \cdot d\mathbf{r} \tag{7.2}$$

is the work done on the particle by the applied force $\mathbf{F}$. The unit of work defined in the International System of Units (SI) is the *Joule*, that is, one *Joule* represents the work done by one Newton force through a displacement of one metre

$$1\,J = 1\,N\,m. \tag{7.3}$$

A diagram representing the applied force on a particle (vertical direction) against the distance moved by the particle (horizontal direction) is in fact a *work* diagram, where the *area below the graph represents the work done*.

Engineering Applications: Analytical and Numerical Calculation with MATLAB, First Edition.
Mihai Dupac and Dan B. Marghitu.
© 2021 John Wiley & Sons Ltd. Published 2021 by John Wiley & Sons Ltd.

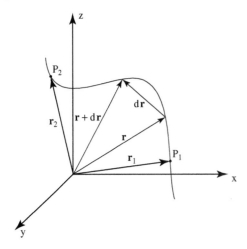

Figure 7.1 Particle moving from position P_1 to P_2 along a defined path.

When the particle is moving from position P_1 to P_2 (Figure 7.1) along a defined path C, the integration along the path is in fact the integration between the initial position P_1 (lower limit) and final position P_2 (upper limit) in Equation (7.1), that is

$$\int_C \mathbf{F} \cdot d\mathbf{r} = \int_{P_1}^{P_2} \mathbf{F} \cdot d\mathbf{r} = \int_{\mathbf{r}_1}^{\mathbf{r}_2} \mathbf{F} \cdot d\mathbf{r}$$

$$= \int_{v_1}^{v_2} \frac{1}{2} m\, d(v^2) = \frac{1}{2} mv^2 \Big|_{v_1}^{v_2}$$

$$= \frac{1}{2} mv_2^2 - \frac{1}{2} mv_1^2 \tag{7.4}$$

where $\mathbf{r}_1$ and respectively $\mathbf{r}_2$ represents the position vector of the particle located at the position P_1 and respectively P_2, and v_1 and v_2 is the particle velocity at the location P_1 and P_2 respectively. The quantity $\int_{\mathbf{r}_1}^{\mathbf{r}_2} \mathbf{F} \cdot d\mathbf{r}$ denoted by

$$U_{1-2} = \int_{\mathbf{r}_1}^{\mathbf{r}_2} \mathbf{F} \cdot d\mathbf{r} \tag{7.5}$$

is the *work done* by the applied force $\mathbf{F}$ (which may vary with position) when the particle is moving from position P_1 to position P_2.

7.2 Kinetic Energy

The quantity denoted by

$$T = \frac{1}{2} m\mathbf{v} \cdot \mathbf{v} = \frac{1}{2} mv^2 \tag{7.6}$$

is the *kinetic energy* of the particle with the mass m and the velocity v.

For the case of independent particles, the total energy can be obtained by adding the kinetic energy of each particle, that is

$$T = \frac{1}{2}\sum_{k=1}^{n} m_k \mathbf{v}_k \cdot \mathbf{v}_k$$

$$= \frac{1}{2}\sum_{k=1}^{n} m_k(\mathbf{v}_O + \mathbf{v}_{k/O}) \cdot (\mathbf{v}_O + \mathbf{v}_{k/O})$$

$$= \frac{1}{2}\sum_{k=1}^{n} m_k \mathbf{v}_O \cdot \mathbf{v}_O + \mathbf{v}_O \cdot \sum_{k=1}^{n} m_k \mathbf{v}_{k/O} + \frac{1}{2}\sum_{k=1}^{n} m_k \mathbf{v}_{k/O} \cdot \mathbf{v}_{k/O}$$

$$= \frac{1}{2}m(\mathbf{v}_O \cdot \mathbf{v}_O) + m(\mathbf{v}_O \cdot \mathbf{v}_{G/O}) + \frac{1}{2}\sum_{k=1}^{n} m_k \mathbf{v}_{k/O} \cdot \mathbf{v}_{k/O} \tag{7.7}$$

where O is the origin of an arbitrary reference frame, and G is the mass centre of the particles.

For the case of a solid body, the body can be represented by n individual particles as in Equation (7.7) where

$$\mathbf{v}_{k/O} = \boldsymbol{\omega} \times \mathbf{r}_{k/O}. \tag{7.8}$$

For the mass centre denoted by G one can write $\mathbf{v}_{G/O} = \boldsymbol{\omega} \times \mathbf{r}_{G/O}$. Replacing Equation (7.8) in Equation (7.7) one can calculate

$$T = \frac{1}{2}m(\mathbf{v}_O \cdot \mathbf{v}_O) + m\mathbf{v}_O \cdot (\boldsymbol{\omega} \times \mathbf{r}_{G/O}) + \frac{1}{2}\sum_{k=1}^{n} m_k(\boldsymbol{\omega} \cdot \mathbf{r}_{k/O}) \cdot (\boldsymbol{\omega} \cdot \mathbf{r}_{k/O})$$

$$= \frac{1}{2}m\mathbf{v}_O \cdot \mathbf{v}_O + m\mathbf{v}_O \cdot (\boldsymbol{\omega} \times \mathbf{r}_{G/O}) + \frac{1}{2}\sum_{k=1}^{n} m_k\boldsymbol{\omega} \cdot [\mathbf{r}_{k/O} \times (\boldsymbol{\omega} \times \mathbf{r}_{k/O})]$$

$$= \frac{1}{2}m\mathbf{v}_O \cdot \mathbf{v}_O + m\mathbf{v}_O \cdot (\boldsymbol{\omega} \times \mathbf{r}_{G/O}) + \frac{1}{2}\boldsymbol{\omega} \cdot \sum_{k=1}^{n} m_k[\mathbf{r}_{k/O} \times (\boldsymbol{\omega} \times \mathbf{r}_{k/O})]$$

$$= \frac{1}{2}m\mathbf{v}_O \cdot \mathbf{v}_O + m\mathbf{v}_O \cdot (\boldsymbol{\omega} \times \mathbf{r}_{G/O}) + \frac{1}{2}\boldsymbol{\omega} \cdot \mathbf{H}_O \tag{7.9}$$

where $\mathbf{H}_O = \sum_{k=1}^{n} m_k[\mathbf{r}_{k/O} \times (\boldsymbol{\omega} \times \mathbf{r}_{k/O})]$ is the angular momentum relative to the centre O of the Cartesian reference frame.

If the centre of the mass G of the rigid body is located at O, the kinetic energy of the rigid body can be expressed for any type of motion by

$$T = \frac{1}{2}m\mathbf{v}_O \cdot \mathbf{v}_O + \frac{1}{2}\boldsymbol{\omega} \cdot \mathbf{H}_O \tag{7.10}$$

or just for pure rotation by

$$T = \frac{1}{2}\boldsymbol{\omega} \cdot \mathbf{H}_O. \tag{7.11}$$

If there is a stationary point S on the body, the kinetic energy of pure rotation can be expressed with respect to the stationary point by

$$T = \frac{1}{2}\boldsymbol{\omega} \cdot \mathbf{H}_S = \frac{1}{2}m(\boldsymbol{\omega} \times \mathbf{r}_{G/S}) \cdot (\boldsymbol{\omega} \times \mathbf{r}_{G/S}) + \frac{1}{2}\boldsymbol{\omega} \cdot \mathbf{H}_G \qquad (7.12)$$

where $\boldsymbol{\omega} \times \mathbf{r}_{G/S}$ is the velocity of the mass centre of the body.

7.3 Work and Power

The location of a particle P when the particle is moving from position P_1 to position P_2 (Figure 7.2) along the trajectory C can be also described by the length s of the trajectory with respect to the initial position P_1. The particle velocity can be expressed by

$$\mathbf{v} = \frac{ds}{dt}\boldsymbol{\xi} \qquad (7.13)$$

where $\boldsymbol{\xi}$ is the tangent vector in a moving Frenet reference frame, and ds is an infinitesimal arc length. From Equation (7.13) an infinitesimal displacement along the particle trajectory C can be calculated as

$$d\mathbf{r} = \mathbf{v}dt = \frac{ds}{dt}\boldsymbol{\xi}dt = \boldsymbol{\xi}ds. \qquad (7.14)$$

From Equations (7.2) and (7.14) one can calculate the work done dU as

$$dU = \mathbf{F} \cdot d\mathbf{r} = \mathbf{F} \cdot \boldsymbol{\xi}ds \qquad (7.15)$$

where the scalar product $\mathbf{F} \cdot \boldsymbol{\xi}$ represents the magnitude of the tangential component of the force action on the particle. When the particle is moving along the trajectory C from the

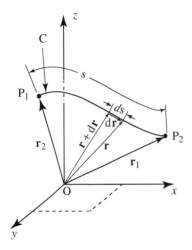

Figure 7.2 Location of a particle P moving from position P_1 to P_2 along a defined path.

position P_1 to position P_2 the work done can be calculated from Equation (7.5) and Equation (7.14) as

$$U_{1-2} = \int_{\mathbf{r}_1}^{\mathbf{r}_2} \mathbf{F} \cdot d\mathbf{r} = \int_{s_1}^{s_2} \mathbf{F} \cdot \boldsymbol{\xi} ds. \tag{7.16}$$

Power, defined as the rate at which work is done, can be computed by

$$P = \mathbf{F} \cdot \frac{d\mathbf{r}}{dt} = \mathbf{F} \cdot \mathbf{v}. \tag{7.17}$$

Since the power is calculated as the ratio between work and time, the unit of power in SI units is the *Watt*, denoted by W, where one Watt is equal to one Newton metre per second, or equivalent to one Joule per second.

7.4 Conservative Forces

Internal forces to a system can be classified as *conservative* or *dissipative*. An internal force **F** is said to be *conservative* if the work done by the force can be expressed as the difference in a scalar function of **r** between the start and the end of the trajectory,

$$U_{1-2} = \int_{P_1}^{P_2} \mathbf{F} \cdot d\mathbf{r} = \int_{\mathbf{r}_1}^{\mathbf{r}_2} \mathbf{F} \cdot d\mathbf{r} = V(\mathbf{r})\Big|_{\mathbf{r}_1}^{\mathbf{r}_2} = V(\mathbf{r}_1) - V(\mathbf{r}_2) \tag{7.18}$$

where P_1 is the initial location represented by the position vector $\mathbf{r}_1$, P_2 is the final location represented by the position vector $\mathbf{r}_2$, and the term $V(\mathbf{r})$ in Equation (7.18) is named *potential*.

One of the main consequences that can be assessed from Equation (7.18) is that the work done by a conservative force between any two locations P_1 and P_2 is independent of the path. Therefore, a *conservative* force does no net work when the particle at which it is applied follows an arbitrary closed path, since the work U_{1-2} only depends on the initial and final potentials.

7.5 Work Done by the Gravitational Force

When the particle of mass m is moving from position P_1 to position P_2 (Figure 7.3) along the trajectory C, the force exerted on the particle can be written as

$$\mathbf{F} = mg\mathbf{J} \tag{7.19}$$

where $Oxyz$ is the associated Cartesian reference frame, m is the particles mass, g is the gravitational acceleration, and $\mathbf{i}, \mathbf{j}$ and $\mathbf{k}$ are the unit vectors associated with the Ox, Oy and Oz axes respectively.

From Equations (7.2) and (7.19) one calculate the work done

$$dU = \mathbf{F} \cdot d\mathbf{r} = \mathbf{G} \cdot d\mathbf{r} = -mg\mathbf{J} \cdot d\mathbf{r}$$
$$= -mg\mathbf{J} \cdot \left(\frac{dx}{dt}\mathbf{i} + \frac{dy}{dt}\mathbf{j} + \frac{dz}{dt}\mathbf{k} \right) dt$$

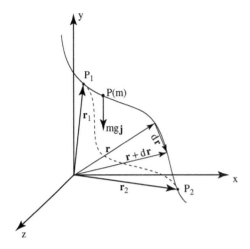

Figure 7.3 Work done by the gravitational force of a particle P moving from position P_1 to P_2 along a defined path.

$$
\begin{aligned}
&= -mg\mathbf{J} \cdot (dx\mathbf{i} + dy\mathbf{J} + dz\mathbf{k}) \\
&= -mgdx\mathbf{J} \cdot \mathbf{i} + mgdy\mathbf{J} \cdot \mathbf{J} + mgdz\mathbf{J} \cdot \mathbf{k} \\
&= -mgdy
\end{aligned}
\tag{7.20}
$$

where $\mathbf{J} \cdot \mathbf{i} = 0$, $\mathbf{J} \cdot \mathbf{J} = 1$ and $\mathbf{J} \cdot \mathbf{k} = 0$.

The work done by the gravitational force when the particle is moving from position P_1 to position P_2 (Figure 7.3) along the trajectory C, can be calculated from Equations (7.5) and (7.20) as

$$
U_{1-2} = \int_{\mathbf{r}_1}^{\mathbf{r}_2} \mathbf{F} \cdot d\mathbf{r} = \int_{y_1}^{y_2} -mgdy = -mgy \Big|_{y_1}^{y_2} = -mgy_2 + mgy_1
\tag{7.21}
$$

where h represents the height, h_1 is the height related to the initial position $\mathbf{r}_1$ of the particle, and h_2 is the height related to the final position $\mathbf{r}_2$ of the particle.

7.6 Work Done by the Friction Force

When a particle of mass m is moving from position P_1 to position P_2 along a trajectory C with friction, the work done by the friction force can be written as

$$
U_{1-2} = \int_{\mathbf{r}_1}^{\mathbf{r}_2} \mathbf{F} \cdot d\mathbf{r} = \int_0^L -\mu mg ds = -\mu mgs \Big|_0^L = -\mu mgL
\tag{7.22}
$$

where L is the length of the trajectory between the position P_1 and P_2, and μ is the coefficient of friction.

7.7 Potential Energy and Conservation of Energy

A change in the kinetic energy can be expressed from Equations (7.4) and (7.5) by

$$U_{1-2} = \int_{P_1}^{P_2} \mathbf{F} \cdot d\mathbf{r} = \int_{\mathbf{r}_1}^{\mathbf{r}_2} \mathbf{F} \cdot d\mathbf{r} = \frac{1}{2}mv_2^2 - \frac{1}{2}mv_1^2 = T_2 - T_1 \tag{7.23}$$

where $T_i, i = 1, 2$, is the kinetic energy of the particle with the velocity $v_i, i = 1, 2$.

A scalar function of the potential V in Equation (7.18) is called *potential energy* and may be expressed as

$$dV = -\mathbf{F} \cdot d\mathbf{r}. \tag{7.24}$$

The work associated with the potential energy in Equation (7.24) is expressed as in Equation (7.18) by

$$U_{1-2} = \int_{\mathbf{r}_1}^{\mathbf{r}_2} \mathbf{F} \cdot d\mathbf{r} = \int_{V_1}^{V_2} dV = V(\mathbf{r})\Big|_{\mathbf{r}_1}^{\mathbf{r}_2} = -[V(\mathbf{r}_2) - V(\mathbf{r}_1)] = -(V_2 - V_1) \tag{7.25}$$

where $V_i, i = 1, 2$ is the potential of the particle at the position $P_i, i = 1, 2$.

An important principle named the *principle of work and energy* relates the *kinetic energy* i.e. energy due to the motion, and *potential energy*, i.e. energy due to the position. The principle of work and energy can be expressed by

$$T_1 + V_1 = T_2 + V_2 \tag{7.26}$$

that is, the sum of the potential and kinetic energy is conserved.

7.8 Work Done and Potential Energy of an Elastic Force

Consider that an elastic force generated by a spring is applied on a particle of mass m. The magnitude of the elastic force can be expressed as $F = -k(r - r_0) = -k\delta$ where r_0 is the spring length in an unstretched state, r is the spring length in a stretched state, k is the spring constant, and $\delta = r - r_0$ is the spring stretch.

Using a system of polar coordinates, defined by the unit vectors $\mathbf{e}_r, \mathbf{e}_\theta$, and the polar angle θ, the vectorial force $\mathbf{F}$ and the value of $d\mathbf{r}$ can be expressed as

$$d\mathbf{r} = \left[\frac{dr}{dt}\mathbf{e}_r + r\frac{d\theta}{dt}\mathbf{e}_\theta\right] dt$$
$$\mathbf{F} = -k(r - r_0)\mathbf{e}_r = -k\delta\mathbf{e}_r. \tag{7.27}$$

Using the definition of potential energy as given by Equation (7.24) one can write

$$dV = -\mathbf{F} \cdot d\mathbf{r}$$
$$= -k(r - r_0)\mathbf{e}_r \cdot \left[\frac{dr}{dt}\mathbf{e}_r + r\frac{d\theta}{dt}\mathbf{e}_\theta\right] dt$$
$$= k(r - r_0)dr$$
$$= k\delta d\delta. \tag{7.28}$$

Integrating Equation (7.28) the potential energy due to the elastic force can be calculated as

$$\int dV = \int k\delta d\delta \Leftrightarrow V = \frac{1}{2}k\delta^2. \tag{7.29}$$

Using Equations (7.23) and (7.24) one can compute the work done on the particle by the elastic force by

$$
\begin{aligned}
U_{1-2} &= \int_{r_1}^{r_2} \mathbf{F} \cdot d\mathbf{r} \\
&= \int_{r_1}^{r_2} \mathbf{F} \cdot \left[\frac{dr}{dt}\mathbf{e}_r + r\frac{d\theta}{dt}\mathbf{e}_\theta\right] dt \\
&= \int_{r_1}^{r_2} -k(r - r_0)\mathbf{e}_r \cdot \left[\frac{dr}{dt}\mathbf{e}_r + r\frac{d\theta}{dt}\mathbf{e}_\theta\right] dt \\
&= -k\int_{r_1}^{r_2} (r - r_0)\mathbf{e}_r \cdot [dr\mathbf{e}_r + rd\theta\mathbf{e}_\theta] \\
&= -k\int_{r_1}^{r_2} (r - r_0)dr \\
&= -k\int_{\delta_1}^{\delta_2} \delta d\delta \\
&= -k\frac{\delta^2}{2}\Big|_{\delta_1}^{\delta_2} \\
&= -k\left(\frac{\delta_2^2}{2} - \frac{\delta_1^2}{2}\right) \\
&= -\frac{1}{2}k(\delta_2^2 - \delta_1^2). \tag{7.30}
\end{aligned}
$$

7.9 Potential Energy Due to the Gravitational Force

7.9.1 Potential Energy Due to the Gravitational Force for a Particle

The weight of (the force acting on) a particle of mass m in a gravitational field can be expressed by Equation (7.19), that is, $F = -mg\mathbf{j}$. Using the definition of potential energy V in Equations (7.24) and (7.27) one can write

$$dV = -\mathbf{F} \cdot d\mathbf{r}$$
$$\Leftrightarrow dV = mg\mathbf{j} \cdot \mathbf{v}dt$$
$$\Leftrightarrow dV = mg\mathbf{j} \cdot \left(\frac{dz}{dt}\mathbf{k} + \frac{dy}{dt}\mathbf{j} + \frac{dx}{dt}\mathbf{i}\right) dt$$
$$\Leftrightarrow dV = mg\mathbf{j} \cdot (dz\mathbf{k} + dy\mathbf{j} + dx\mathbf{i})$$
$$\Leftrightarrow dV = mg\mathbf{j} \cdot dx\mathbf{i} + mg\mathbf{j} \cdot dy\mathbf{j} + mg\mathbf{j} \cdot dz\mathbf{k}$$
$$\Leftrightarrow dV = mg\mathbf{j} \cdot dy\mathbf{j}$$
$$\Leftrightarrow dV = mgdy$$

$$\Leftrightarrow \int dV = \int mgdy$$

$$\Leftrightarrow V = mgy + C \tag{7.31}$$

where $mg\mathbf{j} \cdot d x\mathbf{i} = 0, mg\mathbf{j} \cdot dz\mathbf{k} = 0$ and C is the integration constant. For $C = 0$ one can write

$$V = mgy \tag{7.32}$$

that is, the potential energy due to the gravitational force is a function of position.

Considering a Cartesian reference frame $Oxyz$, one can define the gradient operator ∇ by

$$\nabla() = \frac{\partial()}{\partial x}\mathbf{i} + \frac{\partial()}{\partial y}\mathbf{j} + \frac{\partial()}{\partial z}\mathbf{k}. \tag{7.33}$$

The gradient of the potential energy function $V(x, y, z)$ calculated using

$$\nabla V = \frac{\partial V}{\partial x}\mathbf{i} + \frac{\partial V}{\partial y}\mathbf{j} + \frac{\partial V}{\partial z}\mathbf{k} \tag{7.34}$$

is a vector, called the gradient of V. The differential dV can be calculated using

$$dV = \nabla V \cdot d\mathbf{r}$$
$$= \frac{\partial V}{\partial x} dx + \frac{\partial V}{\partial y} dy + \frac{\partial V}{\partial z} dz \tag{7.35}$$

where $d\mathbf{r} = dx\mathbf{i} + dy\mathbf{j} + dz\mathbf{k}$. Since the potential energy can be written as $dV = -\mathbf{F} \cdot d\mathbf{r}$ from Equation (7.35) one can deduce

$$dV = -\mathbf{F} \cdot d\mathbf{r} = \nabla V \cdot d\mathbf{r}$$
$$\mathbf{F} = -\nabla V. \tag{7.36}$$

7.9.2 Potential Energy Due to the Gravitational Force for a Rigid Body

Using Equation (7.32) of the potential energy of an independent particle k, that is $V_k = m_k g y_k$, and Equation (7.36) written as $\mathbf{F}_k = -\nabla V_k$, the work done on the whole body (sum of the particles) can be computed using

$$\sum_{k=1}^{n} \int_{\mathbf{r}_k^1}^{\mathbf{r}_k^2} \mathbf{F}_i \cdot d\mathbf{r}_i = \sum_{k=1}^{n} -[(V_i)_2 - (V_i)_1] = -(V_2 - V_1). \tag{7.37}$$

For the case of independent particles, the total potential energy due to the gravitational force for a rigid body can be obtained by adding the potential energy of each particle, that is

$$V = \sum_{k=1}^{n} m_i g y_i = mg y_G \tag{7.38}$$

where y_G is the y-coordinate of the mass centre of the rigid body.

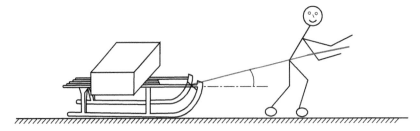

Figure 7.4 Construction worker using a rope and a sledge to pull a heavy block.

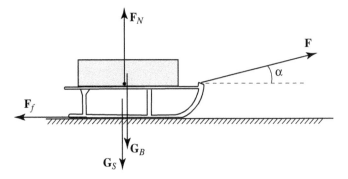

Figure 7.5 Free body diagram of the sledge and the heavy block.

7.10 Examples

Example 7.1

In the winter time the construction worker in Figure 7.4 is using a rope and a sledge of mass m_S to pull a heavy block of mass m_B. The rope is making an angle α with the horizontal as shown on the free-body diagram in Figure 7.5. The friction coefficient between the sledge and the ground is μ_S. How much work is done when:

(a) The sledge is moving a distance d and the acceleration due to the applied force on the sledge is a_S.
(b) The sledge is moving for a time t and the force applied by the worker on the rope is F.

Numerical application:

(a) $d = 4$ m, $m_S = 5$ kg, $m_B = 75$ kg, $g = 9.81$ m s^{-2}, $a_S = 3$ m s^{-2}, $\mu = 0.4$.
(b) $m_S = 5$ kg, $m_B = 75$ kg, $g = 9.81$ m s^{-2}, $t = 5$ s, $\mu = 0.1$, $\alpha = 37°$, $F = 150$ N.

Solution

(a) Considering Newton's law, one can write

$$\mathbf{F} + \mathbf{F}_f + \mathbf{F}_N + \mathbf{G} = m\mathbf{a}. \tag{7.39}$$

Considering motion along the Ox direction one can write

$$F_{Ox} - F_f = ma_S \Leftrightarrow F\cos\alpha - \mu_S F_N = (m_S + m_B)a_S$$
$$\Leftrightarrow F\cos\alpha = \mu_S(m_S + m_B)g + (m_S + m_B)a_S \tag{7.40}$$

where $m = m_S + m_B$ is the total mass (sledge and the block), the friction force magnitude is $F_f = \mu_S F_N = \mu_S mg$, and the magnitude of the Ox component of the force $\mathbf{F}$ is $F_{Ox} = F\cos\alpha$. The input data is declared in MATLAB with:

```
clear all; clc; close all

syms d alpha  mS mB g mu aS F_N F
r_=[d,0,0];

FOx=F*cos(alpha);
FOy=F*sin(alpha);
F_=[FOx,FOy,0];

m=mS+mB;  G_=[0,-m*g,0];
F_N=m*g;  FN_=[0,F_N,0];
Ff=-mu*F_N;  Ff_=[Ff,0,0];
a_=[aS,0,0];
```

Once Newton's law is input in MATLAB, the magnitude of the force $\mathbf{F}$ is calculated with:

```
sumF_  = F_+ Ff_ + G_ +FN_;
F=solve(sumF_(1)-m*a_(1),F);
fprintf('The force F is calculated as\n');
fprintf(' F = %s \n\n', F);
```

The work done by the applied force $\mathbf{F}$ can be computed using

$$W = \mathbf{F} \cdot \mathbf{r}_d = Fd\cos\alpha. \tag{7.41}$$

When the sledge is moving a distance d, from Equations (7.40) and (7.41) one can calculate the work done as

$$W = (F\cos\alpha)d = [\mu_S(m_S + m_B)g + (m_S + m_B)a_S]d. \tag{7.42}$$

The work done is calculated in MATLAB with:

```
W=F*d*cos(alpha);
fprintf('The work done is calculated with\n');
fprintf(' W = %s \n\n', W);
```

The input numerical data are introduced in MATLAB with:

```
lists = {mS,mB,aS,mu,g,d};
listt = {5,75,3,0.4,9.81,4};
```

The numerical results are calculated and printed in MATLAB using:

```
W=subs(W,lists,listt);
fprintf('The numerical value of the work is \n');
fprintf(' W = %g (kJ)\n', W/1000);
```

The numerical result (work) obtained in MATLAB is:

```
The numerical value of the work is
 W = 2.21568 (kJ)
```

(b) Considering Newton's law, one can write

$$\mathbf{F} + \mathbf{F}_f + \mathbf{F}_N + \mathbf{G} = m\mathbf{a}. \tag{7.43}$$

Considering motion along the Ox direction one can calculate the acceleration of the sledge as

$$F_{Ox} - F_f = ma_S \quad \Leftrightarrow \quad F\cos\alpha - F_f = (m_S + m_B)a_S$$

$$\Leftrightarrow \quad a_S = \frac{F\cos\alpha - F_f}{m_S + m_B}. \tag{7.44}$$

The input data is declared in MATLAB with:

```
clear all; clc; close all

syms d alpha   mS mB g mu aS F_N F t
r_ = [d,0,0];

FOx=F*cos(alpha);
FOy=F*sin(alpha);
F_ = [FOx,FOy,0];

m=mS+mB; G_ = [0,-m*g,0];
F_N=m*g; FN_ = [0,F_N,0];
Ff=-mu*F_N; Ff_ = [Ff,0,0];
a_ = [aS,0,0];
```

Once Newton's law is input in MATLAB, the acceleration a_S is calculated with:

```
sumF_ = F_+ Ff_ + G_ +FN_;
aS=solve(sumF_(1)-m*a_(1),aS);
fprintf('The acceleration aS is calculated as\n');
fprintf(' aS = %s \n\n', aS);
```

When the sledge is moving for a time t, the travelled distance can be calculated as

$$d = v_0 t + \frac{1}{2}a_S t^2 = \frac{1}{2}a_S t^2 \tag{7.45}$$

with the initial velocity $v_0 = 0$ since the sledge is initially at rest.

Replacing the calculated acceleration a_S from Equation (7.44) in Equation (7.45) one can obtain

$$d = \frac{1}{2}a_S t^2 = \frac{1}{2}\frac{F\cos\alpha - F_f}{m_S + m_B}t^2. \tag{7.46}$$

The travelled distance is calculated in MATLAB with:

```
v0=0;
veleq=d-v0*t - 1/2*aS*t^2;
d=solve(veleq,d);
fprintf('The traveled distance d is calculated with\n');
fprintf(' d = %s \n\n', d);
```

The work done by the applied force **F** is calculated with

$$W = \mathbf{F} \cdot \mathbf{r}_d = Fd \cos \alpha. \tag{7.47}$$

From Equations (7.47) and (7.46) one can calculate

$$
\begin{aligned}
W &= \frac{F(F \cos \alpha - F_f) \cos \alpha}{2(m_S + m_B)} t^2 \\
&= \frac{F(F \cos \alpha - \mu_S(m_S + m_B)g) \cos \alpha}{2(m_S + m_B)} t^2
\end{aligned}
\tag{7.48}
$$

where the friction force magnitude F_f was calculated using $F_f = \mu_S(m_S + m_B)g$. The work done is calculated in MATLAB with:

```
W=F*d*cos(alpha);
fprintf('The work done is calculated with\n');
fprintf(' W = %s \n\n', W);
```

The input numerical data are introduced in MATLAB with:

```
lists = {mS,mB,t,mu,g,alpha,F};
listt = {5,75,5,0.1,9.81,37*pi/180,150};
```

The numerical results are calculated and printed in MATLAB using:

```
W=subs(W,lists,listt);
fprintf('The numerical value of the work is \n');
fprintf(' W = %g (J)\n', W);
```

The numerical result (work) obtained in MATLAB is:

```
The numerical value of the work is
 W = 773.341 (J)
```

Example 7.2

The worker shown in Figure 7.6 is using a rope and an inclined plane S to lift a block A to a desired height, since this requires less force than lifting it straight up. The rope connecting the block A is parallel to the inclined plane and makes an angle α with the horizontal direction. The friction coefficient between the block A and the inclined plane is μ_S. Knowing the mass m_A of block A and the force **F** applied by the worker to the end of the rope determine the work done by:

(a) The magnitude of the force **F**
(b) The weight of the block

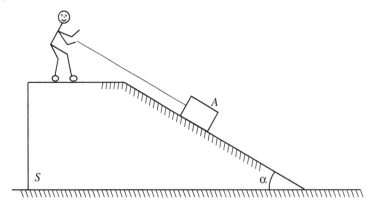

Figure 7.6 Construction worker using a rope and an inclined plane to lift a block to a desired height.

(c) The friction force
(d) The normal force
(e) Total work done

to move the block up the incline plane a distance d.
Numerical application: $m = 25$ kg, $g = 9.81$ m s^{-2}, $\mu_S = 0.25$, $r = 0.1$ m, and $\alpha = 30°$, $F = 300$ N, $d = 4$ m.

Solution
To calculate the work done when the block A is moving up the inclined plane a distance d, one can consider Newton's law (the free-body diagram in Figure 7.7) that is

$$\mathbf{F} + \mathbf{F}_f + \mathbf{F}_N + \mathbf{G} = m\mathbf{a}. \tag{7.49}$$

Considering the Ox (along the inclined plane of motion) and Oy (normal to the plane of motion) directions one can write

$$F_{Ox} - F_f - G_x = ma_x \Leftrightarrow F - F_f - mg \sin \alpha = ma \Leftrightarrow F - \mu F_N - mg \sin \alpha = ma$$

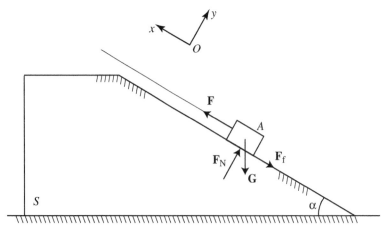

Figure 7.7 Free body diagram of the heavy block located on the inclined plane.

$$F_N - G_y = ma_y \Leftrightarrow F_N - mg\cos\alpha = 0 \Leftrightarrow F_N = mg\cos\alpha \tag{7.50}$$

where $F_f = \mu F_N$ is the magnitude of the friction force acting on block A.
The input data including the calculated normal and friction force is declared in MATLAB with:

```
clear all; clc; close all

syms d alpha m g mu a F_N F
r_=[d,0,0];


FOx=F; FOy=0; F_=[FOx,FOy,0];
G_=[-m*g*sin(alpha),-m*g*cos(alpha),0];
F_N=m*g*cos(alpha); FN_=[0,F_N,0];
Ff=mu*F_N; Ff_=[Ff,0,0];
```

The value of the work done can be expressed as a direct result of the magnitude of the force and the angle it makes with the displacement. One can write:

(a) The work done by the force $\mathbf{F}$

$$W_F = \mathbf{F} \cdot \mathbf{r} = Fd\cos 0 = Fd. \tag{7.51}$$

The work done by the force $\mathbf{F}$ is calculated in MATLAB with:

```
W_F=F*d*cos(alpha);
W_F=subs(W_F,'alpha','0');
fprintf('The work done by the force F is \n');
fprintf(' W_F = %s \n\n', W_F);
```

(b) The work done by the weight of the block

$$\begin{aligned}
W_G = \mathbf{G} \cdot \mathbf{r} &= mgd\cos\left(\alpha + \frac{\pi}{2}\right) \\
&= mgd\left(\cos\alpha\cos\frac{\pi}{2} - \sin\alpha\sin\frac{\pi}{2}\right) \\
&= -mgd\sin\alpha.
\end{aligned} \tag{7.52}$$

The work done by the weight of the block is calculated in MATLAB with:

```
W_G=sum(G_.*r_);
fprintf('The work done by the weight G is \n');
fprintf(' W_G = %s \n\n', W_G);
```

(c) The work done by the friction force

$$W_{F_f} = \mathbf{F}_f \cdot \mathbf{r} = F_f d\cos\pi = -F_f d = -\mu F_N d = -\mu mgd. \tag{7.53}$$

The work done by the friction force F_f is calculated in MATLAB with:

```
W_Ff=sum(Ff_.*r_);
W_Ff=subs(W_Ff,'alpha','pi');
fprintf('The work done by the ');
```

```
fprintf('friction force Ff is \n');
fprintf(' W_Ff = %s \n\n', W_Ff);
```

(d) The work done by the normal force

$$W_{F_N} = \mathbf{F_N} \cdot \mathbf{r} = F_N d \cos\frac{\pi}{2} = 0. \tag{7.54}$$

The work done by the normal force $\mathbf{F_N}$ is calculated in MATLAB with:

```
W_N=sum(FN_.*r_);
fprintf('The work done by the ');
fprintf('normal force F_N is \n');
fprintf(' W_N = %s \n\n', W_N);
```

(e) Total work done

$$W = W_F + W_G + W_{F_f} + W_{F_N} = Fd - mgd\sin\alpha - \mu mgd. \tag{7.55}$$

The total work done is calculated in MATLAB with:

```
W = W_F + W_G + W_Ff + W_N;
fprintf('The total work done is \n');
fprintf(' W = %s \n\n', W);
```

The input numerical data are introduced in MATLAB with:

```
lists = {m,mu,g,d,alpha,F};
listt = {25,0.25,9.81,4,30*pi/180,300};
```

The numerical results (work done by **F**, by the friction force, by the weight of the block, and by the normal force) are calculated and printed in MATLAB using:

```
W_F=subs(W_F,lists,listt);
W_G=eval(subs(W_G,lists,listt));
W_N=subs(W_N,lists,listt);
W_Ff=eval(subs(W_Ff,lists,listt));
fprintf('The numerical value of the ');
fprintf('work due to the force is \n');
fprintf(' W_F = %g (J)\n', W_F);
fprintf('The numerical value of the ');
fprintf('work due to the weight is \n');
fprintf(' W_G = %g (J)\n', W_G);
fprintf('The numerical value of the ');
fprintf('work due to the normal force is \n');
fprintf(' W_N = %g (J)\n', W_N);
fprintf('The numerical value of the ');
fprintf('work due to the friction force is \n');
fprintf(' W_Ff = %g (J)\n', W_Ff);
```

The numerical result for the total work obtained in MATLAB is:

```
W=subs(W,lists,listt);
```

```
fprintf('The numerical value of the total work is \n');
fprintf(' W = %g (J)\n', W);
```

The numerical value of the work due to the force is
 W_F = 1200 (J)
The numerical value of the work due to the weight is
 W_G = -490.5 (J)
The numerical value of the work due to the normal force is
 W_N = 0 (J)
The numerical value of the work due to the friction force is
 W_Ff = -245.25 (J)
The numerical value of the total work is
 W = 464.25 (J)

Example 7.3

The worker shown in Figure 7.8 is using a massless rope with a hook at the end and a frictionless pulley to lift the rectangular block A of mass m to a desired height. Knowing that the block is moving a distance d and the acceleration of the block due to the applied force on the rope is a, determine the work done by:

(a) The worker (applied force)
(b) The weight of the block.

Numerical application: $m = 150$ kg, $d = 3$ m, $g = 9.81$ m s^{-2}, and $a = 0.5$ m s^{-2}.

Solution

To calculate the work done by the worker and the work done by the weight when the block A is moving up a distance d one can consider Newton's law (see the free-body diagram in

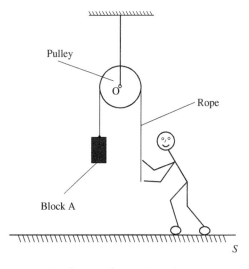

Pulley

Rope

Block A

S

Figure 7.8 Construction worker using a massless rope and frictionless pulley to lift a rectangular block to a desired height.

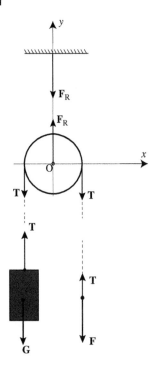

Figure 7.9 Free body diagram of the rectangular block and pulley.

Figure 7.9 for details) that is

$$\mathbf{T} + \mathbf{G} = m\mathbf{a}. \tag{7.56}$$

The input data is declared in MATLAB with:

```
clear all; clc; close all

syms d m g a F
r_=[0,d,0];

FOx=0;
FOy=-F;
F_=[FOx,FOy,0];
T_=-F_;
G_=[0,-m*g,0];
a_=[0,a,0];
```

Considering the Oy direction of motion one can write

$$T_{Oy} - G_{Oy} = ma \Leftrightarrow T - mg = ma \Leftrightarrow F - mg = ma \Leftrightarrow F = mg + ma \tag{7.57}$$

where $T = F$.

Once Newton's law is input in MATLAB, the magnitude of the force **F** is calculated with:

```
sumF_ = T_ + G_;
Fnew=solve(sumF_(2)-m*a_(2),F);
```

```
fprintf('The force F is calculated as\n');
fprintf(' F = %s \n\n', Fnew);
```

One can calculate:

(a) The work done by the worker (applied force **F**)

$$W_F = \mathbf{F} \cdot \mathbf{r} = Fd \cos 0 = Fd. \tag{7.58}$$

From Equations (7.58) and (7.57) one can write

$$W_F = Fd = (mg + ma)d. \tag{7.59}$$

The work done by the worker is calculated in MATLAB with:

```
W_T=sum(T_.*r_);
W_T=subs(W_T,F,Fnew);
fprintf('The work done by the worker ');
fprintf('applied force (tension) is calculated with\n');
fprintf(' W_T = %s \n\n', W_T);
```

(b) The work done by the weight of the block

$$W_G = \mathbf{G} \cdot \mathbf{r} = Gd \cos \pi = -mgd. \tag{7.60}$$

The work done by the weight G is calculated in MATLAB with:

```
W_G=sum(G_.*r_);
fprintf('The work done by the weight ');
fprintf('is calculated with\n');
fprintf(' W_G = %s \n\n', W_G);
```

The input numerical data are introduced in MATLAB with:

```
lists = {m,g,d,a};
listt = {150,9.81,3,0.5};
```

The numerical results are calculated and printed in MATLAB using:

```
W_T=subs(W_T,lists,listt);
W_G=subs(W_G,lists,listt);
fprintf('The numerical value of the work done ');
fprintf('by the worker applied force (tension) is \n');
fprintf(' W_T = %g (J)\n\n', W_T);
fprintf('The numerical value of the work ');
fprintf('done by the weight is \n');
fprintf(' W_G = %g (J)\n', W_G);
```

The numerical results obtained in MATLAB are:

```
The numerical value of the work
done by the worker applied force (tension) is
 W_T = 4639.5 (J)
```

The numerical value of the work done by the weight is
W_G = -4414.5 (J)

Example 7.4

A massless rocket is moving under the action of a variable force $\mathbf{F} = F_x(\mathbf{r})\mathbf{i} + F_y(\mathbf{r})\mathbf{j}$ from an initial position $\mathbf{r}_i = x_i\mathbf{i} + y_i\mathbf{j}$ to a final position $\mathbf{r}_f = x_f\mathbf{i} + y_f\mathbf{j}$. Determine the work done when that the path of the rocket is

(a) A parabola (Figure 7.10) described by the mathematical equation $y(x) = \frac{1}{2}x^2$ where
$F_x(\mathbf{r}) = y$, $F_y(\mathbf{r}) = 3x$.
Numerical application: $x_i = 0$ m, $y_i = 0$ m, $x_f = 2$ m, $y_f = 2$ m.
(b) Described by the mathematical equation $y(x) = 9x - x^3$ (Figure 7.11), where $F_x(\mathbf{r}) = y$ and $F_y(\mathbf{r}) = 2$.

Numerical application: $x_i = 0$ m, $y_i = 0$ m, $x_f = 1$ m, $y_f = 8$ m.

Solution
Using the formula for the work in a two-dimensional case one can calculate

$$W = W_x + W_y$$

$$= \int_{x_i}^{x_f} F_x(\mathbf{r})dx + \int_{y_i}^{y_f} F_y(\mathbf{r})dy$$

$$= \int_{x_i}^{x_f} ydx + \int_{y_i}^{y_f} 3xdy$$

$$= \int_{x_i}^{x_f} \frac{1}{2}x^2dx + \int_{y_i}^{y_f} 3x^2dx$$

$$= \frac{1}{2}\int_{x_i}^{x_f} x^2dx + 3\int_{y_i}^{y_f} x^2dx$$

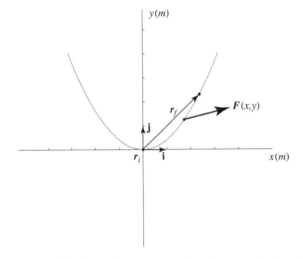

Figure 7.10 A massless rocket moving along a parabola under the action of a variable force.

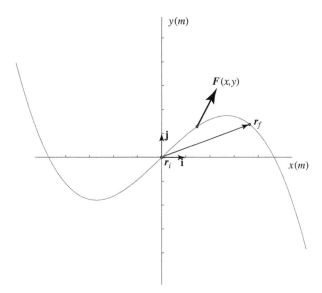

Figure 7.11 A massless rocket moving along a path described by the mathematical equation $y(x) = 9x - x^3$ under the action of a variable force.

$$= \frac{x^3}{6} \Big|_{x_i}^{x_f} + x^3 \Big|_{y_i}^{y_f}$$

$$= \left(\frac{x_f^3}{6} - \frac{x_i^3}{6} \right) + (y_f^3 - y_i^3) \tag{7.61}$$

where $dy = 2\frac{1}{2}xdx = xdx$.

The input data is declared in MATLAB with:

```
clear all; clc; close all
syms x y xi yi xf yf

y = 1/2*x^2;
Fx=y; Fy=3*x;
F=[Fx,Fy];
```

The *Ox* component of the work is calculated in MATLAB with:

```
Wx=int(Fx,xi,xf);
fprintf('The work Wx ');
fprintf('with respect to x is\n');
fprintf('Wx = %s   \n\n',Wx);
```

The *Oy* component of the work is calculated in MATLAB with:

```
dy=diff(y,x);
Wy=int(Fy*dy,yi,yf);
fprintf('The work Wy ');
```

```
fprintf('with respect to y is\n');
fprintf('Wy = %s   \n\n',Wy);
```

The total work is calculated in MATLAB with:

```
W=Wx+Wy;
fprintf('The total work W is calculated as\n');
fprintf(' W = Wx + Wy = %s \n\n', W);
```

The input numerical data are introduced in MATLAB with:

```
% numerical results
lists = {xi,yi,xf,yf};
listn = {0,0,2,2};
```

The numerical results are calculated and printed in MATLAB using:

```
Wx = eval(subs(Wx,lists,listn));
Wy = eval(subs(Wy,lists,listn));
fprintf('The numerical values of the work components are \n');
fprintf('Wx = %f [J] \n',Wx);
fprintf('Wy = %f [J]\n\n',Wy);


W = eval(subs(W,lists,listn));
fprintf('The numerical values of the total work is \n');
fprintf('W = %f [J]\n\n',W);
```

The numerical results obtained in MATLAB are:

```
The numerical values of the work components are
Wx = 1.333333 [J]
Wy = 8.000000 [J]


The numerical values of the total work is
W = 9.333333 [J]
```

(b) Using the formula for the work in a two-dimensional case one can write

$$W = W_x + W_y$$
$$= \int_{x_i}^{x_f} F_x(\mathbf{r})dx + \int_{y_i}^{y_f} F_y(\mathbf{r})dy$$
$$= \int_{x_i}^{x_f} ydx + \int_{y_i}^{y_f} 2dy$$
$$= \int_{x_i}^{x_f} (9x - x^3)dx + 2\int_{y_i}^{y_f} (9 - 3x^2)dx$$
$$= 9\int_{x_i}^{x_f} xdx - \int_{x_i}^{x_f} x^3 dx + 18\int_{y_i}^{y_f} dx - 6\int_{y_i}^{y_f} x^2 dx$$
$$= 9\frac{x^2}{2}\Big|_{x_i}^{x_f} - \frac{x^4}{4}\Big|_{x_i}^{x_f} + 18x\Big|_{y_i}^{y_f} - 6\frac{x^3}{3}\Big|_{y_i}^{y_f}$$

$$= \frac{9}{2}(x_f^2 - x_i^2) - \left(\frac{x_f^4}{4} - \frac{x_i^4}{4}\right) + 18(y_f - y_i) - 2(y_f^3 - y_i^3) \qquad (7.62)$$

where $dy = (9 - 3x^2)dx$.

The input data is declared in MATLAB with:

```
clear all; clc; close all
syms x y xi yi xf yf


y = 9*x - x^3;
Fx=y; Fy=2;
F=[Fx,Fy];
```

The *Ox* component of the work is calculated in MATLAB with:

```
Wx=int(Fx,xi,xf);
fprintf('The work Wx ');
fprintf('with respect to x is\n');
fprintf('Wx = %s   \n\n',expand(Wx));
```

The *Oy* component of the work is calculated in MATLAB with:

```
dy=diff(y,x);
Wy=int(Fy*dy,yi,yf);
fprintf('The work Wy ');
fprintf('with respect to y is \n');
fprintf('Wy = %s   \n\n',Wy);
```

The total work is calculated in MATLAB with:

```
W=Wx+Wy;
fprintf('The total work W is calculated as\n');
fprintf(' W = Wx + Wy = %s \n\n', W);
```

The input numerical data are introduced in MATLAB with:

```
% numerical results
lists = {xi,yi,xf,yf};
listn = {0,0,1,8};
```

The numerical results are calculated and printed in MATLAB using:

```
Wx = eval(subs(Wx,lists,listn));
Wy = eval(subs(Wy,lists,listn));
fprintf('The numerical values of the work are \n');
fprintf('Wx = %f [J] \n',Wx);
fprintf('Wy = %f [J]\n\n',Wy);
```

```
W = eval(subs(W,lists,listn));
fprintf('The numerical values of the total work is \n');
fprintf('W = %f [J]\n\n',W);
```

The numerical results obtained in MATLAB are:

```
The numerical values of the work are
Wx = 4.250000 [J]
Wy = -880.000000 [J]

The numerical values of the total work is
W = -875.750000 [J]
```

Example 7.5

To speed up the work at a construction site two workers are using a frictionless ramp of height h to slide down some bricks and roll down some spherical shells, as shown in Figure 7.12. While the worker at the top of the ramp is supposed to release the bricks and respectively the shells, the worker at the bottom of the ramp is suppose to catch them. Using the *principle of work and energy* calculate:

(a) The velocity of a brick at the bottom of the ramp.
 Numerical application: $m = 4$ kg, $h = h_{b_i} = 4$ m, $g = 9.81$ m s^{-2}.
(b) The velocity of a spherical shell at the bottom of the ramp.

 Numerical application: $m = 4$ kg, $h_{s_i} = h = 4$ m, $g = 9.81$ m s^{-2}.

Solution

(a) Applying the *principle of work and energy* for the change in height and speed of the brick (Figure 7.13) one can write

$$W_b = (K_{b_f} - K_{b_i}) + (U_{b_f} - U_{b_i}) \tag{7.63}$$

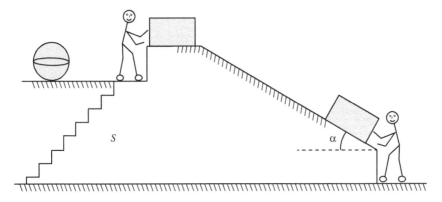

Figure 7.12 Construction worker using a frictionless ramp to slide down bricks and spherical shells.

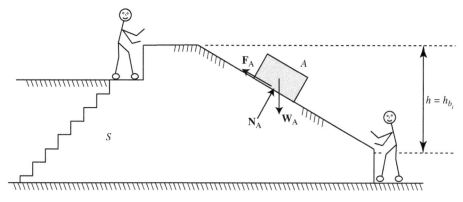

Figure 7.13 Free body diagram of a brick on the inclined plane.

where W_b is the work done by the brick, K_{b_f} is the final kinetic energy of the brick, K_{b_i} is the initial kinetic energy of the brick, U_{b_f} is the final potential energy of the brick, and U_{b_i} is the initial potential energy of the brick.

The input data is declared in MATLAB with:

```
clear all; clc; close all
syms m h_bi v_bi v_bf g

h_bf=0; v_bi=0;
K_bf=1/2*m*v_bf^2; K_bi=1/2*m*v_bi^2;
U_bf=m*g*h_bf; U_bi=m*g*h_bi;
W=0;
E=(K_bf-K_bi)+(U_bf-U_bi);
```

Since the ramp is frictionless, the work W_b is zero therefore from Equation (7.63) one can write

$$K_{b_f} - K_{b_i} = U_{b_i} - U_{b_f}$$

$$\Leftrightarrow \quad \frac{1}{2}mv_{b_f}^2 - \frac{1}{2}mv_{b_i}^2 = mgh_{b_i} - mgh_{b_f}$$

$$\Leftrightarrow \quad \frac{1}{2}mv_{b_f}^2 = mgh_{b_i}$$

$$\Leftrightarrow \quad v_{b_f}^2 = \sqrt{2gh_{b_i}} \qquad (7.64)$$

where the final height is $h_{b_f} = 0$, the initial height $h_{b_i} = h$ is the height of the ramp, the initial velocity of the brick is $v_{b_i} = 0$ and its final velocity at the bottom of the ramp is v_{b_f}.

Using the *principle of work and energy* in MATLAB, one can calculate the velocity of the brick at the end (bottom) of the ramp with:

```
v_bf=solve(W-E,v_bf);
fprintf('The velocity v_bf is \n');
fprintf(' v_bf = %s \n\n', v_bf(1));
```

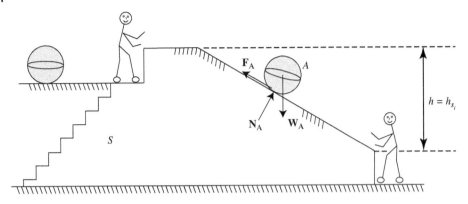

Figure 7.14 Free body diagram of a spherical shell on the inclined plane.

The input numerical data are introduced in MATLAB with:

```
% numerical results
lists = {g,h_bi};
listn = {9.81,4};
```

The numerical result (velocity of the brick) is calculated and printed in MATLAB using:

```
v_bf = abs(eval(subs(v_bf,lists,listn)));
fprintf('The velocoity v_bf is \n');
fprintf('v_bf = %f [m/s^2]\n\n',v_bf(1));
```

The numerical result (brick velocity) obtained in MATLAB is:

```
The velocoity v_bf is
v_bf = 8.858894 [m/s^2]
```

(b) The spherical shell rolling down the ramp (Figure 7.14) acquires both translational K_{Ts} and rotational K_{Rs} kinetic energy. Using the *principle of work and energy* for the spherical shell one can write

$$W_s = (K_{Ts_f} - K_{Ts_i}) + (K_{Rs_f} - K_{Rs_i}) + (U_{s_f} - U_{s_i}) \qquad (7.65)$$

where W_s is the work done by the spherical shell, K_{Ts_f} is the final translational kinetic energy of the shell, K_{Ts_i} is the initial translational kinetic energy of the shell, K_{Rs_f} is the final rotational kinetic energy of the shell, K_{Rs_i} is the initial rotational kinetic energy of the shell, U_{s_f} is the final potential energy of the shell, and U_{s_i} is the initial potential energy of the shell.

The input data is declared in MATLAB with:

```
clear all; clc; close all
syms m g v_Tsf R h v_Rsf


v_Tsi=0; v_Rsi=0; I=2/3*m*R^2; h_sf=0; h_si=h;
omega_Rsf=v_Tsf/R; omega_Rsi=v_Rsi/R;
K_Tsf=1/2*m*v_Tsf*v_Tsf; K_Tsi=1/2*m*v_Tsi*v_Tsi;
```

```
K_Rsf=1/2*I*omega_Rsf*omega_Rsf;
K_Rsi=1/2*I*omega_Rsi*omega_Rsi;
U_sf=m*g*h_sf; U_si=m*g*h_si;
W=0;
E=(K_Tsf-K_Tsi)+(K_Rsf-K_Rsi)+(U_sf-U_si);
```

Since the ramp is frictionless, the work W_s is zero therefore from Equation (7.65) one can write

$$(K_{Ts_f} - K_{Ts_i}) + (K_{Rs_f} - K_{Rs_i}) = U_{s_f} - U_{s_i}$$

$$\Leftrightarrow \left(\frac{1}{2}mv_{Ts_f}^2 - \frac{1}{2}mv_{Ts_i}^2\right) + \left(\frac{1}{2}I\omega_{Rs_f}^2 - \frac{1}{2}I\omega_{Rs_i}^2\right) = mgh_{s_i} - mgh_{s_f}$$

$$\Leftrightarrow \frac{1}{2}mv_{Ts_f}^2 + \frac{1}{2}I\omega_{Rs_f}^2 = mgh_{s_i}$$

$$\Leftrightarrow \frac{1}{2}mv_{Ts_f}^2 + \frac{1}{2}\frac{2}{3}mR^2\frac{v_{Ts_f}^2}{R^2} = mgh_{s_i}$$

$$\Leftrightarrow \frac{1}{2}mv_{Ts_f}^2 + \frac{1}{3}mv_{Ts_f}^2 = mgh_{s_i}$$

$$\Leftrightarrow v_{Ts_f} = \sqrt{\frac{6}{5}gh_{s_i}} = \sqrt{1.2gh_{s_i}} \tag{7.66}$$

where the final height is $h_{s_f} = 0$, the initial height $h_{s_i} = h$ is the height of the ramp, the initial translational velocity of the spherical shell is $v_{Ts_i} = 0$, the final translational velocity at the bottom of the ramp is v_{Ts_f}, the initial rotational velocity of the spherical shell is $v_{Rs_i} = 0$, the final rotational velocity at the bottom of the ramp is v_{Rs_f}, the angular velocity of the spherical shell is $\omega_{Rs_f} = \dfrac{v_{Ts_f}}{R}$, the moment of inertia for the spherical shell is $I = \frac{2}{3}mR^2$, and R is the radius of the spherical shell.

Using the *principle of work and energy* in MATLAB one can calculate the velocity of the spherical shell at the bottom of the ramp with:

```
v_Tsf=solve(-W+E,v_Tsf);
fprintf('The velocity v_Tsf is \n');
fprintf(' v_Tsf = %s \n\n', v_Tsf(2));
```

The input numerical data are introduced in MATLAB with:

```
% numerical results
lists = {g,h};
listn = {9.81,4};
```

The numerical result (spherical shell velocity) is calculated and printed in MATLAB using:

```
v_Tsf = abs(eval(subs(v_Tsf,lists,listn)));
fprintf('The velocoity v_Tsf is \n');
fprintf('v_Tsf = %f [m/s^2]\n\n',v_Tsf(1));
```

The numerical result obtained in MATLAB is:

```
The velocoity v_Tsf is
v_Tsf = 6.862070 [m/s^2]
```

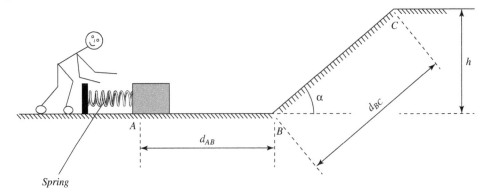

Figure 7.15 Construction worker using a spring to shoot a heavy block along a built track.

Example 7.6

To speed up the work at a construction site a worker is using a spring with a spring constant k to shoot a heavy block of mass m along a built track. The track is composed of a horizontal part AB of length d_{AB} and an inclined ramp BC of length d_{BC} making an angle α with the horizontal direction as shown in Figure 7.15. The friction coefficient between the track and the block is μ. The spring is compressed a distance Δx_i by the worker before the releasing the block. Using the *principle of work and energy* calculate:

(a) The work done by friction
(b) The velocity of a block at the top of the ramp.

Numerical application: $\mu = 0.4$, $m = 20$ kg, $d_{AB} = 7$ m, $d_{BC} = 5$ m, $\alpha = 30°$, $g = 9.81$ m s^{-2}, $\Delta_{x_i} = 0.2$ m, and $k = 15$.

Solution

The work done by the friction can be calculated as

$$
\begin{aligned}
W &= W_{AB} + W_{BC} \\
&= -F_f d_{AB} - F_f d_{BC} \\
&= -\mu mg d_{AB} - \mu mg d_{BC} \cos \alpha.
\end{aligned}
\tag{7.67}
$$

The input data is declared in MATLAB with:

```
clear all; clc; close all
syms m g d_AB d_BC alpha mu h k Delta_xi vf

N_AB=m*g;
N_BC=m*g*cos(alpha);
F_fAB=mu*N_AB;
F_fBC=mu*N_BC;
W_AB = -F_fAB*d_AB;
W_BC = -F_fBC*d_BC;
```

The work done by the friction is calculated in MATLAB with:

```
W=W_AB+W_BC;
fprintf('The total work W done by the friction is \n');
fprintf(' W = W_AB + W_BC = %s \n\n', W);
```

Applying the *principle of work and energy* to the block released by the spring one can write

$$W = (K_f - K_i) + (U_{sf} - U_{si}) + (U_f - U_i) \tag{7.68}$$

where W is the work done, $K_f = \frac{1}{2}mv_f^2$ is the final kinetic energy, $K_i = 0$ is the initial kinetic energy, $U_{si} = \frac{1}{2}k\Delta x_i$ is the initial energy stored in the spring, $U_{sf} = 0$ because of the equilibrium position, U_f and U_i are the final and the initial potential energy of the block respectively.

The kinetic energy, potential energy, work done, and energy stored in the spring is entered in MATLAB with:

```
Kf=1/2*m*vf^2; Ki=0;
h_f=h;
Uf=m*g*h_f; Ui=0;
Usf=0; Usi=1/2*k*Delta_xi;
E=(Kf-Ki)+(Usf-Usi)+(Uf-Ui);
```

Replacing the work done by the friction (Equation (7.67)) in Equation (7.68) one can calculate the velocity of the block, that is

$$-\mu mgd_{AB} - \mu mgd_{BC}\cos\alpha = \frac{1}{2}mv_f^2 - \frac{1}{2}k(\Delta x_i)^2 + mgh_f$$

$$\Leftrightarrow \quad -2\mu gd_{AB} - 2\mu gd_{BC}\cos\alpha = v_f^2 - \frac{k}{m}(\Delta x_i)^2 + 2gh_f$$

$$\Leftrightarrow \quad v_f = \sqrt{\frac{k}{m}(\Delta x_i)^2 - 2gh_f - 2\mu gd_{AB} - 2\mu gd_{BC}\cos\alpha}$$

$$\Leftrightarrow \quad v_f = \sqrt{\frac{k}{m}(\Delta x_i)^2 - 2gd_{BC}\sin\alpha - 2\mu gd_{AB} - 2\mu gd_{BC}\cos\alpha} \tag{7.69}$$

where $h_f = d_{BC}\sin\alpha$.

The velocity of the block is calculated in MATLAB using:

```
listsh={h}; listnh={d_BC*sin(alpha)};
vf=solve(W-E,vf);
vf = subs(vf,listsh,listnh);
fprintf('The velocity vf is \n');
fprintf(' vf = %s \n\n', vf(1));
```

The input numerical data are introduced in MATLAB with:

```
% numerical results
lists = {mu,m,d_AB,d_BC,g,alpha,Delta_xi,k};
listn = {0.4,20,7,5,9.81,30*pi/180,0.2,15};
```

The numerical results (work done by the friction and velocity of the heavy block) are calculated and printed in MATLAB using:

```
W_AB = eval(subs(W_AB,lists,listn));
W_BC = eval(subs(W_BC,lists,listn));
fprintf('The numerical values ');
fprintf('of the work W_AB and W_BC are \n');
fprintf('W_AB = %f [J] \n',W_AB);
fprintf('W_BC = %f [J]\n\n',W_BC);

W = eval(subs(W,lists,listn));
fprintf('The numerical value of the work ');
fprintf('done by the friction is \n');
fprintf('W = %f [J]\n\n',W);

vf = abs(eval(subs(vf,lists,listn)));
fprintf('The velocity vf is \n');
fprintf('vf = %f [m/s^2]\n\n',vf(1));
```

The numerical results obtained in MATLAB are:

```
The numerical values of the work W_AB and W_BC are
W_AB = -549.360000 [J]
W_BC = -339.828368 [J]

The numerical value of the work done by the friction is
W = -889.188368 [J]

The velocity vf is
vf = 11.739627 [m/s^2]
```

Example 7.7

A sphere, P, of mass m is at a distance y_1 above the ground. The sphere is stationary and it is struck at a velocity v_1 straight upward, as shown in Figure 7.16(a). The duration of the impact is dt. Find:

(a) The maximum hight above the ground the sphere travels
(b) The velocity of the sphere just before it collides the ground
(c) The average power transferred to the sphere during the initial impact
(d) Repeat (a) and (b) using Newton's equation of motion.

Numerical application:

```
m = 0.5;     % (kg)
y1= 0.5;     % (m)
v1= 5;       % (m/s)
dt= 0.002;   % (s)
g = 9.81;    % (m/s^2)
```

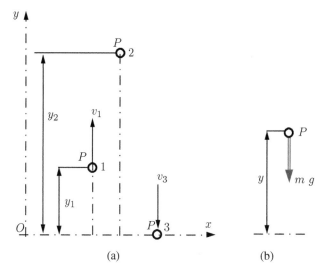

Figure 7.16 Sphere under impact.

Solution

(a) The maximum height above the ground of the sphere is determined from the conservation of energy at the initial position 1 and the position of maximum height 2. For position 1 the kinetic energy is $T_1 = mv_1^2/2$ and the potential energy is $V_1 = mg\, y_1$. For position 2 the kinetic energy is $T_2 = mv_2^2/2 = 0$ because at maximum height the velocity of the sphere is zero, $v_2 = 0$. The potential energy is $V_2 = mg\, y_2 = V_2 = mg\, y_{max}$. The conservation of the total energy gives

$$T_1 + V_1 = T_2 + V_2 \quad \text{or} \quad mv_1^2/2 + mg\, y_1 = mg\, y_2. \tag{7.70}$$

The maximum hight is calculated in MATLAB with:

```
% conservation of energy
T1 = m*v1^2/2;
V1 = m*g*y1;

v2 = 0;
T2 = m*v2^2/2;
V2 = m*g*y2;

% E = T1+V1 = T2+V2
E = (T1+V1)-(T2+V2);
y2n = vpasolve(E, y2);
fprintf('y2=ymax=%6.3g(m)\n',y2n)

% y2=ymax=  1.77(m)
```

(b) The velocity of the sphere just before it collides with the ground is determined from the conservation of energy at the initial position 1 and the final position 3 when $y_3 = 0$. For position 3 the kinetic energy is $T_3 = mv_3^2/2$ and the potential energy is $V_1 = mg\, y_3 = 0$. The conservation of the total energy between 1 and 3 is

$$T_1 + V_1 = T_3 + V_3 \quad \text{or} \quad mv_1^2/2 + mg\, y_1 = mv_3^2/2. \tag{7.71}$$

The velocity before it contacts the ground in MATLAB is calculated with:

```
T3 = m*v3^2/2;
y3 = 0;
V3 = m*g*y3;
```

```
% E = T1+V1 = T3+V3
E = (T1+V1)-(T3+V3);
v3n = vpasolve(E, v3);
fprintf('v3=vmax=%6.3g(m/s)\n',v3n(1))
```

```
% v3=vmax=  -5.9(m/s)
```

(c) The average power transferred to the sphere during the initial impact of 0.002 s is

```
v0 = 0; %(m/s)
U01 = m*v1^2/2 - m*v0^2/2;
% power
P = U01/dt;
fprintf('power P=%6.3f(W)=%6.3f(kW)\n',P,P/1000)
```

```
% power P=3125.000(W)= 3.125(kW)
```

(d) The free-body diagram of the sphere is shown in Figure 7.16(b). The Newton's equation of motion is $md^2y/dt^2 = -mg$ or in MATLAB:

```
syms t y(t)
% m a = -m g => d2y/dt2 = -g
eq = diff(y(t),t,t)+g;
```

To solve the ordinary differential equation the following MATLAB commands are used:

```
[V,S] = odeToVectorField(eq)
% converting scalar differential equations
% to coupled first order systems
% V =
%       Y[2]
%  -981/100
%
% S =
%   y -> Y[1]
%   Dy -> Y[2]
syms Y
```

```
eom_ = matlabFunction(V,'vars', {t,Y})
% generate a MATLAB  function from a sym object
% eom_ = function_handle with value:
% @(t,Y)[Y(2);-9.81e+2./1.0e+2]
```

For the upward motion the initial conditions are:

```
% upward motion
timeu_ = [0 10];
% initial condition
x0u_ = [y1 v1];
% y(0) = y1
% dy(0)/dt = v1
```

An option is created the ordinary differential equation using odeset:

```
optiond = ...
odeset('RelTol',1e-3,'MaxStep',1e-3,'Events',@eventu);
```

The function eventu is developed to stop the integration when velocity is zero, at the maximum height:

```
function [value, isterminal, direction] = eventu(t,x)
% velocity zero v=dy=0
value = x(2);
isterminal = 1;
direction = 0;
end
%value(i) describes the ith event.
%isterminal(i) = 1 integration ends when event is true
%otherwise, it is 0.
%direction(i) = 0 if all zeros are determined.
```

The ordinary differential equation is numerically solved using ode45:

```
[tu, xsu_, tum, xsum] = ...
ode45(eom_,timeu_,x0u_,optionu);
```

The time and maximum height for the upward motion are:

```
ymax = xsum(1);
vtop = xsum(2);

yup_ = xsu_(:,1);
vup_ = xsu_(:,2);

fprintf('tup = %6.3g (s) \n',tum);
fprintf('ymax = %6.3g (m) \n',ymax);

% tup =   0.51 (s)
% ymax =   1.77 (m)
```

For the downward motion when velocity down negative the following MATLAB commands are used:

```
% downward motion
% velocity down negative

timed_ = [tum 10];
xd_ = [ymax 0]; % initial conditions

optiond = ...
odeset('RelTol',1e-3,'MaxStep',1e-3,'Events',@eventd);

[td, xsd_, tdm, xsdm] = ...
ode45(eom_, timed_, xd_, optiond);

ttotal = tdm;
tdown = ttotal-tum;
yd = xsdm(1);
vd = xsdm(2);
ydown_ = xsd_(:,1);
vdown_ = xsd_(:,2);

fprintf('\n');
fprintf('ttotal = %6.3g (s) \n',ttotal);
fprintf('tdown = ttotal-tup = %6.3g (s) \n',tdown);
fprintf('vd = %6.3g (m/s) \n',vd);

% ttotal =    1.11 (s)
% tdown = ttotal-tup =   0.601 (s)
% vd =    -5.9 (m/s)
```

The event was when the distance from the ground is zero:

```
function [value, isterminal, direction] = eventd(t,x)
% displacement zero y=0
value = x(1);
isterminal = 1;
direction = 0;
end
```

The kinetic energies of the sphere for upward and downward motions are:

```
Tu  = 1/2*vup_.*vup_;
Td  = 1/2*vdown_.*vdown_;
```

The distance from the ground, the velocity, and the kinetic energy of the sphere are depicted in Figure 7.17. The velocity as a function of the ground distance is shown in Figure 7.18. The MATLAB program for the previous two figures is

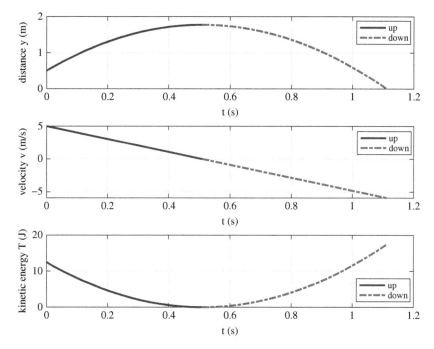

Figure 7.17 The distance from the ground, the velocity, and the kinetic energy of the sphere.

```
figure(1)
subplot(3,1,1)
plot(tu,yup_,'b-','LineWidth',2)
hold on
plot(td,ydown_,'r-.','LineWidth',2)
hold on
legend('up','down')
xlabel('t (s)'); ylabel('distance y (m)');
grid
subplot(3,1,2)
plot(tu,vup_,'b-','LineWidth',2)
hold on
plot(td,vdown_,'r-.','LineWidth',2)
hold on
legend('up','down')
xlabel('t (s)'); ylabel('velocity v (m/s)');
grid
subplot(3,1,3)
plot(tu,Tu,'b-','LineWidth',2)
hold on
plot(td,Td,'r-.','LineWidth',2)
hold on
```

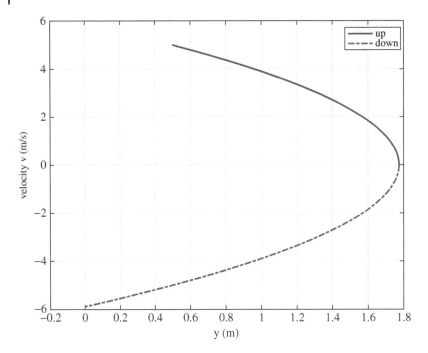

Figure 7.18 The velocity of the sphere as a function of the ground distance.

```
legend('up','down')
xlabel('t (s)'); ylabel('kinetic energy T (J)');
grid

figure(2)
plot(yup_,vup_,'b-','LineWidth',2)
hold on
plot(ydown_,vdown_,'r-.','LineWidth',2)
legend('up','down')
xlabel('y (m)'); ylabel('velocity v (m/s)');
grid
```

Example 7.8

A constant force F acts on a block as is shown in Figure 7.19(a). The block is initially at rest and moves on a horizontal surface. The weight of the block is G. The coefficient of kinetic friction between the surface and the block is μ_k. Find the velocity of the block at the distance d from the initial position using the principle of work and energy and the Newton's equation of motion.

Numerical application:

```
G = 30;    % (lb) weight
```

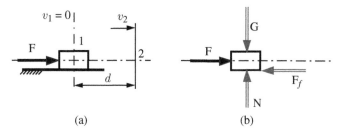

Figure 7.19 (a) Force acting on a block initially resting on a horizontal surface, (a) Free body diagram of the block.

```
F = 10;    % (lb) constant force
muk=0.2;   % coefficient of kinetic friction
d = 3;     % (ft)
g = 32.2; % (ft/s^2)
```

Solution

The free-body diagram of the block is shown in Figure 7.19(b) where N is the reaction of the surface on the block and $F_f = \mu_k N$ is the friction force. The equilibrium equation on the vertical axis gives: $N = G$. The principle of work and energy can be written as

$$U_{12} = T_2 + T_1, \tag{7.72}$$

where $T_2 = mv_2^2/2$ is the kinetic energy of the block with the mass $m = G/g$ at the distance d and $T_1 = 0$ is the initial kinetic energy $(v_1 0)$. The work between the positions 1 and 2 is $U_{12} = (F - F_f)\, d$. The velocity v_2 at the distance d is calculated in MATLAB with:

```
syms v2
N = G;
Ff= muk*N;
v1 = 0;
T1 = (G/g)*v1^2/2;
T2 = (G/g)*v2^2/2;
U12 = (F - Ff)*d;

PWE = U12-(T2-T1);
v2n = vpasolve(PWE, v2);
fprintf('v2=%6.3g(ft/s)\n',v2n(2))
```

```
v2=   5.08(ft/s)
```

The same result is obtained using Newton's second law. The ordinary differential equation of motion

$$\frac{G}{m}\frac{d^2y}{dt^2} = F - \mu_k\, G, \text{ with } y(0) = \frac{dy}{dt}(0) = 0, \tag{7.73}$$

is solved in MATLAB with:

```
syms t y(t)
```

```
% m a = F-Ff => (G/g)d2y/dt2 = F-Ff
eq = (F - Ff) - (G/g)*diff(y(t),t,t);

[V,S] = odeToVectorField(eq);
syms Y
eom_ = matlabFunction(V,'vars', {t,Y});
time_ = [0 1.75];
x0_ = [0 0]; % initial condition
option = ...
odeset('RelTol',1e-3,'MaxStep',1e-3,'Events',@event);
[tu_, xsu_, td, xsum] = ...
ode45(eom_,time_,x0_,option);
yd = xsum(1);
vd = xsum(2);
yt_ = xsu_(:,1);
vt_ = xsu_(:,2);
fprintf('t = %6.3g (s) \n',td);
fprintf('d = %6.3g (ft) \n',yd);
fprintf('v = %6.3g (ft/s) \n',vd);
```

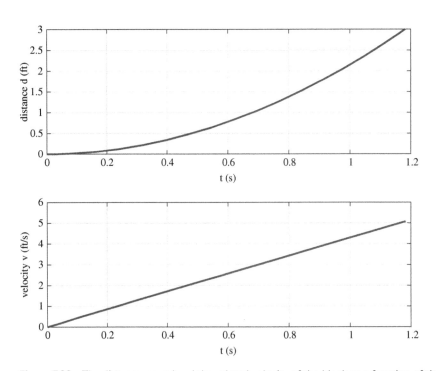

Figure 7.20 The distance moved and the related velocity of the block as a function of time.

```
%t =    1.18 (s)
%d =       3 (ft)
%v =    5.08 (ft/s)

function [value,isterminal,direction]=event(t,x)
% distance  y - d = 0
global d
value = x(1) - d;
isterminal = 1;
direction = 0;
end
```

The distance and the velocity of the block are depicted in Figure 7.20.

8

Simple Machines

A machine can be defined as a device that makes work easier by converting energy form one form to another form, that is, it utilizes the energy received to do some useful work.

A simple machine can be defined as a mechanical device that modifies the direction and magnitude of a force or effort and makes work easier. A simple machine has no moving (or a few) moving parts, and uses a small amount of force over distance (by changing the magnitude, speed, or direction of force) to overcome larger resistance, thus making work easier by using less mechanical effort for moving an object.

There are generally accepted (defined) five classical simple machines namely the:

(1) Lever
(2) Inclined plane
(3) Screw and wedge
(4) Pulley
(5) Wheel and axle.

The above simple machines are generally used to lift loads hence they are also called simple lifting machines. When a set of simple machines are working together this is named a compound machine. If a simple machine does not dissipate energy, the machine energy is conserved, as it has 100% efficiency, and it is called an ideal machine.

The idea of a simple machine was introduced in the third century BC by the Greek mathematician, scientist and philosopher Archimedes of Syracuse. His remark "give me a place to stand on, and I will move the Earth," relates to the idea of the mechanical advantage of a lever.

8.1 Load and Effort, Mechanical Advantage, Velocity Ratio and Efficiency of a Simple Machine

8.1.1 Load and Effort

A simple machine is driven by a *input force* named *effort* to overcome a considerable resistance (*output force*) named *load*. The input force, e.g., *the effort*, is formally denoted by F_E and the larger resistance, e.g., *the load* to be overcome is denoted by F_L.

Engineering Applications: Analytical and Numerical Calculation with MATLAB, First Edition.
Mihai Dupac and Dan B. Marghitu.
© 2021 John Wiley & Sons Ltd. Published 2021 by John Wiley & Sons Ltd.

8.1.2 Mechanical Advantage

Equilibrium of a simple machine implies the balance of the moments of unequal applied forces, thus the work done by the effort should be equal to the work done by the load. If an increase in the effort is considered the equilibrium could be preserve by decreasing the distance traversed by the load. The proportion of output force, e.g., *the load*, to the input force, e.g., *the effort*, is defined as *mechanical advantage* (M_A) of the simple machine, that is,

$$M_A = \frac{F_L}{F_E} \tag{8.1}$$

where F_L is the load, and F_E is the applied effort.

Since for an ideal machine the input power P_{in} and output power P_{out} (with power defined as the product of the force and the velocity) must be equal, one we can write

$$P_{in} = P_{out} \Leftrightarrow F_E v_{in} = F_L v_{out} \Leftrightarrow \frac{F_L}{F_E} = \frac{v_{in}}{v_{out}} \tag{8.2}$$

where v_{in} and v_{out} are the velocity of the effort and respectively of the load at the application point. From Equations (8.1) and (8.2) one can write

$$M_A = \frac{F_L}{F_E} = \frac{v_{in}}{v_{out}} = V_r \tag{8.3}$$

that is, $M_A = V_r$ for an ideal machine.

8.1.3 Velocity Ratio and Efficiency

The *velocity ratio* (V_r) of a machine is defined as the proportion of the distance travelled by effort to the distance travelled by the load, that is,

$$V_r = \frac{d_E}{d_L} \tag{8.4}$$

where d_E is the distance traveled by the effort and d_L is the distance traveled by the load.

All the machines lose energy to friction and heat (deformation and wear) during operation. The mechanical efficiency of the simple machine (a dimensionless number often denoted by the Greek letter η) is calculated as the proportion of output power P_{out} to input power P_{in} by

$$\eta = \frac{P_{out}}{P_{in}} = \frac{F_L v_{out}}{F_E v_{in}} = M_A \frac{v_{out}}{v_{in}} = \frac{M_A}{V_r} \tag{8.5}$$

and measure the effectiveness of the machine in transforming the input power to output power.

Rearranging Equation (8.5) one can write

$$M_A = \eta \frac{v_{in}}{v_{out}} = \eta V_r \tag{8.6}$$

The mechanical efficiency of the simple machine could be defined alternatively as the proportion between the work done by the load to the work done by the effort, that is

$$\eta = \frac{F_L d_L}{F_E d_E} = \frac{M_A}{V_r} \tag{8.7}$$

where F_L is the load, d_L is the distance travelled by load, F_E is the effort and d_E is the distance travelled by effort.

8.2 Effort and Load of an Ideal Machine

Since for an ideal machine the input and output power are equal, the efficiency is 100% or 1, that is, $\eta = 100\%$ or 1. In real applications (non-ideal machines) the efficiency is less than 1, that is, $\eta < 1$. Replacing η by 1 in Equation (8.6), i.e. considering an ideal machine, one can obtain $M_A = V_r$, that is, the mechanical advantage is equal to the velocity ratio, hence an ideal machine could be regarded as the baseline for performance assessment of real machines. From $V_r = M_A = \dfrac{F_L}{F_E}$, an ideal effort can be written as

$$F_{E_{ideal}} = \frac{F_L}{V_r}.$$

(8.8)

Rearranging Equation (8.5) one can write $\eta V_r = M_A = \dfrac{F_L}{F_E}$, thus an actual effort can be written as

$$F_{E_{actual}} = \frac{F_L}{V_r} \frac{1}{\eta}.$$

(8.9)

The loss in effort of a non-ideal machine can be calculated by

$$F_{E_{loss}} = F_{E_{actual}} - F_{E_{ideal}} = \frac{1}{\eta} \frac{F_L}{V_r} - \frac{F_L}{V_r} = \left(\frac{1}{\eta} - 1\right) \frac{F_L}{V_r}.$$

(8.10)

For a given effort F_E, the load $F_{L_{ideal}}$ that can be lifted by an ideal machine can be calculated as

$$F_{L_{ideal}} = F_E V_r.$$

(8.11)

In a non-ideal machine, the amount of load (named actual load) that can be lifted by a given effort F_E can be calculated by

$$F_{L_{actual}} = \eta F_E V_r.$$

(8.12)

The decrease in the amount of load of a non-ideal machine can be calculated by

$$F_{L_{decrease}} = F_{L_{ideal}} - F_{L_{actual}} = F_E V_r - \eta F_E V_r = F_E V_r (1 - \eta).$$

(8.13)

8.3 The Lever

The lever is a simple machine. A lever consists of a straight rod resting on a pivot (fulcrum) and an input force F_{in} (force provided to do the work) and an output force F_{out} (force pushing against) act at the two ends of the rod, as shown in Figure 8.1. The levers can be classified as first order, second order and third order and differ based on the location of the effort, fulcrum (pivot) and load. A first order lever has the pivot between the effort and load, a second order lever has the load between the effort and pivot, and a third order lever has the effort between the load and the pivot.

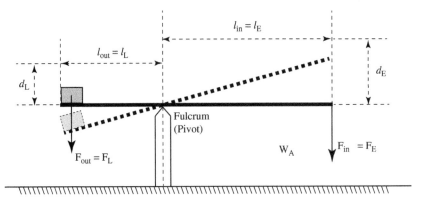

Figure 8.1 Lever represented as a straight rod resting on a pivot (fulcrum).

For a first order lever at equilibrium, the input torque should be equal to the output torque, that is

$$F_{in}l_{in} = F_{out}l_{out} \Leftrightarrow F_E l_E = F_L l_L \tag{8.14}$$

where $l_{in} = l_E$ is the effort arm, i.e. the distance between the effort and the fulcrum, and $l_{out} = l_L$ is the load arm, i.e. the distance between the load and the fulcrum.

The actual mechanical advantage is therefore

$$M_{A_{actual}} = \frac{F_{out}}{F_{in}} = \frac{F_L}{F_E} = \frac{l_{in}}{l_{out}} = \frac{l_E}{l_L}. \tag{8.15}$$

Since the friction is negligible, the actual and the ideal mechanical advantage are identical, that is

$$M_{A_{ideal}} = M_{A_{actual}} = M_A = \frac{F_L}{F_E} = \frac{l_E}{l_L}. \tag{8.16}$$

Using the proportion of the edges of the similar triangles in Figure 8.1 one can calculate

$$V_r = \frac{d_E}{d_L} = \frac{l_E}{l_L} \tag{8.17}$$

where d_E is the distance travelled by the effort and d_L is the distance travelled by the load.

8.4 Inclined Plane (Wedge)

The inclined plane – a surface making an angle against a horizontal surface – is one of the simplest types of simple machine that produces mechanical advantage. An inclined plane is used to raise heavy objects that are otherwise very hard to hoist vertically. It consists of an inclined surface having an inclined length (slope) l making an angle α with the horizontal direction (surface) and a vertical elevation (height) h, as shown in Figure 8.2.

Load F_L is to be lifted to the height H when an effort F_E is applied. The velocity ratio can be computed as

$$V_r = \frac{d_E}{d_L} = \frac{l}{h} = \frac{l}{l \sin \alpha} = \frac{1}{\sin \alpha}. \tag{8.18}$$

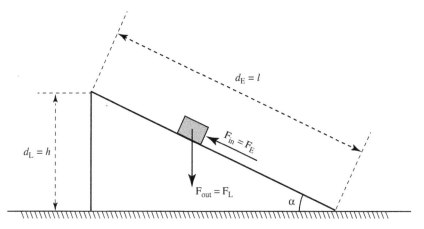

Figure 8.2 Inclined plane making an angle α with the horizontal direction.

Assuming an ideal machine, i.e. $M_A = V_r$, from Equation (8.18) the ideal mechanical advantage can be calculated as

$$M_{A_{ideal}} = \frac{1}{\sin \alpha}. \tag{8.19}$$

The actual advantage (a non-ideal machine) can be calculated with

$$M_{A_{actual}} = \frac{F_L}{F_E} \tag{8.20}$$

and due to the inclined plane acting friction, $M_{A_{actual}} < M_{A_{ideal}}$.

8.5 Screws

A screw – one of the classical simple machines – is a mechanism that converts rotational motion to linear motion. A screw thread could be regarded as a uniform inclined plane section wrapped around a cylindrical surface on a helical path. The common profile of a thread is the V thread (Figure 8.3(a)), but other cross section profiles such as square thread (Figure 8.3(b)) of saw-teeth (Figure 8.3(c)) are used.

The terminology of screw threads include the *lead* (axial advance), the *pitch* (consecutive threads distance), the *major* and *minor diameter*, *pitch diameter*, and *thread angle*, as shown in Figure 8.3. The slope of screw thread measured with respect to lead is called the *lead angle*. Except for single-threaded screws, lead and pitch distances are different.

8.6 Simple Screwjack

A screwjack is a type of power screw used to lift heavy loads and (or) to retrieve precise positioning along the screw axis. A screwjack is made of a nut attached to a pedestal, a screw containing the screw head [where the weight to be lifted (or lowered) is placed[and a lever (handle) attached to the screw head, as shown in Figure 8.4.

Consider a screwjack with a square-threaded screw having the pitch diameter d_m.

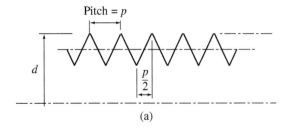

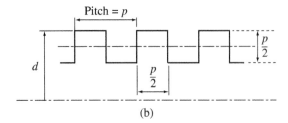

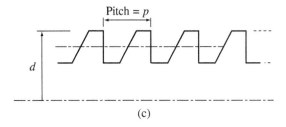

Figure 8.3 Screw thread (a) V profile, (b) square profile, and (c) saw-teeth profile.

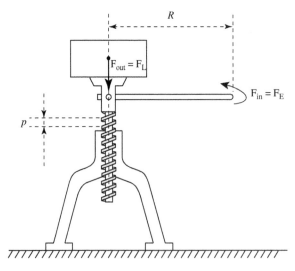

Figure 8.4 A simple screwjack.

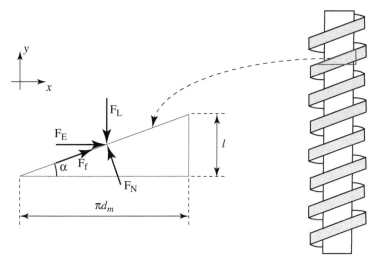

Figure 8.5 Equilibrium of forces along the horizontal and vertical axes of the thread – motion impending upwards.

8.6.1 Motion Impending Upwards

Consider a load F_L to be raised by the screwjack by the application of an external force F_E (which is considered to be positive). The lead angle of the thread is denoted by α, the lead is denoted by l, the normal force acting on the inclined plane is F_N, the coefficient of friction is denoted by μ, and the friction force F_f acting against to the motion is calculated as $F_f = \mu F_N$. The equilibrium of forces (Figure 8.5) along the horizontal and vertical axes gives

$$F_E = F_N \sin \alpha + \mu F_N \cos \alpha$$
$$F_N \cos \alpha = F_L + \mu F_N \sin \alpha. \tag{8.21}$$

Eliminating F_N in Equation (8.21) one can obtain

$$F_E = F_L \frac{\sin \alpha + \mu \cos \alpha}{\cos \alpha - \mu \sin \alpha} \tag{8.22}$$

or equivalent

$$F_E = F_L \frac{\sin (\alpha + \alpha_s)}{\cos (\alpha + \alpha_s)} = W \tan (\alpha + \alpha_s) = F_L \frac{\tan \alpha + \tan \alpha_s}{1 - \tan \alpha \tan \alpha_s} \tag{8.23}$$

where $\mu = \tan \alpha_s$. The actual effort $F_{E_{actual}}$ at the circumference of the screw is $F_{E_{actual}} = F_E = F_L \tan (\alpha + \alpha_s)$, and the ideal effort $F_{E_{ideal}}$ when the friction force is zero can be written as $F_{E_{ideal}} = F_L \tan \alpha$.

Dividing Equation (8.22) by $\cos \alpha$ one may obtain

$$F_E = \frac{F_L \left(l\pi d_m + \mu \right)}{1 - \mu l\pi d_m} \tag{8.24}$$

where $\tan \alpha = \dfrac{l}{\pi d_m}$.

The actual moment needed to raise the load in the presence of friction is

$$T_{\text{actual}} = F_E \frac{d_m}{2} = \frac{F_L d_m}{2} \frac{l\pi d_m + \mu}{1 - \mu l\pi d_m} \tag{8.25}$$

The moment, T_{ideal}, required to raise the load when $\mu = 0$ (no friction), is calculated from Equation (8.25) as

$$T_{\text{ideal}} = \frac{F_L l}{2\pi} \tag{8.26}$$

The screw efficiency η can be computed with

$$\eta = \frac{T_{\text{ideal}}}{T_{\text{actual}}} = \frac{F_L l}{2\pi T_{\text{actual}}} \tag{8.27}$$

The ideal mechanical advantage of the screwjack can be computed by

$$M_{A_{\text{ideal}}} = \frac{F_L}{F_E} = \frac{2\pi R}{p}. \tag{8.28}$$

Since in the presence of friction $M_{A_{\text{actual}}} < M_{A_{\text{ideal}}}$ the screwjack cannot be considered an efficient machine.

8.6.2 Motion Impending Downwards

Consider next a load F_L to be lowered by the application of an external force F_E (which is considered to be negative). The lead angle of the screwjack thread is denoted by α, the normal force acting on the inclined plane is F_N, the coefficient of friction is denoted by μ, and the friction force F_f is calculated as $F_f = \mu F_N$. The equilibrium of forces (Figure 8.6) along the horizontal and vertical axes can be written as

$$\mu F_N \cos \alpha = F_E + F_N \sin \alpha$$
$$F_L = \mu F_N \sin \alpha + F_N \cos \alpha = 0. \tag{8.29}$$

From Equation (8.29) one can obtain

$$F_E = F_L \frac{\mu \cos \alpha - \sin \alpha}{\cos \alpha + \mu \sin \alpha}. \tag{8.30}$$

Using $\tan \alpha = \dfrac{l}{\pi d_m}$ and dividing Equation (8.30) by $\cos \alpha$ one may obtain

$$F_E = \frac{F_L (\mu - l\pi d_m)}{1 + \mu l\pi d_m}. \tag{8.31}$$

The moment needed for lowering the load while overcoming the friction is

$$M_{\text{actual}} = F_E \frac{d_m}{2} = \frac{F_L d_m}{2} \frac{\mu - l\pi d_m}{1 + \mu l\pi d_m}. \tag{8.32}$$

For a large lead or a low friction (coefficient) the load will lower without any external effort.

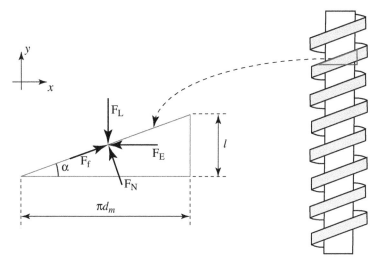

Figure 8.6 Equilibrium of forces along the horizontal and vertical axes of the thread – motion impending downwards.

8.6.3 Efficiency While Hoisting Load

The actual effort $F_{E_{actual}}$ at the circumference of the screw is $F_{E_{actual}} = F_E = F_L \tan(\alpha + \alpha_s)$, and the ideal effort $F_{E_{ideal}}$ when the friction force is zero can be written as $F_{E_{ideal}} = F_L \tan \alpha$. Thus the efficiency of the machine can be calculated as

$$\eta = \frac{F_{E_{ideal}}}{F_{E_{actual}}} = \frac{F_L \tan \alpha}{F_L \tan(\alpha + \alpha_s)} = \frac{\tan \alpha}{\tan(\alpha + \alpha_s)}. \tag{8.33}$$

8.7 Differential Screwjack

A differential screwjack is used in precision equipment where slow movement or finer adjustment is necessary. It consists of two different threads having two different pitches p_1 and p_2, one threaded both on inside and outside and the other threaded on the outside only, as shown in Figure 8.7.

During a single rotation of the lever, the load F_L gets raised by a distance $(p_2 - p_1)$ and the effort F_E travel a complete rotation that is a distance $2\pi R$.

The velocity ratio (V_r) of the differential screwjack can be calculated using

$$V_r = \frac{d_E}{d_L} = \frac{2\pi R}{p_2 - p_1}. \tag{8.34}$$

The mechanical advantage M_A of the differential screwjack can be calculated as

$$M_A = \frac{F_L}{F_E}. \tag{8.35}$$

The efficiency η of the differential screwjack is calculated with

$$\eta = \frac{M_A}{V_r} = \frac{F_L(p_2 - p_1)}{2\pi R F_E}. \tag{8.36}$$

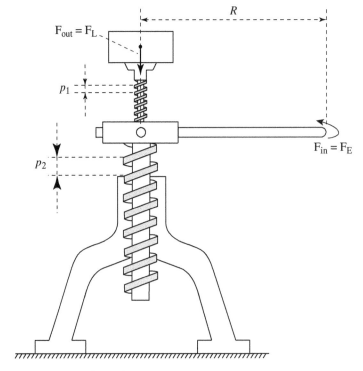

Figure 8.7 A differential screwjack.

8.8 Pulleys

A pulley – wheel mounted on an axle – is another of the six simple machines used for transmission of power and hoisting loads. The driving element of a pulley is usually a rope, but cables and belts are also used.

A pulley system consisting of one or more pulleys and used to gain mechanical advantage could be classified depending on the arrangement of pulleys. There are three main types of basic pulley systems: one that changes the direction of the force named fixed (the drum is secured to a single point), one that changes the magnitude of the force named movable (the drum moves when the load is moving), and one that changes both the direction and the magnitude of the force named compound (fixed and movable pulleys forms a block and tackle).

A pulley can be arranged in many different ways in order to accomplish different tasks. Complex pulley systems can move very heavy loads, but while the magnitude of applied effort is greatly reduced, the distance travelled by the load (to accomplish the work) is greatly increased. Some complex pulley systems are presented next.

8.8.1 First-order Pulley System

For a first-order pulley system consisting of a number of n pulleys, one pulley is fixed and the rest of the $n - 1$ pulleys are movable. In a first-order pulley system the effort F_E is applied

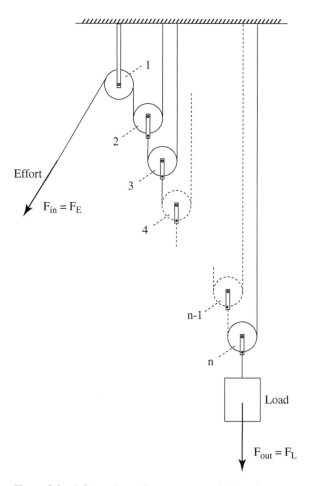

Figure 8.8   A first order pulley system consisting of a number of n pulleys.

through a rope wrapped over the fixed pulley, while the load F_L is suspended at the end of the nth (last) pulley. A first-order pulley system consisting of one fixed pulley and n movable pulleys is shown in Figure 8.8.

From the vertical force equilibrium of pulley n, one can write

$$2^n F_E = F_L \Leftrightarrow F_E = \frac{F_E}{2^n}. \tag{8.37}$$

The mechanical advantage can be calculated using

$$M_A = \frac{F_L}{F_E} = \frac{F_L}{\dfrac{F_L}{2^n}} = 2^n = 2^{\text{number of movable pulleys}}. \tag{8.38}$$

Considering an ideal machine, the velocity ratio for a number of n movable pulleys can be written as

$$V_r = M_A = 2^n. \tag{8.39}$$

8.8.2 Second-order Pulley System

For a second-order pulley system two pulley blocks are considered: a top block fastened to a support (fixed) and a movable bottom block, as shown in Figure 8.9. The load F_L is suspended at the bottom pulley located at the end of bottom block, and the effort F_E is applied through the top pulley of the top block through the rope. The rope is connected to the bottom pulley of the top block, then successively wrapped around each and every pulley and finally exiting at the top pulley of the top block.

Considering force equilibrium one can write

$$2nF_E = F_L \Leftrightarrow F_E = \frac{F_L}{2n}. \tag{8.40}$$

The mechanical advantage of the system can be computed with

$$M_A = \frac{F_L}{F_E} = \frac{F_L}{\dfrac{F_L}{2n}} = 2n = 2 \cdot \text{(number of movable pulleys in the bottom block)}.$$

$$\tag{8.41}$$

Considering it as an ideal machine, the velocity ratio can be calculated as

$$V_r = M_A = 2n = 2 \cdot \text{(number of movable pulleys in the bottom block).} \tag{8.42}$$

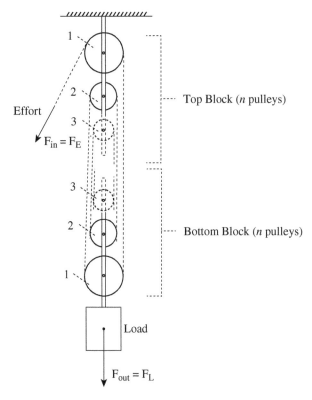

Figure 8.9 A second order pulley system consisting of a two pulley blocks.

For an ideal second-order pulley system having n pulleys in the bottom block one can write

$$V_r = M_A = 2n. \tag{8.43}$$

8.8.3 Third-order Pulley System

For a third-order pulley system consisting of n pulleys, one pulley is fixed and the rest of $n - 1$ pulleys are movable, as shown in Figure 8.10. The load F_L is suspended at the bottom of the rigid base, and the effort F_E is applied through the movable bottom pulley by the rope number n. There are n ropes in a third-order pulley system consisting of a number of n pulleys, each one connecting the rigid base with another pulley while passing over another pulley.

Considering the forces acting on rigid base where the load is attached one can write

$$\left(2^n - 1\right) F_E = F_L \Leftrightarrow F_E = \frac{F_L}{2^n - 1}. \tag{8.44}$$

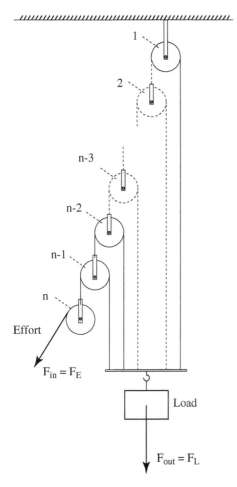

Figure 8.10 A third order pulley system consisting of a number of n pulleys.

The mechanical advantage of the system can be calculated as

$$M_A = \frac{F_L}{F_E} = \frac{F_L}{\frac{F_L}{2^n - 1}} = 2^n - 1 = 2^{\text{number of pulleys in the system}} - 1. \qquad (8.45)$$

For an ideal third order pulley system, the velocity ratio can be calculated using

$$V_r = M_A = 2^n - 1 = 2^{\text{number of pulleys in the system}} - 1. \qquad (8.46)$$

8.9 Differential Pulley

A differential pulley, shown in Figure 8.11, is a special type of pulley used to lift heavy objects by a small distance. The differential pulley (also named the Weston differential pulley) was invented by the American mechanical engineer Thomas Aldridge Weston.

A differential pulley consist of an upper and a lower block. The upper block – which is fixed – consists of two pulleys of unequal diameter coaxially attached to each other. The lower block – which is movable – consists of a single pulley holding the load F_L to be lifted. A rope (or chain) is wrapped around the pulleys of the upper block and lower block as shown in Figure 8.11.

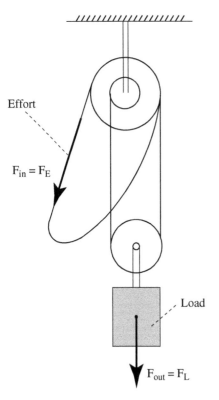

Figure 8.11 A differential pulley (also named a Weston differential pulley).

For an ideal machine, i.e. $V_r = M_A$, the equilibrium of forces about the axis of the pulleys (Figure 8.11) can be written as

$$R\frac{F_L}{2} = F_E R + r\frac{F_L}{2} \Leftrightarrow (R-r)\frac{F_L}{2} = F_E R \Leftrightarrow \frac{F_L}{F_E} = \frac{2R}{R-r} \tag{8.47}$$

where r and R are the radii of the smaller and larger pulley respectively.

Using Equation (8.47) the mechanical advantage can be calculated as

$$M_A = \frac{F_L}{F_E} = \frac{2R}{R-r} = \frac{2D}{D-d} \tag{8.48}$$

where d and D are the diameters of the smaller and larger pulley respectively.

Since the system was considered as an ideal machine, the velocity ratio is

$$V_r = M_A = \frac{2R}{R-r} = \frac{2D}{D-d}. \tag{8.49}$$

8.10 Wheel and Axle

The wheel and axle is a simple machine where the effort wheel is rigidly connected to an axle and coaxially mounted on a shaft. An effort F_E is applied by a rope wrapped around the wheel in order to raise (or to lower) a load F_L suspended by another rope wrapped over the axle in opposite direction to the rope around the wheel (Figure 8.12).

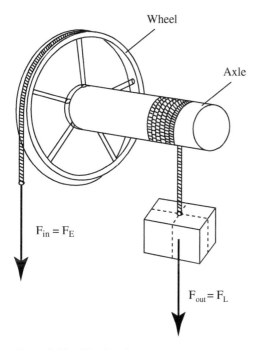

Wheel

Axle

$F_{in} = F_E$

$F_{out} = F_L$

Figure 8.12   Wheel and axle.

Considering R and r the radii of the wheel and respectively the axle, the distance traversed by the effort (wheel) on a full rotation is $2\pi R$ and the distance travelled by the load (axle) is $2\pi r$. The velocity ratio of the wheel and axle can be calculated as

$$V_r = \frac{d_E}{d_L} = \frac{2\pi R}{2\pi r} = \frac{R}{r}.$$ (8.50)

For an ideal machine, i.e. no-friction condition (100% efficiency), the mechanical advantage is

$$M_A = V_r = \frac{R}{r}.$$ (8.51)

In a real machine, the work done by the effort and the work done by the load are

$$W_E = 2\pi R F_{E_{actual}}$$
$$W_L = 2\pi r W$$ (8.52)

where $F_{E_{actual}}$ is the effort required to lift load F_L.

The actual efficiency of the wheel and axle can be calculated as

$$\eta_{actual} = \frac{W_L}{W_E} = \frac{F_L \cdot 2\pi r}{F_{E_{actual}} \cdot 2\pi R} = \frac{F_L r}{F_{E_{actual}} R} = M_{A_{actual}} \frac{r}{R}.$$ (8.53)

From Equation (8.53) the actual mechanical advantage can be written as

$$M_{A_{actual}} = \eta_{actual} \frac{R}{r}.$$ (8.54)

8.11 Wheel and Differential Axle

A wheel and differential axle has a different axle configuration and a better mechanical advantage when compared to single wheel and axle.

As shown in Figure 8.13 a wheel and differential axle consists of two axles of different diameters – a larger axle of diameter d_1 and a smaller axle of diameter d_2 – coaxially fitted with the wheel of diameter D. The effort F_E is applied through a wrapped rope wrapped around the wheel. Another rope wrapped around the two axles carries a load F_L with the help of a movable pulley. The rope on the wheel and smaller axle are wrapped in the same direction, whereas that on the larger axle is in the opposite direction.

For a single rotation of wheel and axles the distance travelled by the effort F_E is πD, the length of rope unwrapped on the smaller axle is πd_2, the length of rope wrapped on the larger axle is πd_1, and the net length of rope is $\pi (d_1 - d_2)$. Thus the load F_L gets lifted by a distance $\pi \dfrac{d_1 - d_2}{2}$.

The velocity ratio of the wheel and differential axle is calculated with

$$V_r = \frac{\pi D}{\pi \dfrac{d_1 - d_2}{2}} = \frac{2D}{d_1 - d_2}.$$ (8.55)

Considering the free-body diagram of wheels and differential axle and taking moment equilibrium at point O, one can write

$$F_E \frac{D}{2} + \frac{F_L}{2} \frac{d_2}{2} = \frac{F_L}{2} \frac{d_1}{2} \Leftrightarrow F_E D = \frac{F_L (d_1 - d_2)}{2}.$$ (8.56)

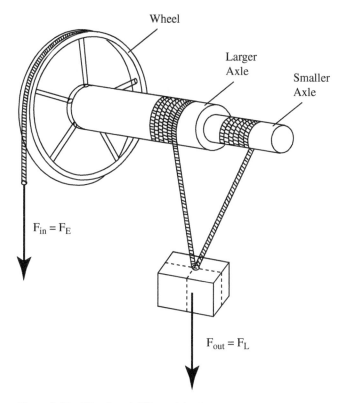

Figure 8.13 Wheel and differential axle.

The mechanical advantage can be calculated from Equations (8.1) and (8.56) as

$$M_A = \frac{F_L}{F_E} = \frac{2D}{d_1 - d_2}.$$

(8.57)

Since the velocity ratio in Equation (8.55) and the mechanical advantage in Equation (8.57) are equal, the efficiency is 100%, that is, an ideal machine with no friction.

When friction is considered, the actual effort is denoted by P_{actual}. For a full rotation the work done by the effort is $W_E = F_{E_{actual}} \pi D$ and the work done by the load is $W_L = F_L \frac{\pi}{2}(d_1 - d_2)$. The efficiency of the machine can be calculated using

$$\eta = \frac{W_L}{W_E} = \frac{F_L \frac{\pi}{2}(d_1 - d_2)}{F_{E_{actual}} \pi D} = \frac{F_L}{F_{E_{actual}}} \frac{d_1 - d_2}{2D} = M_{A_{actual}} \frac{d_1 - d_2}{2D}.$$

(8.58)

From Equation (8.58) the actual mechanical advantage can be written as

$$M_{A_{actual}} = \eta_{actual} \frac{2D}{d_1 - d_2}.$$

(8.59)

8.12 Examples

Example 8.1

A worker is using the clip pliers (class 1 lever) shown in Figure 8.14 to cut a thin metallic wire. The dimensions of the pliers are d_L (handle length) and d_E (blade length). Calculate the exerted force at the joint/fulcrum A, the force that the pliers applies at the metallic wire and the mechanical advantage when a force F is applied at both ends of the pliers handle for cutting the wire. Numerical application: $F = 4$ N, $d_E = 0.04$ m, $d_L = 0.14$ m.

Solution

Considering the free-body diagram in Figure 8.15 one can write the equilibrium equations (sum of forces about the horizontal and vertical direction) as

$$\sum F_x = 0 \Leftrightarrow F_{Ax} = 0,$$
$$\sum F_y = 0 \Leftrightarrow -F_D + F_{Ay} - F = 0. \tag{8.60}$$

The equilibrium equations (sum of the moments) about D can be written as

$$\sum M_D = 0 \Leftrightarrow F(d_L + d_E) - F_{Ay}d_E = 0, \tag{8.61}$$

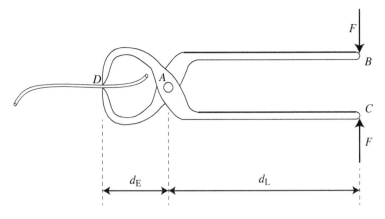

Figure 8.14 Clip pliers (class 1 lever).

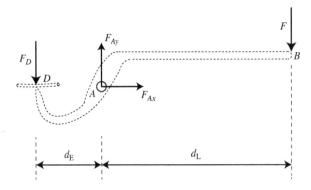

Figure 8.15 The free body diagram of the clip pliers (class 1 lever).

or about A

$$\sum M_A = 0 \Leftrightarrow F_D d_E - F d_L = 0,$$ (8.62)

The equilibrium equations are written and printed in MATLAB with:

```
clear all; clc; close all
syms d_L d_E F F_Ax F_Ay F_D

eqFx=F_Ax;
eqFy=-F_D + F_Ay - F;
eqM=F*(d_L+d_E)-F_Ay*d_E;
fprintf('The equilibrium equations are \n')
fprintf('%s = 0 \n\n',eqFx)
fprintf('%s = 0 \n\n',eqFy)
fprintf('%s = 0 \n\n',eqM)
```

Solving the equilibrium Equations (8.60) and (8.61) one can write

$$F_{Ay} = F\frac{(d_L + d_E)}{d_E},$$

$$F_D = F_{Ay} - F = F\frac{(d_L + d_E)}{d_E} - F = F\frac{d_L}{d_E}.$$ (8.63)

In MATLAB the equilibrium equations are solved with:

```
sol=solve(eqFx,eqFy,eqM,'F_Ax','F_Ay','F_D');
fprintf('The force at the joint A is \n')
fprintf('F_Ax = %s \n\n',sol.F_Ax)
fprintf('F_Ay = %s \n\n',sol.F_Ay)
fprintf('F_D = %s \n\n',sol.F_D)
```

One can calculate the mechanical advantage using

$$MA = \frac{d_L}{d_E},$$ (8.64)

and in MATLAB with:

```
MA=d_L/d_E;
fprintf('Mechanical Advantage is calculated as \n')
fprintf('MA = %s \n\n',MA)
```

The input numerical data are introduced in MATLAB with:

```
% numerical results
lists = {F,d_E,d_L};
listn = {4,0.04,0.14};
```

The numerical results are calculated and printed in MATLAB using:

```
F_Ax = eval(subs(sol.F_Ax,lists,listn));
F_Ay = eval(subs(sol.F_Ay,lists,listn));
```

```
F_D = eval(subs(sol.F_D,lists,listn));
MAn=eval(subs(MA,lists,listn));
fprintf('The numerical value of the force F_Ax is \n ')
fprintf('F_Ax = %f [N]\n\n',F_Ax);
fprintf('The numerical value of the force F_Ay is \n ')
fprintf('F_Ay = %f [N]\n\n',F_Ay);
fprintf('The numerical value of the force F_D is \n ')
fprintf('F_D = %f [N]\n\n',F_D);
fprintf('The mechanical advantage is \n')
fprintf('MA = %f \n\n',MAn);
```

The MATLAB numerical results for the friction force and mechanical advantage are:

```
The numerical value of the force F_Ax is
  F_Ax = 0.000000 [N]

The numerical value of the force F_Ay is
  F_Ay = 18.000000 [N]

The numerical value of the force F_D is
  F_D = 14.000000 [N]

The mechanical advantage is
MA = 3.500000
```

Example 8.2

A worker is applying a force F to the top of a wedge to split a log (Figure 8.16). The applied force F causes the wedge to be about to move further into the log. Knowing the angle α of the wedge and the coefficient of static friction μ between the wedge and the log, determine the friction forces and the normal forces acting on the sides of the wedge.

Numerical application: $F = 600$ N, $\mu = 0.45$, $\alpha = 10°$.

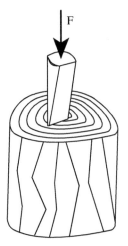

F

Figure 8.16 A wedge and a log.

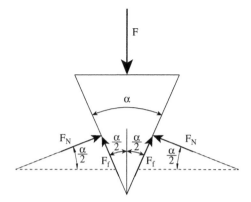

Figure 8.17 Free body diagram of the wedge.

Solution

(a) Considering the free-body diagram in Figure 8.17 one can write the equilibrium equation (sum of forces about the vertical direction) as

$$\sum F_y = 0 \Leftrightarrow 2F_f \cos\frac{\alpha}{2} + 2F_N - F = 0 \tag{8.65}$$

where F_N is the normal force acting on the side of the wedge and F_f is the fiction force that opposes the motion of the wedge.

The friction force can be written as

$$F_f = \mu F_N. \tag{8.66}$$

The equilibrium and friction force equations are introduced in MATLAB with:

```
clear all; clc; close all
syms alpha F_f mu F_N F

theta=alpha/2
eqFf = F_f-mu*F_N;
eqFy = 2*F_f*cos(theta) + 2*F_N - F;
fprintf('The Oy equilibrium equation is \n');
fprintf('%s = 0 \n\n',eqFy);
fprintf('The friction force equation is \n');
fprintf('%s = 0 \n\n',eqFf);
```

From Equations (8.65) and (8.66) one can calculate

$$F_f = \frac{1}{2}\frac{\mu F}{\mu \cos\frac{\alpha}{2} + 1} \quad \text{and} \quad F_N = \frac{1}{2}\frac{F}{\mu \cos\frac{\alpha}{2} + 1}. \tag{8.67}$$

The equations are solved in MATLAB with:

```
sol=solve(eqFf,eqFy,'F_f','F_N');
fprintf('The friction force and nornal force is \n');
fprintf('F_f = %s \n\n',sol.F_f);
fprintf('F_N = %s \n\n',sol.F_N);
```

The input numerical data are introduced in MATLAB with:

```
lists = {alpha, mu, F};
listn = {10*pi/180, 0.45, 600};
```

The numerical results are calculated and printed in MATLAB using:

```
F_f = eval(subs(sol.F_f, lists, listn));
F_N = eval(subs(sol.F_N, lists, listn));
fprintf('The numerical value of the friction force F_f is \n ')
fprintf('F_Ax = %f [N] \n\n', F_f);
fprintf('The numerical value of the normal force F_N is \n ')
fprintf('F_Ay = %f [N] \n\n', F_N);
```

The MATLAB numerical results for the friction force and mechanical advantage are:

```
The numerical value of the friction force F_f is
 F_Ax = 93.213529 [N]


The numerical value of the normal force F_N is
 F_Ay = 207.141176 [N]
```

Example 8.3

A worker is using the crowbar (class 1 lever) shown in Figure 8.18 to pull out a nail. The external force F_A applied by the worker on the crowbar at point A has a horizontal direction as shown in Figure 8.19. The height d_A of the point A (vertical component of the distance moved by the effort), the direction α of the force F_B applied to the nail, and the horizontal component d_B of the distance moved by the load are known. Calculate the value of the external force the worker should apply to pull out the nail if the output force due to the nail is F_B. What is the mechanical advantage of the lever when the force F is applied as shown. Numerical application: $F_B = 85$ N, $d_B = 0.05$ m, $d_A = 0.45$ m, $\alpha = 45°$.

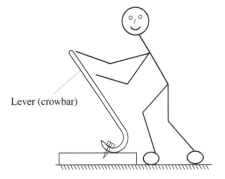

Lever (crowbar)

Figure 8.18 A worker using a crowbar (class 1 lever).

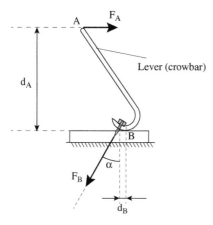

Figure 8.19 Free body diagram of the crowbar.

Solution
The torque due to the applied external force F_A and the reaction force the nail exerts back on the puller (equal and opposite to F_B) must be equal to each other in order to satisfy equilibrium. In fact, for the nail to actually move, the torque due to the applied external force must be slightly greater than the torque due to the reaction force. One can write

$$F_A d_A = F_B d_B \Leftrightarrow F_A d_A = F_B d_{B_h} \cos \alpha \tag{8.68}$$

where the distance moved by the load d_B is calculated by $d_B = d_{B_h} \cos \alpha$. The equilibrium equations are written MATLAB using:

```
clear all; clc; close all
syms d_A d_Bh F_B F_A alpha

d_B=d_Bh*cos(alpha);
eqM=F_A*d_A-F_B*d_B;
fprintf('The equilibrium equation is \n');
fprintf('%s = 0 \n\n',eqM);
```

From Equation (8.68) one can calculate the value of the external force as

$$F_A = \frac{F_B d_B \cos \alpha}{d_A} \tag{8.69}$$

The external force is calculated in MATLAB with:

```
sol=solve(eqM,'F_A');
fprintf('The external force the worker should apply is \n');
fprintf('F_A = %s \n\n',sol);
F_A=sol;
```

The mechanical advantage can be calculated using

$$\text{MA} = \frac{F_B}{F_A} = \frac{F_B}{\dfrac{F_B d_B \cos \alpha}{d_A}} = \frac{d_A}{d_B \cos \alpha} \tag{8.70}$$

and in MATLAB with:

```
MA=F_B/F_A;
fprintf('Mechanical Advantage is calculated as \n');
fprintf('MA = %s \n\n',MA);
```

The input numerical data are introduced in MATLAB with:

```
% numerical results
lists = {F_B,d_B,d_A, alpha};
listn = {85,0.05,0.45,45*pi/180};
```

The numerical results are calculated and printed in MATLAB using:

```
F_An = eval(subs(F_A,lists,listn));
MAn=eval(subs(MA,lists,listn));
fprintf('The numerical value of the applied force is \n');
fprintf('F_A = %f (N)\n\n',F_An);
fprintf('The mechanical advantage is \n');
fprintf('MA = %f \n\n',MAn);
```

The obtained MATLAB numerical results are:

```
The numerical value of the applied force is
F_A = 9.444444 (N)

The mechanical advantage is
MA = 9.000000
```

Example 8.4

The wheelbarrow (class 2 lever) shown in Figure 8.20 is used by a worker to carry bricks to a specific location. The combined mass of the bricks and wheelbarrow is m_{BW}, the horizontal component of the distance moved by the effort is d_{Eh}, and the horizontal component of the distance moved by the load is d_{Lh} as shown in Figure 8.21. Calculate:

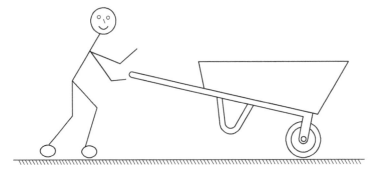

Figure 8.20 A worker using a wheelbarrow (class 2 lever).

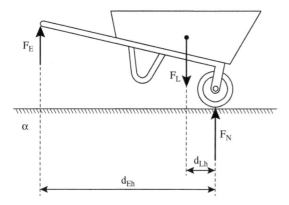

Figure 8.21 Free body diagram of the wheelbarrow.

(a) The upward force the worker should exert to support the combined mass of the wheel-barrow and the bricks

(b) The force the wheelbarrow exerts on the ground

(c) The mechanical advantage of the wheelbarrow.

Numerical application: $m_{BW} = 40$ kg, $g = 9.81$ m s^{-2}, $d_{Eh} = 1.2$ m, $d_{Lh} = 0.15$ m, $\alpha = 30°$.

Solution

(a) The upward force F_E the worker should exert to support the combined mass of the wheelbarrow and the bricks (Figure 8.21) can be calculated from

$$F_E d_E = F_L d_L \Leftrightarrow F_E d_{Eh} \tan \alpha = F_L d_{Lh} \tan \alpha \Leftrightarrow F_E d_{Eh} = F_L d_{Lh}. \tag{8.71}$$

The equation of the upward force is written in MATLAB with:

```
clear all; clc; close all
syms d_Lh d_Eh F_E F_L F_N m_BW g alpha

d_L=d_Lh*tan(alpha);
d_E=d_Eh*tan(alpha);
eqM=F_E*d_E-F_L*d_L;
fprintf('The equilibrium equation is \n');
fprintf('%s = 0 \n\n',eqM);
```

From Equation (8.71) one can calculate the value of the upward force as

$$F_E = \frac{F_L d_{Lh}}{d_{Eh}}. \tag{8.72}$$

The upward force F_E is calculated in MATLAB with:

```
sol_eqM=solve(eqM,'F_E');
fprintf('The external force the worker should apply is \n');
fprintf('F_E = %s \n\n',sol_eqM);
F_E=sol_eqM;
```

(b) From the free-body diagram shown in Figure 8.21 one can calculate

$$F_E + F_N = F_L. \tag{8.73}$$

The equilibrium equation for the normal force is written in MATLAB with:

```
eqN=F_E+F_N-F_L;
fprintf('The equilibrium equation for normal force is \n');
fprintf('%s = 0 \n\n',eqN);
```

From Equation (8.73) one can calculate the value of the normal force as

$$F_N = F_L - F_E \tag{8.74}$$

and in MATLAB with:

```
sol_eqN=solve(eqN,'F_N');
fprintf('The external force the worker should apply is \n');
fprintf('F_N = %s \n\n',sol_eqN);
F_N=sol_eqN;
```

(c) The mechanical advantage can be calculated by

$$MA = \frac{F_L}{F_E} = \frac{F_L}{\dfrac{F_L d_{Lh}}{d_{Eh}}} \tag{8.75}$$

In MATLAB the mechanical advantage MA is calculated with:

```
MA=F_L/F_E;
fprintf('Mechanical Advantage is calculated as \n');
fprintf('MA = %s \n\n',MA);
```

The input numerical data are introduced in MATLAB with:

```
% numerical results
lists = {m_BW,d_Eh,d_Lh, alpha,g};
listn = {40,1.2,0.15,30*pi/180,9.81};
```

The numerical results are calculated and printed in MATLAB using:

```
F_L=m_BW*g;
F_L=eval(subs(F_L,lists,listn));
F_En = eval(subs(F_E,lists,listn));
F_Nn = eval(subs(F_N,lists,listn));
MAn=eval(subs(MA,lists,listn));
fprintf('The numerical value of the upward force is \n');
fprintf('F_E = %f (N)\n\n',F_En);
fprintf('The numerical value of the normal force is \n');
fprintf('F_N = %f (N)\n\n',F_Nn);
fprintf('The mechanical advantage is \n');
fprintf('MA = %f \n\n',MAn);
```

The obtained MATLAB numerical results are:

```
The numerical value of the upward force is
F_E = 49.050000 (N)
```

The numerical value of the normal force is
F_N = 343.350000 (N)

The mechanical advantage is
MA = 8.000000

Example 8.5

The worker shown in Figure 8.22 is using a rope and an inclined plane to lift a block A to a desired height, since this requires less force than lifting it straight up. The rope connecting to the block A is parallel to the inclined plane and makes an angle α with the horizontal direction. The friction coefficient between the block A and the inclined plane S is μ_S. Knowing that the mass m_A of block A determine:

(a) The minimum force F_{Down} the worker must apply to the end of the rope for the block A not to slide down the inclined plane.
(b) The acceleration of the block A when the block is moving down the inclined plane due to an applied force on the rope of one third of the minimum calculated force F_{Down}.
(c) The minimum force F_{Up} the worker must apply to the end of the rope for the block A to start moving up the inclined plane.
(d) The acceleration of the block A when the block is moving up the inclined plane due to an applied force on the rope twice of the calculated minimum force F_{Up}.

Numerical application: $m_A = 7.5$ kg, $g = 9.81$ m s^{-2}, $\mu_S = 0.4$, $r = 0.1$ m, and $\alpha = \dfrac{\pi}{6}$.

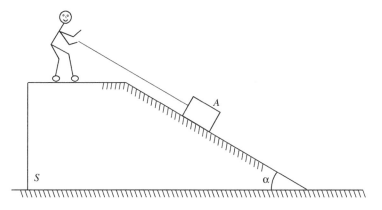

Figure 8.22   Worker using a rope and an inclined plane to lift a block.

Solution

(a) As shown in Figure 8.23, the rope making an angle α with the horizontal direction is connecting with the block A. To calculate the minimum force the worker must apply to the end of the rope for the block A not to slide down the inclined plane one can consider that the block A is moving down the inclined plane. Considering the equilibrium of block A shown in the free body diagram in Figure 8.23, one can write

$$\begin{cases} N_A - W_A \cos \alpha = 0 \\ F_{Down} - W_A \sin \alpha + F_A = 0 \end{cases} \Leftrightarrow \begin{cases} N_A - W_A \cos \alpha = 0 \\ F_{Down} - W_A \sin \alpha + \mu_S N_A = 0 \end{cases} \tag{8.76}$$

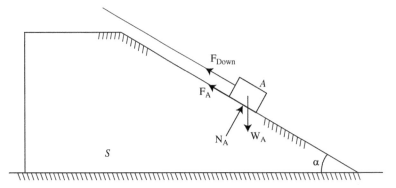

Figure 8.23 Free body diagram of the block (block A not to slide down the inclined plane).

where $F_A = \mu_S N_A$ is the friction force acting on block A, calculated in MATLAB with:

```
clear all; clc; close all
syms N_A W_A F_Down F_Up alpha mu_S m_A a_Down a_Up g

F_A=mu_S*N_A;
fprintf('The friction force acting ')
fprintf('on block A is \n')
fprintf('F_A = %s \n\n',F_A)
```

The equilibrium equations shown in Equation (8.76) are written and printed in MATLAB with:

```
eq1a=N_A-W_A*cos(alpha);
eq2a=F_Down-W_A*sin(alpha)+F_A;
fprintf('The equilibrium equations are \n')
fprintf('%s = 0 \n',eq1a);
fprintf('%s = 0 \n',eq2a);
```

From Equation (8.76) one can obtain

$$F_{Down} = W_A \sin \alpha - \mu_S W_A \cos \alpha. \tag{8.77}$$

The minimum force the worker must apply to the end of the rope for the block A not to slide down is calculated in MATLAB with:

```
sol_a=solve(eq1a,eq2a,'F_Down', 'N_A');
fprintf('The minimum force the worker must apply ')
fprintf('at the end of the rope \n')
fprintf('for the block not to slide ')
fprintf('down the inclined plane \n')
fprintf('F_Down = %s \n\n',sol_a.F_Down)
```

(b) When the block A is sliding down the inclined plane under the action of a force equal to $\dfrac{F_{Down}}{3}$ one can write

$$N_A - W_A \cos \alpha = 0$$
$$\frac{F_{Down}}{3} - W_A \sin \alpha + \mu_S N_A = m_A a_{Down} . \tag{8.78}$$

From Equation (8.78) one can calculate the acceleration of the block A as

$$a_{\text{Down}} = \frac{\frac{F_{\text{Down}}}{3} - W_A \sin \alpha + \mu_S W_A \cos \alpha}{m_A}. \tag{8.79}$$

The acceleration of the block A when the applied force on the rope is equal to $\frac{F_{\text{Down}}}{3}$ is calculated and printed in MATLAB with:

```
eq1b=N_A-W_A*cos(alpha);
eq2b=(sol_a.F_Down)/3-W_A*sin(alpha)+F_A -m_A*a_Down;
sol_b=solve(eq1b,eq2b,'a_Down', 'N_A');
fprintf('The acceleration of the block ')
fprintf('down the inclined plane is \n')
fprintf('a_Down = %s \n\n',sol_b.a_Down)
```

(c) To calculate the minimal force the worker must apply to the end of the rope for the block A to start moving up the inclined plane one can consider the equilibrium of block A – shown in the free body diagram in Figure 8.24 – that is

$$\begin{cases} N_A - W_A \cos \alpha = 0 \\ F_{\text{Up}} - W_A \sin \alpha - F_A = 0 \end{cases} \Leftrightarrow \begin{cases} N_A - W_A \cos \alpha = 0 \\ F_{\text{Up}} - W_A \sin \alpha - \mu_S N_A = 0 \end{cases} \tag{8.80}$$

where $F_A = \mu_S N_A$ is the friction force acting on block A. The equilibrium of the block A is written in MATLAB with:

```
eq1c=N_A-W_A*cos(alpha);
eq2c=F_Up-W_A*sin(alpha)-F_A;
fprintf('The equilibrium equations are \n')
fprintf('%s = 0 \n',eq1c);
fprintf('%s = 0 \n',eq2c);
```

From Equation (8.80) one can calculate the minimal force the worker must apply at the end of the rope for the block A to start moving up the inclined plane as

$$T_A = W_A \sin \alpha + \mu_S W_A \cos \alpha. \tag{8.81}$$

The minimal force the worker must apply at the end of the rope for the block A to start moving up the inclined plane is calculated in MATLAB with:

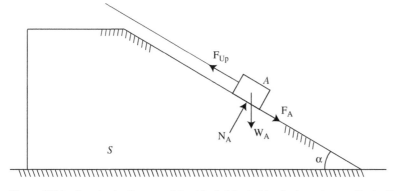

Figure 8.24 Free body diagram of the block (block A to start moving up the inclined plane).

```
sol_c=solve(eq1c,eq2c,'F_Up', 'N_A');
fprintf('The minimal force the worker must apply ')
fprintf('at the end of the rope \n');
fprintf('for the block to start moving ');
fprintf('up the inclined plane is \n');
fprintf('F_Up = %s \n\n',sol_c.F_Up);
```

(d) When the block A is moving up the inclined plane under the action of a force equal to $1.5F_{Up}$ one can write

$$N_A - W_A \cos \alpha = 0$$

$$1.5 * F_{Up} - W_A \sin \alpha - \mu_S N_A = m_A a_{Up} \qquad (8.82)$$

From Equation (8.82) one can calculate the acceleration of the block A

$$a_{Up} = \frac{1.5F_{Up} - W_A \sin \alpha - \mu_S W_A \cos \alpha}{m_A} \qquad (8.83)$$

The acceleration of the block A when the applied force on the rope is $\frac{3}{2}F_{Up}$ is calculated and printed in MATLAB with:

```
eq1d=N_A-W_A*cos(alpha);
eq2d=1.5*sol_c.F_Up-W_A*sin(alpha)-F_A - m_A*a_Up;
sol_d=solve(eq1d,eq2d,'a_Up', 'N_A');
fprintf('The acceleration of the block ')
fprintf('up the inclined plane is \n')
fprintf('a_Up = %s \n\n',sol_d.a_Up)
```

The input numerical data are introduced in MATLAB with:

```
% numerical results
lists = {m_A,g,mu_S,alpha};
listn = {7.5,9.81,0.4,pi/6};
```

The numerical results are calculated and printed in MATLAB using:

```
W_A=m_A*g;
W_A=eval(subs(W_A,lists,listn));
F_Down = eval(subs(sol_a.F_Down,lists,listn));
a_Down = eval(subs(sol_b.a_Down,lists,listn));
F_Up = eval(subs(sol_c.F_Up,lists,listn));
a_Up = eval(subs(sol_d.a_Up,lists,listn));
fprintf('The minimal force the worker must apply ')
fprintf('for the block not to slide ')
fprintf('down the inclined plane is \n')
fprintf('F_Down = [%f] (N)\n\n',F_Down);
fprintf('The acceleration of the block ')
fprintf('down the inclined plane ');
fprintf('when F_Down is [%f] (N) is \n',F_Down/3);
fprintf('a_Down = [%f] (m/s^2)\n\n',a_Down);
fprintf('The minimal force the worker must apply ')
```

```
fprintf('for the block to start moving ');
fprintf('up the inclined plane is \n');
fprintf('F_Up = [%f] (N)\n\n',F_Up);
fprintf('The acceleration of the block ')
fprintf('up the inclined plane when F_Up is [%f] (N) is \n',
1.5*F_Up);
fprintf('a_Up = [%f] (m/s^2)\n',a_Up);
```

The MATLAB numerical results for the forces are:

```
The minimal force the worker must apply
for the block not to slide down the inclined plane is
F_Down = [11.300372] (N)

The acceleration of the block
down the inclined plane when F_Down is [3.766791] (N) is
a_Down = [-1.004478] (m/s^2)

The minimal force the worker must apply
for the block to start moving up the inclined plane is
F_Up = [62.274628] (N)

The acceleration of the block
up the inclined plane when F_Up is [93.411941] (N) is
a_Up = [4.151642] (m/s^2)
```

Example 8.6

The worker shown in Figure 8.25 is using a massless rope with a hook at the end, a friction-less massless pulley and a cable placed around the rectangular block A to lift it to a desired

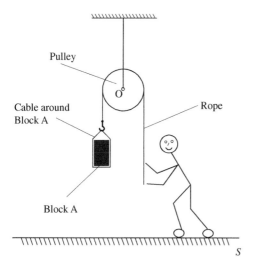

Figure 8.25 Worker using a rope and a frictionless pulley to lift a block to a desired height.

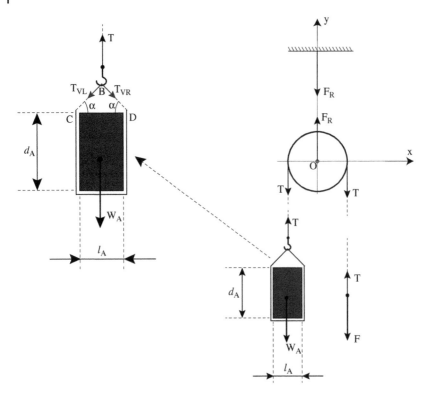

Figure 8.26 Free body diagram of the block and pulley.

height. The block, having length l_A and width d_A, is supported by a cable with length s_C in two different positions, a vertical one shown in Figure 8.26 and a horizontal one shown in Figure 8.27. Knowing the mass m_A of the block, determine:

(a) The force F the worker must apply to the end of the rope to keep the block A stationary and the mechanical advantage of the pulley.
(b) The force exerted on the ceiling by the supporting rod holding the pulley.
(c) The force acting in the cable for the two different positions shown in Figure 8.26 and Figure 8.27

Numerical application: $m_A = 7.5$ kg, $g = 9.81$ m s^{-2}, $\mu_S = 0.4$, $r = 0.1$ m, and $\alpha = \dfrac{\pi}{6}$.

Solution

(a) To calculate the force the worker must apply to the end of the rope for the block to be stationary (in equilibrium) one can consider the equilibrium equations. The equilibrium of the block A and applied worker force F can be written using the free-body diagram shown in Figure 8.26 or Figure 8.27 by

$$\sum \mathbf{F}_x = 0 \Leftrightarrow 0 = 0$$

$$\sum \mathbf{F}_y = 0 \Leftrightarrow T - W_A = 0 \text{ and } T - F = 0 \tag{8.84}$$

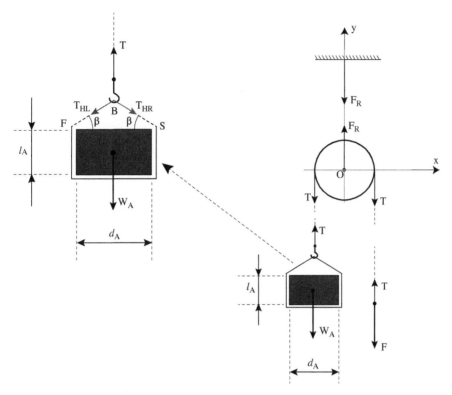

Figure 8.27 Free body diagram of the block and pulley.

where T is the tension in the rope. The equilibrium of the block A is written in MATLAB with:

```
clear all; clc; close all
syms m g T F F_R W_A alpha beta sinalpha sinbeta cosalpha
cosbeta
syms d_BC d_A l_A s_A T_VL T_VR T_HL T_HR d_FB m_A

W_A=m_A*g;
eqFy1=T-W_A;
eqFy2=T-F;
fprintf('The equilibrium equations are \n');
fprintf('%s = 0 \n',eqFy1);
fprintf('%s = 0 \n\n',eqFy2);
```

From Equation (8.84) the force the worker must apply to the rope for the block to be stationary can be computed as

$$T = W_A = m_A g \text{ and } T = F \Leftrightarrow F = m_A g \tag{8.85}$$

and in MATLAB with:

```
sol_eqFy12=solve([eqFy1,eqFy2],[F,T]);
fprintf('The force the worker should apply is \n');
fprintf('F = %s \n\n',sol_eqFy12.F);
F=sol_eqFy12.F;
```

The pulley shown in Figure 8.25 has an mechanical advantage of 1 since the effort applied by the worker is equal to the block weight (load force).

(b) To calculate the force exerted on the ceiling by the supporting rod holding the pulley one can consider the equilibrium equations for the pulley, weight of block A and applied worker force F using the free body diagram shown in Figure 8.26 or Figure 8.27 by

$$\sum F_x = 0 \Leftrightarrow 0 = 0$$
$$\sum F_y = 0 \Leftrightarrow F_R - W_A - F = 0. \tag{8.86}$$

The equilibrium equations for the pulley can be written in MATLAB with:

```
eqCeiling=F_R-W_A-F;
fprintf('The equilibrium equation for the ceiling is \n');
fprintf('%s = 0 \n\n',eqCeiling);
```

From Equation (8.86) the force exerted on the ceiling is

$$F_R = W_A + F = m_A g + m_A g = 2m_A g. \tag{8.87}$$

In MATLAB, the force exerted on the ceiling by the supporting rod is calculated with:

```
sol_eqCeiling=solve(eqCeiling,'F_R');
fprintf('The force acting on the ceiling is \n');
fprintf('F_R = %s \n\n',sol_eqCeiling);
F_R=sol_eqCeiling;
```

(c) To calculate the force acting in the cable for the vertical position of the block shown in Figure 8.26 one can calculate the angle α between the cable and the horizontal direction using either

$$\cos \alpha = \frac{\dfrac{d_{CD}}{2}}{d_{BC}} = \frac{d_{CD}}{2d_{BC}} = \frac{d_{CD}}{2\left(\dfrac{s_A - 2d_A - l_A}{2}\right)} = \frac{l_A}{s_A - 2d_A - l_A} \tag{8.88}$$

or

$$\sin \alpha = \frac{h_{BCD}}{d_{BC}} = \frac{\sqrt{d_{BC}^2 - \left(\dfrac{d_{CD}}{2}\right)^2}}{d_{BC}} = \sqrt{1 - \left(\dfrac{d_{CD}}{2d_{BC}}\right)^2}$$
$$= \sqrt{1 - \left(\dfrac{l_A}{s_A - 2d_A - l_A}\right)^2} \tag{8.89}$$

where $s_A = d_{BC} + d_{BD} + 2d_A + l_A$ is the length of the cable around the block having the height d_A and the width l_A (Figure 8.26), and $d_{BC} = d_{BD}$.

Considering the equilibrium equations for the block A shown in Figure 8.26 one can write

$$\sum F_x = 0 \Leftrightarrow -T_{VL} \cos \alpha + T_{VR} \cos \alpha = 0$$

$$\sum F_y = 0 \Leftrightarrow T - T_{VL} \sin \alpha - T_{VR} \sin \alpha = 0. \tag{8.90}$$

Solving Equation (8.90) one can obtain

$$T_{VL} = T_{VR} = \frac{T}{2 \sin \alpha} = \frac{T}{2\sqrt{1 - \left(\dfrac{l_A}{s_A - 2d_A - l_A}\right)^2}}. \tag{8.91}$$

The force acting in the cable when the block A is placed in a vertical position can be calculated in MATLAB with:

```
d_BD=d_BC;
d_CD=l_A;
eqCableV=-s_A+d_BC+d_BD+2*d_A+l_A;
d_BC=solve(eqCableV,'d_BC');
h_BCD=sqrt((d_BC)^2 -(d_CD/2)^2);
sinalpha=h_BCD/d_BC;
SumFxV=-T_VL*cosalpha+T_VR*cosalpha;
SumFyV=T-T_VL*sinalpha-T_VR*sinalpha;
solFxFyV=solve([SumFxV,SumFyV],'T_VL','T_VR');
fprintf('The tension T_VL in the cable is \n\n');
pretty(solFxFyV.T_VL);
```

To calculate the force acting in the cable for the horizontal position of the block shown in Figure 8.27 one can calculate the angle β between the cable and the horizontal direction using either

$$\cos \beta = \frac{d_{FS}}{2d_{FB}} = \frac{d_A}{s_A - 2l_A - d_A} \tag{8.92}$$

or

$$\sin \beta = \frac{h_{BFS}}{d_{FB}} = \frac{\sqrt{d_{FB}^2 - \left(\dfrac{d_{FS}}{2}\right)^2}}{d_{FB}} = \sqrt{1 - \left(\dfrac{d_{FS}}{2d_{FB}}\right)^2}$$

$$= \sqrt{1 - \left(\dfrac{d_A}{s_A - 2l_A - d_A}\right)^2} \tag{8.93}$$

where $s_A = d_{FB} + d_{BS} + 2l_A + d_A$ is the length of the cable around the block having the height l_A and the width d_A (Figure 8.27), and $d_{FB} = d_{BS}$.

Considering the equilibrium equations for the block A shown in Figure 8.27 one can write

$$\sum F_x = 0 \Leftrightarrow -T_{HL} \cos \beta + T_{HR} \cos \beta = 0$$

$$\sum F_y = 0 \Leftrightarrow T - T_{HL} \sin \beta - T_{HR} \sin \beta = 0. \tag{8.94}$$

Solving Equation (8.94) one can obtain

$$T_{HL} = T_{HR} = \frac{T}{2 \sin \beta} = \frac{T}{2\sqrt{1 - \left(\dfrac{d_A}{s_A - 2l_A - d_A}\right)^2}}. \tag{8.95}$$

The force acting in the cable when the block *A* is placed in a horizontal position can be calculated in MATLAB with:

```
d_BS=d_FB;
d_FS=d_A;
eqCableH=-s_A+d_FB+d_BS+d_A+2*l_A;
d_FB=solve(eqCableH,'d_FB');
h_BFS=sqrt((d_FB)^2 -(d_FS/2)^2);
sinbeta=h_BFS/d_FB;
SumFxH=-T_HL*cosbeta+T_HR*cosbeta;
SumFyH=T-T_HL*sinbeta-T_HR*sinbeta;
solFxFyH=solve([SumFxH,SumFyH],'T_HL','T_HR');
fprintf('The tension T_HL in the cable is \n\n');
pretty(solFxFyH.T_HL);
```

The input numerical data are introduced in MATLAB with:

```
% numerical results
lists = {m_A,g,d_A,l_A,s_A};
listn = {7.5,9.81,3,2.5,12.5};
```

The numerical results are calculated and printed in MATLAB using:

```
F = eval(subs(F,lists,listn));
T=F;
F_R = eval(subs(F_R,lists,listn));
T_VL = eval(subs(solFxFyV.T_VL,lists,listn));
T_HL = eval(subs(solFxFyH.T_HL,lists,listn));

fprintf('The numerical value of the force F is \n');
fprintf('F = %f (N)\n\n',F);
fprintf('The tension T in the rope is \n');
fprintf('T = %f (N)\n\n',F);
fprintf('The numerical value of the force exerted on the
ceiling is \n');
fprintf('F_R = %f (N)\n\n',F_R);
fprintf('The tension T_VL in the cable is \n');
fprintf('T_VL = %f (N) \n\n',T_VL);
fprintf('The tension T_HL in the cable is \n');
fprintf('T_HL = %f (N) \n\n',T_HL);
```

The MATLAB numerical results for the force *F* worker must apply for equilibrium, the tension *T* in the rope, force F_R exerted on the ceiling and tension T_{VL} and T_{HL} in the cable are:

```
The numerical value of the force F is
F = 73.575000 (N)

The tension T in the cables is
T = 73.575000 (N)
```

The numerical value of the force exerted on the ceiling is
F_R = 147.150000 (N)

The tension T_VL in the cable is
T_VL = 47.125716 (N)

The tension T_HL in the cable is
T_HL = 49.355610 (N)

Example 8.7
The worker shown in Figure 8.28 is using a massless rope and a frictionless $n = 3$ pulley system to lift a block A to a desired height. Knowing the mass m_A of the block, determine:

(a) The force F the worker must apply to the end of the rope to keep the block A stationary.
(b) The force ratio and movement ratio when the effort applied by the worker is twice the value of the force F required to keep the block stationary.
(c) The efficiency of the system.

  Numerical application: $m_A = 6.5$ kg and $g = 9.81$ m s².

Solution
(a) To calculate the force the worker must apply to the end of the rope for the block to be stationary (in equilibrium) one can consider the equilibrium equations. The equilibrium of the weight of block A and applied worker force F can be written using the free body diagram shown in Figure 8.29 by

$$\sum F_x = 0 \Leftrightarrow 0 = 0$$

$$\sum F_y = 0 \Leftrightarrow F = T, \ T = T_3, \ T_3 = T_2, \ T_2 = T_1 \text{ and } W_A = T_1 + T_2 + T_3 \quad (8.96)$$

where T_1, T_2 and T_3 are the tensions in the ropes.

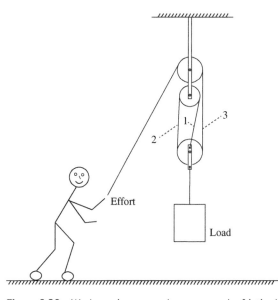

Figure 8.28 Worker using a massless rope and a frictionless pulley system to lift a block.

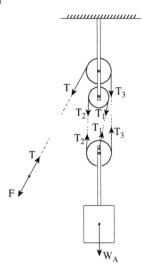

Figure 8.29 Free body diagram of the block and the pulley system.

The equilibrium equations are written in MATLAB with:

```
clear all; clc; close all
syms m_A g F T T1 T2 T3

W_A=m_A*g;
eqFy1=F-T;
eqFy2=T2-T1;
eqFy3=T3-T2;
eqFy4=T-T3;
eqFy5=W_A-T1-T2-T3;
fprintf('The equilibrium equations are \n');
fprintf('%s = 0 \n',eqFy1);
fprintf('%s = 0 \n',eqFy2);
fprintf('%s = 0 \n',eqFy3);
fprintf('%s = 0 \n',eqFy4);
fprintf('%s = 0 \n\n',eqFy5);
```

From Equation (8.96) the force the worker must apply to the rope for the block to be stationary can be computed as

$$F = \frac{W_A}{3} = \frac{m_A g}{3} \tag{8.97}$$

and in MATLAB the computation is performed with:

```
sol_eqFy15=solve([eqFy1,eqFy2,eqFy3,eqFy4,eqFy5],[F,T,T1,T2,T3]);
fprintf('The force the worker should apply ');
fprintf('to the rope for the block to be stationary is \n');
fprintf('F = %s \n\n',sol_eqFy15.F);
F=sol_eqFy15.F;
```

(b) From the definition of the force ratio, e.g., $FR = \dfrac{Load}{Effort}$ one can write

$$FR = \frac{m_A g}{2F} = \frac{3m_A g}{2m_A g} = \frac{3}{2} \tag{8.98}$$

where $m_A g$ is the weight of the block A and $2F$ is the effort applied by the worker. The force ratio is calculated in MATLAB with:

```
Load=W_A;
Effort=2*F;
FR=Load/Effort;
fprintf('The force ratio is ');
fprintf('FR = %s \n\n',FR);
```

By definition, the movement ratio MR for a system having n pulleys is n, that is, for the pulley shown in Figure 8.29, the movement ratio is $n = 3$.

The movement ratio is calculated in MATLAB with:

```
d_Effort=3;
d_Load=1;
MR=d_Effort/d_Load;
fprintf('The movement ratio is ');
fprintf('MR = %d \n\n',MR);
```

(c) From the definition of efficiency one can calculate

$$E = \frac{FR}{MR} 100\% = \frac{3}{2n} 100\% \tag{8.99}$$

The efficiency of the system is calculated in MATLAB with:

```
Efficiency=FR*100/MR;
fprintf('The efficiency of the pulley is \n');
fprintf('Efficiency = %s%% \n\n',Efficiency);
```

The input numerical data are introduced in MATLAB with:

```
% numerical results
lists = {m_A,g};
listn = {6.5,9.81};
```

The numerical results are calculated and printed in MATLAB using:

```
F = eval(subs(F,lists,listn));

fprintf('The numerical value of the force F is \n');
fprintf('F = %f [N] \n\n',F);
fprintf('The numerical value of the force ratio is \n');
fprintf('FR = %s \n\n',FR);
fprintf('The numerical value of the movement ratio is \n');
fprintf('MR = %d \n\n',MR);
```

```
fprintf('The numerical value of the efficiency of the
pulley is \n');
fprintf('Efficiency = %s%% \n\n',Efficiency);
```

The MATLAB numerical results for the force F worker must apply for equilibrium, force ratio FR, movement ratio MR and efficiency of the system are:

```
The numerical value of the force F is
F = 21.255000 [N]
```

```
The numerical value of the force ratio is
FR = 3/2
```

```
The numerical value of the movement ratio is
MR = 3
```

```
The numerical value of the efficiency of the pulley is
Efficiency = 50%
```

Example 8.8

The worker shown in Figure 8.30 is using a massless rope and a frictionless $n = 4$ pulley system to lift a block A to a desired height. Knowing the mass m_A of the block, determine:

(a) The minimum force F the worker must apply to the end of the rope to keep the block A stationary.
(b) The force ratio when the effort applied by the worker is twice the value of the force F required to keep the block stationary.

Numerical application: $m_A = 6.5$ kg and $g = 9.81$ m s^{-2}.

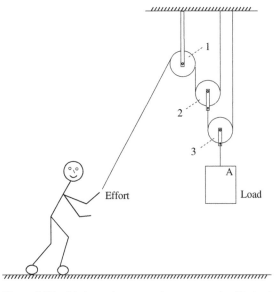

Figure 8.30 Worker using a massless rope and a frictionless pulley system to lift a block.

Solution

(a) To calculate the force the worker must apply to the end of the rope for the block to be stationary (in equilibrium) one can consider the equilibrium equations. The equilibrium of the weight of block A and applied worker force F can be written using the free body diagram shown in Figure 8.31 by

$$\sum F_x = 0 \Leftrightarrow 0 = 0$$
$$\sum F_y = 0 \Leftrightarrow F = T, \ T = T_1, \ T_1 = T_2,$$
$$T_3 = T_2 + T_1, \ T_3 = T_4, \ W_A = T_3 + T_4 \qquad (8.100)$$

where T_1, T_2, T_3 and T_4 are the tensions in the ropes. The equilibrium equations are written in MATLAB with:

```
clear all; clc; close all
syms m_A g F T T1 T2 T3 T4

W_A=m_A*g;
eqFy0=F-T;
eqFy1=T1-T;
eqFy2=T2-T1;
eqFy3=T3-T2-T1;
eqFy4=T4-T3;
eqFy5=W_A-T4-T3;
fprintf('The equilibrium equations are \n');
```

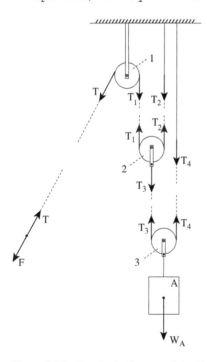

Figure 8.31 Free body diagram of the block and pulley system.

```
fprintf('%s = 0 \n',eqFy0);
fprintf('%s = 0 \n',eqFy1);
fprintf('%s = 0 \n',eqFy2);
fprintf('%s = 0 \n',eqFy3);
fprintf('%s = 0 \n',eqFy4);
fprintf('%s = 0 \n\n',eqFy5);
```

From Equation (8.100) one can write

$$W_A = T_3 + T_4 = 2T_3 = 2\left(T_1 + T_2\right) = 4T_1 = 4T = 4F \tag{8.101}$$

or equivalent, the force the worker must apply to the rope for the block to be stationary can be

$$F = \frac{W_A}{4} = \frac{m_A g}{4}. \tag{8.102}$$

In MATLAB the computation of the force F is performed with:

```
sol_eqFy05=solve([eqFy0,eqFy1,eqFy2,eqFy3,eqFy4,eqFy5],
[F,T,T1,T2,T3,T4]);
fprintf('The minimum force the worker should apply ');
fprintf('to the rope for the block to be stationary is \n');
fprintf('F = %s \n\n',sol_eqFy05.F);
F=sol_eqFy05.F;
```

(b) From the definition of the force ratio, e.g., FR $= \dfrac{\text{Load}}{\text{Effort}}$ one can write

$$\text{FR} = \frac{m_A g}{2F} = \frac{4m_A g}{2m_A g} = \frac{4}{2} = 2 \tag{8.103}$$

where $m_A g$ is the weight of the block A and $2F$ is the effort applied by the worker. The force ratio is calculated in MATLAB with:

```
Load=W_A;
Effort=2*F;
FR=Load/Effort;
fprintf('The force ratio is ');
fprintf('FR = %s \n\n',FR);
```

The input numerical data are introduced in MATLAB with:

```
% numerical results
lists = {m_A,g};
listn = {6.5,9.81};
```

The numerical results are calculated and printed in MATLAB using:

```
F = eval(subs(F,lists,listn));
fprintf('The numerical value of the force F is \n');
fprintf('F = %f [N]\n\n',F);
fprintf('The numerical value of the force ratio is \n');
fprintf('FR = %s \n\n',FR);
```

The MATLAB numerical results for the force F worker must apply for equilibrium and force ratio FR of the system are:

```
The numerical value of the force F is
F = 15.941250 [N]

The numerical value of the force ratio is
FR = 2
```

Example 8.9

A worker is using an horizontal bar to turn the screw of a square threaded jack (Figure 8.32) to move up a rectangular block of weight $W = mg$. Knowing the mass m of the block, the length l of the horizontal bar, the mean radius r and the lead L_S of the single threaded screw, determine:

(a) The force F the worker must exert to the end of the bar to move the block up a distance $h = p$ when friction is neglected.
(b) The mechanical advantage of the jack when the force F is applied.
(c) The minimal torque M required to move up the block when the friction coefficient between the screw and the supporting thread is μ. Determine if the block remains stationary when the torque is removed.

Numerical application: $m_A = 115$ kg and $g = 9.81$ m s^{-2}.

Solution

(a) To move the block a distance h equal to the pitch p of the screw the worker must perform a complete rotation of the horizontal bar about the screw axis. The force F the worker must apply to the end of the bar to move the block up a distance h can be calculated from

$$Fd = Wh \Leftrightarrow 2\pi Fl = Wp \tag{8.104}$$

where $d = 2\pi l$ is the distance to be moved in order to advance the block a distance h equal to the pitch p of the screw.

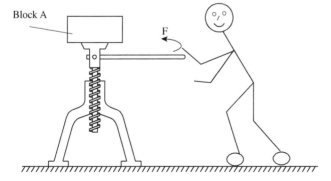

Block A   F

Figure 8.32 Worker using a horizontal bar to turn the screw of a square threaded jack.

Equation (8.104) is written in MATLAB using:

```
clear all; clc; close all
syms m g l p F r mu

W=m*g;
d=2*pi*l;
h=p;
eqF=F*d-W*h;
fprintf('The equations to calculate force F is \n');
fprintf('%s = 0 \n\n',eqF);
```

From Equation (8.104) one can calculate the force F using

$$F = \frac{Wp}{2\pi l} \tag{8.105}$$

and in MATLAB with:

```
sol_eqF=solve(eqF,F);
fprintf('The force the worker should apply to move ');
fprintf('the block up a distance h=p is \n');
fprintf('F = %s \n\n',sol_eqF);
F=sol_eqF;
```

(b) The mechanical advantage of the system can be calculated using

$$MA = \frac{\text{Load}}{\text{Effort}} = \frac{2\pi l}{p} \tag{8.106}$$

and in MATLAB with:

```
d_Load=2*pi*l;
d_Effort=p;
MA=d_Load/d_Effort;
fprintf('The mechanical advantage is \n');
fprintf('MA = %s \n\n',MA);
```

(c) For a screw motion (moving up) opposite to the direction of the load the minimal torque M required to turn up the screw can be calculated using

$$M = Wr\tan(\theta + \varphi)$$
$$= Wr\tan\left(\tan^{-1}\left(\frac{L_S}{2\pi r}\right) + \tan^{-1}(\mu)\right)$$
$$= Wr\tan\left(\tan^{-1}\left(\frac{p}{2\pi r}\right) + \tan^{-1}(\mu)\right) \tag{8.107}$$

where r is the mean radius of the thread, L_S is the lead of the screw, $L_S = p$ since the pitch and lead are equal for single start screws, $\theta = \tan^{-1}\left(\frac{p}{2\pi r}\right)$, and $\varphi = \tan^{-1}(\mu)$ is the angle of friction. The torque is calculated in MATLAB with:

```
L_S=p;
theta=atan(L_S/(2*pi*r));
```

```
phi=atan(mu);
M=W*r*tan(theta+phi);
fprintf('The minimal torque required to move up the block ');
fprintf('when the static coefficient of friction is mu can be
calculated as \n');
fprintf('M = %s \n\n',M);
```

The screw is self-locking if

$$\mu \geq \tan(\theta) = \frac{p}{2\pi r}. \tag{8.108}$$

The self-locking condition is introduced in MATLAB with:

```
fprintf('The self-locking condition of the screw is \n');
fprintf('mu >   %s \n\n',tan(theta));
```

The input numerical data are introduced in MATLAB with:

```
% numerical results
lists = {m,g,mu,p,r,l};
listn = {450,9.81,0.25,0.01,0.008,0.75};
```

The numerical results are calculated and printed in MATLAB using:

```
F = eval(subs(sol_eqF,lists,listn));
MA = eval(subs(MA,lists,listn));
M = eval(subs(M,lists,listn));
theta = eval(subs(theta,lists,listn));
mu = eval(subs(mu,lists,listn));
fprintf('The numerical value of the force F is \n');
fprintf('F = %f [N]\n\n',F);
fprintf('The numerical value of the mechanical advantage
is \n');
fprintf('MA = %f \n\n',MA);
fprintf('The numerical value of the minimal torque required
to move up the block is \n');
fprintf('M = %f [N][m]\n\n',M);
fprintf('The self-locking condition \n');
fprintf('%f > %f \n\n',mu,tan(theta));
fprintf('is TRUE so the block is not going down under its
own weight \n');
```

The MATLAB numerical results for the force F, mechanical advantage M_A, torque M and force are:

```
The numerical value of the force F is
F = 9.367860 [N]


The numerical value of the mechanical advantage is
MA = 471.238898
```

The numerical value of the minimal torque required to move
up the load is
```
M = 16.684725  [N] [m]
```

The self-locking condition
```
0.250000 > 0.198944
```

is TRUE so the block is not going down under its own weight

Example 8.10

The worker shown in Figure 8.33 is using a lever A to press on the exterior of a pulley C to slow down and stop the block B at a desired height. The lever A, which can rotate about the fixed point D, has length d_L. The pulley radius is r_p, the friction radius (where the lever acts) is r_c, and the friction coefficient between the lever and the pulley is μ. Knowing the weight W_B of the block B, determine:

(a) The equilibrium equations for the lever A, pulley C and for the cantilever beam D.
(b) The minimum force the worker must apply to the end of the lever A (normal to the lever) such as the whole system is in equilibrium, i.e. the block B is not moving.

Numerical application: $\mu = 0.45$, $\mu_{AB} = 0.6$, $g = 9.81$ m s^{-2}, $m_B = 4$ kg, $d_F = 0.4$ m, $d_{Ff} = 0.2$ m, $d_{R_F} = 0.002$ m, $r_p = 0.1$ m, $r_c = 0.15$ m, $d_D = 0.3$ m and $\alpha = 36.87°$.

Solution

The equilibrium of the block B (Figure 8.34) can be written as

$$\sum F_{\text{Block}} = 0 \Leftrightarrow W_B + T = 0. \tag{8.109}$$

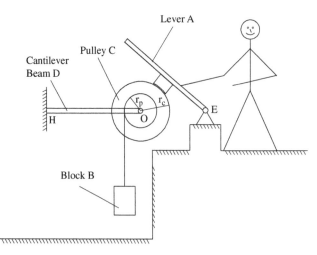

Figure 8.33 Worker using a mechanical system (lever pressing a pulley) to slow down and stop a block at a desired height.

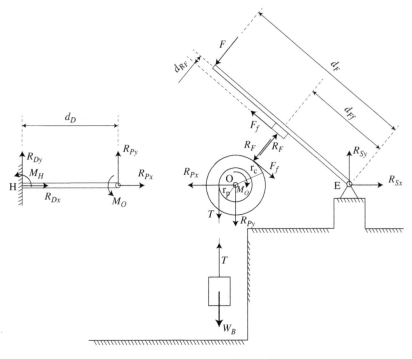

Figure 8.34 Free body diagram of the lever, pulley and block.

Considering the fixed reference frame Oxy shown in Figure 8.34 the equilibrium of the block B can be written as

$$\sum F_y = 0 \Leftrightarrow T - W_B = 0$$
$$\sum F_x = 0 \Leftrightarrow 0 = 0 \qquad (8.110)$$

or equivalent

$$T = W_B. \qquad (8.111)$$

The equilibrium of the block is written in MATLAB with

```
clear all; clc; close all
syms alpha T F F_f R_F R_Sx R_Sy d_Ff d_F d_RF
syms m_B g mu r_c r_p R_Px R_Py M_O R_Dx R_Dy M_H d_D Fnew

W_B = m_B*g;
T_  = [0 T 0];
W_B_  = [0 -W_B 0];

eq_block = T_  + W_B_;
fprintf('The block equilibrium equations are \n')
fprintf('%s = 0 \n\n',char(eq_block(2)))
```

```
solB=solve(eq_block(2),'T');
fprintf('Solving the block equations: \n')
fprintf('The rope tension T is \n')
fprintf('T = %s \n\n',char(simplify(solB)))
```

The equilibrium of the lever A used to press on the exterior of a pulley C can be written as

$$\sum \mathbf{F}_{Lever} = 0 \Leftrightarrow \mathbf{F} + \mathbf{R}_F + \mathbf{F}_f + \mathbf{R}_S = 0. \tag{8.112}$$

In the fixed reference frame Oxy the equilibrium equations of the lever A can be written as

$$\sum F_x = 0 \Leftrightarrow R_{Sx} + (R_F - F)\sin\alpha - F_f\cos\alpha = 0$$
$$\sum F_y = 0 \Leftrightarrow R_{Sy} + (R_F - F)\cos\alpha + F_f\sin\alpha = 0 \tag{8.113}$$

where the friction force F_f is expressed as $F_f = \mu R_F$.

The equilibrium (moment) of the lever A about point E can be written as

$$\sum M_E = 0 \Leftrightarrow Fd_F - F_f d_{F_f} - R_F d_{R_F} = 0. \tag{8.114}$$

From Equations (8.113) and (8.114) one can write

$$R_{Sx} + (R_F - F)\sin\alpha - F_f\cos\alpha = 0$$
$$R_{Sy} + (R_F - F)\cos\alpha + F_f\sin\alpha = 0$$
$$Fd_F - F_f d_{Ff} - R_F d_{R_F} = 0. \tag{8.115}$$

The equilibrium of the lever A (sum of forces and sum of moments) is written in MATLAB with:

```
F_f=mu*R_F;
R_F_  = [R_F*sin(alpha) R_F*cos(alpha) 0];
F_f_  = [-F_f*cos(alpha) F_f*sin(alpha) 0];
R_S_  = [R_Sx R_Sy 0];
F_    = [-F*sin(alpha) -F*cos(alpha) 0];
d_Ff_ = [d_Ff 0 0];
d_F_  = [d_F 0 0];
d_RF_ = [d_RF 0 0];
R_Fp_ = [0 R_F 0];
F_fp_ = [0 F_f 0];
F_p_  = [0 -F 0];

eq_leverF = R_F_ + R_S_ + F_ + F_f_;
eq_leverM = cross(F_fp_,d_Ff_) + ...
     cross(F_p_,d_F_) + cross(R_Fp_,d_RF_);
fprintf('The piston equilibrium equations: \n')
fprintf('%s = 0 \n',char(eq_leverF(1)))
fprintf('%s = 0 \n',char(eq_leverF(2)))
fprintf('%s = 0 \n\n',char(eq_leverM(3)))
```

The lever equilibrium equations are solved in MATLAB with:

```
solL=solve(eq_leverF(1),eq_leverF(2),...
    eq_leverM(3),'R_Sx', 'R_Sy','R_F');
fprintf('Solving the piston equations: \n')
fprintf('The reaction R_Sx is \n')
fprintf('R_Sx = %s \n\n',char(solL.R_Sx))
fprintf('The reaction R_Sx is \n')
fprintf('R_Sy = %s \n\n',char(solL.R_Sy))
fprintf('The reaction R_F is \n')
fprintf('R_F = %s \n\n',char(solL.R_F))
```

The equilibrium of the pulley C can be written as

$$\sum \mathbf{F}_{\text{Pulley}} = 0 \Leftrightarrow \mathbf{T} + \mathbf{F}_f + \mathbf{R}_F + \mathbf{R}_{Py} + \mathbf{R}_{Px} = 0. \tag{8.116}$$

In the fixed reference frame Oxy the equilibrium equations can be written as

$$\sum F_x = 0 \Leftrightarrow F_f \cos \alpha - R_{Px} - R_F \sin \alpha = 0$$
$$\sum F_y = 0 \Leftrightarrow -T - R_{Py} - R_F \cos \alpha - F_f \sin \alpha = 0 \tag{8.117}$$

or equivalent

$$F_f \cos \alpha - R_{Px} - R_F \sin \alpha = 0$$
$$-T - R_{Py} - R_F \cos \alpha - F_f \sin \alpha = 0. \tag{8.118}$$

The (moment) equilibrium of the pulley C about its centre O can be written as

$$\sum M_O = 0 \Leftrightarrow F_f r_c - T r_p + M_O = 0. \tag{8.119}$$

From Equations (8.118) and (8.119) one can write

$$F_f \cos \alpha - R_{Px} - R_F \sin \alpha = 0$$
$$-T - R_{Py} - R_F \cos \alpha - F_f \sin \alpha = 0$$
$$F_f r_c - T r_p + M_O = 0. \tag{8.120}$$

The equilibrium of the pulley C (sum of forces and sum of moments) is written in MATLAB with:

```
R_P_  = [-R_Px -R_Py 0];
r_c_  = [r_c 0 0];
r_p_  = [r_p 0 0];
R_F_= - R_F_;
F_f_  = - F_f_;
F_fp_ = - F_fp_;
M_O_  = [0 0 M_O];

eq_pulleyF = T_ + F_f_ + R_F_ + R_P_;
eq_pulleyM = cross(F_fp_,r_c_) + ...
    cross(T_,r_p_) + M_O_;
fprintf('The pulley equilibrium equations are: \n')
```

```
fprintf('%s = 0 \n',char(eq_pulleyF(1)))
fprintf('%s = 0 \n',char(eq_pulleyF(2)))
fprintf('%s = 0 \n\n',char(eq_pulleyM(3)))
```

The pulley equilibrium equations are solved in MATLAB with:

```
solP=solve(eq_pulleyF(1),eq_pulleyF(2),...
    eq_pulleyM(3),'R_Px', 'R_Py','M_O');
solP.R_Px=subs(solP.R_Px,R_F,solL.R_F);
solP.R_Py=subs(solP.R_Py,{R_F,T},{solL.R_F,solB});
solP.M_O=subs(solP.M_O,{R_F,T},{solL.R_F,solB});
fprintf('Solving the pulley equations: \n')
fprintf('The reaction R_Px is \n')
fprintf('R_Px = %s \n\n',char(solP.R_Px))
fprintf('The reaction R_Py is \n')
fprintf('R_Py = %s \n\n',char(solP.R_Py))
fprintf('The moment M_O at O is \n')
fprintf('M_O = %s \n\n',char(solP.M_O))
```

From Equations (8.115) and (8.120) the minimum force the worker must apply to the end of the lever A such as the whole system is in equilibrium, i.e. the block B is not moving, could be calculated. In MATLAB the moment equation is solved using:

```
solF=solve(solP.M_O,'F');
fprintf('The minimum force the worker must apply to the
lever is \n')
fprintf('F = %s \n\n',char(solF))
```

The equilibrium of the cantilever beam D can be written as

$$\sum \mathbf{F}_{Beam} = 0 \Leftrightarrow \mathbf{R}_{Dx} + \mathbf{R}_{Dy} + \mathbf{R}_{Px} + \mathbf{R}_{Py} = 0. \tag{8.121}$$

In the fixed reference frame Oxy the equilibrium equations can be written as

$$\sum F_x = 0 \Leftrightarrow R_{Dx} + R_{Px} = 0$$
$$\sum F_y = 0 \Leftrightarrow R_{Dx} + R_{Px} = 0. \tag{8.122}$$

The equilibrium (sum of moments) of the cantilever beam D about point H can be written as

$$\sum M_H = 0 \Leftrightarrow M_O + M_H + d_D R_{Py} = 0. \tag{8.123}$$

From Equations (8.122) and (8.123) one can write

$$R_{Dx} + R_{Px} = 0$$
$$R_{Dx} + R_{Px} = 0$$
$$M_O + M_H + d_D R_{Py} = 0. \tag{8.124}$$

The equilibrium of the cantilever beam D is written in MATLAB with:

```
R_D_ = [-R_Dx -R_Dy 0];
d_D_ = [d_D 0 0];
M_H_ = [0 0 M_H];
```

```
eq_beamF = R_D_ + R_P_;
eq_beamM = cross(R_D_,d_D_) + M_H_ + M_O_;
fprintf('The beam equilibrium equations are: \n')
fprintf('%s = 0 \n',char(eq_beamF(1)))
fprintf('%s = 0 \n',char(eq_beamF(2)))
fprintf('%s = 0 \n\n',char(eq_beamM(3)))
```

The cantilever beam equilibrium equations are solved in MATLAB with:

```
solBM=solve(eq_beamF(1),eq_beamF(2),...
    eq_beamM(3),'R_Dx', 'R_Dy','M_H');
solBM.R_Dx=subs(solBM.R_Dx,R_Px,solP.R_Px);
solBM.R_Dy=subs(solBM.R_Dy,R_Py,solP.R_Py);
solBM.M_H=subs(solBM.M_H,{M_O,R_Py},...
    {solP.M_O,solP.R_Py});
fprintf('Solving the beam equations: \n')
fprintf('The reaction R_Dx is \n')
fprintf('R_Dx = %s \n\n',char(solBM.R_Dx))
fprintf('The reaction R_Dy is \n')
fprintf('R_Dy = %s \n\n',char(solBM.R_Dy))
fprintf('The moment M_H at H is \n')
fprintf('M_H = %s \n\n',char(solBM.M_H))
```

The input numerical data are introduced in MATLAB with:

```
lists = {mu,g,m_B,alpha,d_F,d_Ff,d_RF,r_p,r_c,d_D};
listn = {0.45,9.81,4,atan(0.15/0.2),0.4,0.2,0.002,0.1,0.15,0.3};
```

The numerical value of the minimum force the worker must apply to the end of the lever is calculated and printed in MATLAB with:

```
Fnew=eval(subs(solF,lists,listn));
fprintf('The minimum force the worker must apply ')
fprintf('to the end of the lever $A$ is \n')
fprintf('F = %f [N]\n\n',Fnew)
```

The numerical value of the reactions R_{Sx}, R_{Sy} and R_F at equilibrium (piston) are calculated and printed in MATLAB with:

```
R_Sx=subs(solL.R_Sx,lists,listn);
R_Sx=eval(subs(R_Sx,{'lists',F},{'listn',Fnew}));
R_Sy=subs(solL.R_Sy,lists,listn);
R_Sy=eval(subs(R_Sy,{'lists',F},{'listn',Fnew}));
R_F=subs(solL.R_F,lists,listn);
R_F=eval(subs(R_F,{'lists',F},{'listn',Fnew}));
fprintf('Solving the piston equations: \n')
fprintf('The reaction R_Sx, R_Sy and R_F at equilibrium are \n')
fprintf('R_Sx = %f [N]\n',R_Sx);
fprintf('R_Sy = %f [N]\n',R_Sy);
fprintf('R_F = %f [N]\n\n',R_F);
```

The numerical value of the pulley reactions R_{Px} and R_{Py} are calculated and printed in MAT-LAB with:

```
R_Px=subs({solP.R_Px,solF},lists,listn);
R_Px=eval(subs(R_Px(1),{'lists',F},{'listn',Fnew}));
R_Py=subs({solP.R_Py,solF},lists,listn);
R_Py=eval(subs(R_Py(1),{'lists',F},{'listn',Fnew}));
M_O=subs({solP.M_O,solF},lists,listn);
M_O=eval(subs(M_O(1),{'lists',F},{'listn',Fnew}));
fprintf('Solving the pulley equations: \n');
fprintf('The moment M_O and reactions R_Px and R_Py at
equilibrium are \n');
fprintf('M_O = %f [Nm]\n',M_O);
fprintf('R_Px = %f [N]\n',R_Px);
fprintf('R_Py = %f [N]\n\n',R_Py);
```

The numerical value of the moment M_H and cantilevered beam reactions R_{Dx} and R_{Dy} are calculated and printed in MATLAB with:

```
R_Dx=subs({solBM.R_Dx,solF},lists,listn);
R_Dx=eval(subs(R_Dx(1),{'lists',F},{'listn',Fnew}));
R_Dy=subs({solBM.R_Dy,solF},lists,listn);
R_Dy=eval(subs(R_Dy(1),{'lists',F},{'listn',Fnew}));
M_H=subs({solBM.M_H,solF},lists,listn);
M_H=eval(subs(M_H(1),{'lists',F},{'listn',Fnew}));
fprintf('Solving the beam equations: \n')
fprintf('The moment M_H and reaction R_Dx, R_Dy at H are \n')
fprintf('M_H = %f [Nm]\n',M_H);
fprintf('R_Dx = %f [N]\n',R_Dx);
fprintf('R_Dy = %f [N]\n\n',R_Dy);
```

The obtained MATLAB numerical results are:

```
The minimum force the worker must apply to the end of the lever
$A$ is
F = 13.370667 [N]

Solving the piston equations:
The reaction R_Sx, R_Sy and R_F at equilibrium are
R_Sx = -0.598187 [N]
R_Sy = -51.842879 [N]
R_F = 58.133333 [N]

Solving the pulley equations:
The moment M_O and reactions R_Px and R_Py at equilibrium are
M_O = 0.000000 [Nm]
R_Px = -7.477338 [N]
R_Py = -24.068131 [N]
```

Solving the beam equations:
The moment M_H and reaction R_Dx, R_Dy at H are
M_H = -7.220439 [Nm]
R_Dx = 7.477338 [N]
R_Dy = 24.068131 [N]

Example 8.11

The L bracket 1 shown in Figure 8.35(a) supports block 2. A single bolt at A holds bracket 1 against the smooth fixed wall 0. Bracket 1 has a mass m_1 and has its mass centre at C. Block 2 has a concentrated mass m_2 and has its mass centre at D. Under the weight of block 2 the the bracket makes an angle θ with the vertical wall, as shown in Figure 8.35(b). The following dimensions are given: $OA = AB = a$ and $BC = CD = l$. Determine the reaction forces on the bracket for $\theta = 0°$ and $\theta = 3°$. Numerical application: $m_1 = 20$ kg, $m_2 = 5$ kg, $a = 50$ mm, $l = 250$ mm, $g = 9.81$ m s^{-2}.

Solution

The symbolic input data is given by:

```
syms a l theta m1 m2 g
sd = {m1, m2, a, l, g};
% m1 =  20;   % (kg)
% m2 =   5;   % (kg)
% a  =  50;   % (mm)
```

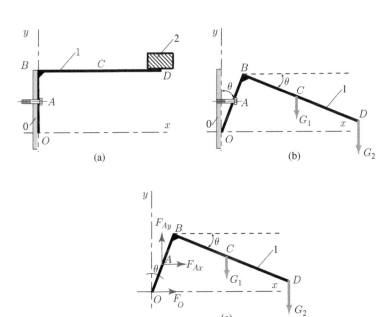

(a)

(b)

(c)

Figure 8.35 (a) L bracket and single bolt supporting a block, (b) angle θ between the bracket and the wall under the weight of the block, and (c) free body diagram of the bracket, bolt and block.

```
% l = 250;    % (mm)
% g = 9.81;  % (m/s^2)
% corresponding numerical data list
nd = {20, 5, 50*10^-3, 250*10^-3, 9.81};
```

Figure 8.35(c) shows the free-body diagram of bracket 1. The reaction force of the smooth wall on bracket 1 at the origin O is

$$\mathbf{F}_O = F_O \, \mathbf{\imath},$$

and the reaction force of the bolt on bracket 1 at A is

$$\mathbf{F}_A = F_{Ax} \, \mathbf{\imath} + F_{Ay} \, \mathbf{\jmath},$$

or with MATLAB:

```
syms FO FAx FAy
FO_ = [FO, 0, 0];  % at O
FA_ = [FAx, FAy, 0];  % at A
```

The gravity forces are introduced as:

```
G1_ = [0, -m1*g, 0];  % at C
G2_ = [0, -m2*g, 0];  % at D
```

The position vectors of the force application points are O, A, C, D and are calculated with:

```
% position vectors
% origin
rO_ = [0, 0, 0];
% position vector of A
rA_ = [a*sin(theta), a*cos(theta), 0];
% position vector of B
rB_ = 2*rA_;
rBC_ = [l*cos(theta), -l*sin(theta), 0];
% position vector of C
rC_ = rB_ + rBC_;
% position vector of D
rD_ = rB_ + 2*rBC_;
```

The reaction force F_O is calculated from the moment equilibrium equation of all forces with respect to A or with MATLAB:

```
% sum of moments on 1 about A
% MA_ = AO_xFO_+AC_xG1_+AD_xG2_
MA_ = cross(rO_-rA_, FO_)+...
      cross(rC_-rA_, G1_)+...
      cross(rD_-rA_, G2_);
```

The reaction F_O is found with the solve function:

```
FOs = solve(MA_(3),FO);
```

```
FO0  = subs(FOs,theta,0);
FO3  = subs(FOs,theta,3*pi/180);

FO0n = subs(FO0,sd,nd);
FO3n = subs(FO3,sd,nd);

fprintf('0 deg => FO = %s = %6.3f (N) \n',FO0,FO0n)
fprintf('3 deg => FO = %6.3f (N) \n',FO3n)
```

and the results are:

```
% 0 deg => FO = (g*l*m1 + 2*g*l*m2)/a = 1471.500 (N)
% 3 deg => FO = 1484.353 (N)
```

The reaction of the bolt on the bracket, F_A, is calculated from the force equilibrium for the link 1:

```
% sum of forces on 1
% FO_+G1_+G2_+FA_ = 0
% FA_ = -(FO_+G1_+G2_)
FA_ = -(FO_+G1_+G2_);

FA0_ = subs(FA_,theta,0);
FA3_ = subs(FA_,theta,3*pi/180);

FA0_ = subs(FA_,FO,FO0);
FA3_ = subs(FA_,FO,FO3);

FA0n_ = subs(FA0_,sd,nd);
FA3n_ = subs(FA3_,sd,nd);
```

and the results are:

```
% 0 deg =>
% FA_ = [-(g*l*m1 + 2*g*l*m2)/a,g*m1 + g*m2,0]
% FA_ = [-1471.500,245.250,0] (N)
% 3 deg =>
% FA_ = [-1484.353,245.250,0] (N)
```

Example 8.12

The link in Figure 8.36(a) is supported by smooth radial-thrust bearings at O. A moment M_m is applied to the link as shown. At the end B acts force F_B in a plane parallel to the x–z plane. This force F_B makes an angle θ with a parallel to the x axis. Determine the force F_B, the joint reaction forces and the reaction moments at O for equilibrium of the link. Numerical application: $l = 6$ m, $a = 2$ m, $\theta = \pi/4$, $M_m = 100$ N m.

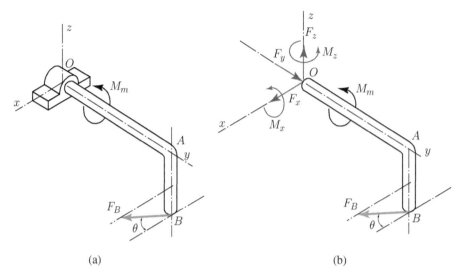

(a)   (b)

Figure 8.36 (a) Link supported by a smooth radial-thrust bearing, and (b) free body diagram of the link.

Solution

The symbolic input data is given by:

```
% symbolic input data
syms a l theta Mm
sd = {1, a, theta, Mm};
% l = 6; % (m)
% a = 2; % (m)
% theta = pi/4;
% Mm = 100; % (N m)
% corresponding numerical data
nd = {6, 2, pi/4, 100};
```

The position vector of the force F_B (Figure 8.36(b)) is calculated is:

```
% position vectors
rO_ = [0, 0,  0]; % origin
% position vector of A
rA_ = [0, 1,  0];
% position vector of B
rB_ = [0, 1, -a];
```

The external moment (Figure 8.36(b)) is along the y axis:

```
Mm_ = [0 Mm 0];
```

The joint at origin O has only one degree of freedom, rotation about the y axis. There are five constraints: three joint reaction forces, F_x, F_y, F_z, and two reaction moments M_x and M_z:

```
syms Fx Fy Fz Mx Mz
```

```
F_ = [Fx, Fy, Fz];
M_ = [Mx, 0, Mz];
```

The unknown force F_B is written in a vectorial form:

```
syms  FB
FB_ = [FB*cos(theta), 0, FB*sin(theta)];
```

The equilibrium force and moment equations for the link are:

```
% equilibrium
% sum of forces
SF_ = F_ + FB_;
SFx = SF_(1); %(1)
SFy = SF_(2); %(2)
SFz = SF_(3); %(3)
% sum of moments about O
% MO_ = rB_xFB_ + M + Mm_
MO_ = cross(rB_, FB_) + M_ + Mm_;
MOx = MO_(1); %(4)
MOy = MO_(2); %(5)
MOz = MO_(3); %(6)
```

There are six equations (1)–(6) with six unknowns and the system can be solved using the MATLAB:

```
sol=solve(SFx,SFy,SFz,MOx,MOy,MOz,...
          Fx, Fy, Fz, Mx, Mz, FB);

FBs = sol.FB;
Fxs = sol.Fx;
Fys = sol.Fy;
Fzs = sol.Fz;
Mxs = sol.Mx;
Mzs = sol.Mz;

FBn = subs(FBs,sd,nd);
Fxn = subs(Fxs,sd,nd);
Fyn = subs(Fys,sd,nd);
Fzn = subs(Fzs,sd,nd);
Mxn = subs(Mxs,sd,nd);
Mzn = subs(Mzs,sd,nd);

fprintf('FB = %s = %6.3f (N) \n',FBs,FBn)
fprintf('Fx = %s = %6.3f (N) \n',Fxs,Fxn)
fprintf('Fy = %s = %6.3f (N) \n',Fys,Fyn)
fprintf('Fz = %s = %6.3f (N) \n',Fzs,Fzn)
fprintf('Mx = %s = %6.3f (N m) \n',Mxs,Mxn)
fprintf('Mz = %s = %6.3f (N m) \n',Mzs,Mzn)
```

The results are:

```
% FB = Mm/(a*cos(theta))  =  70.711  (N)
% Fx = -Mm/a = -50.000  (N)
% Fy = 0 =   0.000  (N)
% Fz = -(Mm*sin(theta))/(a*cos(theta))  =  -50.000  (N)
% Mx = -(Mm*l*sin(theta))/(a*cos(theta))  =  -300.000  (N m)
% Mz = (Mm*l)/a = 300.000  (N m)
```

References

1 E.A. Avallone, T. Baumeister, A. Sadegh, *Marks' Standard Handbook for Mechanical Engineers*, 11th Edition, McGraw-Hill Education, New York, 2007.
2 A. Bedford, W. Fowler, *Dynamics*, Addison Wesley, Menlo Park, CA, 1999.
3 A. Bedford, W. Fowler, *Statics*, Addison Wesley, Menlo Park, CA, 1999.
4 F.P. Beer and E.R. Johnston Jr., *Vector Mechanics for Engineers: Statics and Dynamics*, McGraw-Hill, New York, (1996).
5 F.P. Beer, E.R. Johnston, D.F. Mazurek, *Vector Mechanics for Engineers: Statics*, 10th edn.1115 (McGraw-Hill Publishing Company, New York, 2012.
6 Ferdinand P. Beer, E. Russell Johnston, John T. DeWolf, David Mazurek, *Statics and Mechanics of Materials*, McGraw Hill, 1st Edition, 2020.
7 P.P. Benham, R.J. Crawford, C.G. Armstrong, *Mechanics of Engineering Materials*, Prentice Hall, 1996.
8 V.B. Bhandari, *Design of Machine Elements*, Tata McGraw-Hill, 2010.
9 S.S. Bhavikatti, *Mechanics of Solids*, Daryaganj, Delhi, NEW AGE International, 2010.
10 John Bird, *Engineering Mathematics*, 8th ed Paperback, Routledge, 2017.
11 John Bird, Carl Ross, *Mechanical Engineering Principles*, 4th ed Paperback, Routledge, 2019.
12 P. Boresi and R.J. Schmidt, *Engineering Mechanics*, Brooks/Cole Thomson Learning, 2001.
13 P. Boresi, O.M. Sidebottom, F.B. Selly, and J. O. Smith, *Advanced Mechanics of Materials*, 3rd edition, John Wiley & Sons, New York, 1978.
14 G. Boothroyd, C. Poli, *Applied Engineering Mechanics: Statics and Dynamics (Mechanical Engineering)*, CRC Press, 1980.
15 R. Budynas, K.J. Nisbett, *Shigley's Mechanical Engineering Design*, 9th Edition, McGraw-Hill, New York, 2013.
16 Thomas Burns, Applied Statics and Strength of Materials, Delmar; Revised Edition, 2009.
17 S.J. Chapman, MATLAB Programming for Engineers, Thomson, 2007.
18 Peter R. N. *Childs, Mechanical Design Engineering Handbook*, Butterworth-Heinemann; 2nd edition, 2018.
19 J.A. Collins, H.R. Busby, G.H. Staab, *Mechanical Design of Machine Elements and Machines*, 2nd Edition, John Wiley & Sons, 2009.

Engineering Applications: Analytical and Numerical Calculation with MATLAB, First Edition.
Mihai Dupac and Dan B. Marghitu.
© 2021 John Wiley & Sons Ltd. Published 2021 by John Wiley & Sons Ltd.

20 A. Ertas, J.C. Jones, *The Engineering Design Process*, John Wiley & Sons, New York, 1996.

21 D. M. Etter and D.C. Kuncicky, *Introduction to MATLAB for Engineers and Scientists*, Prentice Hall, Upper Saddle River, NJ, 1996.

22 A.S. Hall, A.R. Holowenko, H.G. Laughlin, *Schaum's Outline of Machine Design*, McGraw-Hill, New York, 2013.

23 B.G. Hamrock, B. Jacobson, S.R. Schmid, *Fundamentals of Machine Elements*, McGraw-Hill, New York, 1999.

24 R.C. Hibbeler, *Engineering Mechanics*, Prentice Hall, Upper Saddle River, NJ, 2010.

25 R.C. Hibbeler, *Statics and Mechanics of Materials*, Pearson, 2016.

26 A. Hristev, Probleme de fizica, Editura Prometeu, 1991.

27 R.C. Juvinall, K.M. Marshek, *Fundamentals of Machine Component Design*, 5th Edition, John Wiley & Sons, New York, 2010.

28 Thomas R. Kane, David A. Levinson, *Dynamics Theory and Applications*, McGraw-Hill, 2005.

29 K. Lingaiah, *Machine Design Databook*, 2nd Edition, McGraw-Hill Education, New York, 2003.

30 The MathWorks: http://www.mathworks.com/.

31 Filipe Marques, Paulo Flores, J.C. Pimenta Claro, Hamid M. Lankarani, A survey and comparison of several friction force models for dynamic analysis of multibody mechanical systems, *Nonlinear Dynamics*, (2016), 86, 1407 1443.

32 D. B. Marghitu, *Mechanical Engineer's Handbook*, Academic Press, San Diego, CA, 2001.

33 D.B. Marghitu, M.J. Crocker, *Analytical Elements of Mechanisms*, Cambridge University Press, Cambridge, 2001.

34 D.B. Marghitu, *Kinematic Chains and Machine Component Design*, Elsevier, Amsterdam, 2005.

35 D.B. Marghitu, M. Dupac, N.H. Madsen, *Statics with MATLAB*, Springer, New York, NY, 2013.

36 D.B. Marghitu, M. Dupac, *Advanced Dynamics: Analytical and Numerical Calculations with MATLAB*, Springer, New York, NY, 2012.

37 D.B. Marghitu, *Mechanisms and Robots Analysis with MATLAB*, Springer, New York, NY, 2009.

38 C.R. Mischke, "Prediction of Stochastic Endurance Strength," *Transaction of ASME, Journal Vibration, Acoustics, Stress, and Reliability in Design*, Vol. 109 (1), pp. 113–122, 1987.

39 J.L. Meriam, L.G. Kraige, *Engineering Mechanics: Statics and Dynamics*, John Wiley & Sons, New York, 2007.

40 R.L. Mott, *Machine Elements in Mechanical Design*, Prentice Hall, Upper Saddle River, NJ, 1999.

41 W.A. Nash, *Strength of Materials, Schaum's Outline Series*, McGraw-Hill, New York, 1972.

42 R.L. Norton, *Machine Design*, Prentice-Hall, Upper Saddle River, NJ, 1996.

43 R.L. Norton, *Design of Machinery*, McGraw-Hill, New York, 1999.

44 W.C. Orthwein, *Machine Component Design*, West Publishing Company, St. Paul, 1990.

45 D. Planchard and M. Planchard, *SolidWorks 2013 Tutorial with Video Instruction*, SDC Publications, 2013.

46 E. Pennestri, V. Rossi, P. Salvini, P.P. Valentini, *Review and comparison of dry friction force models, Nonlinear Dynamics*, 83, 1785 1801, 2016.

47 C.A. Rubin, *The Student Edition of Working Model*, Addison–Wesley Publishing Company, Reading, MA, 1995.

48 A. Ruina, R. Pratap, *Introduction to Statics and Dynamics* (Oxford University Press, NewYork, 2002).

49 A. Ruina and R. Pratap, *Introduction to Statics and Dynamics* (Oxford University Press, 2002).

50 I.H. Shames, *Engineering Mechanics - Statics and Dynamics*, Prentice-Hall, Upper Saddle River, NJ, 1997.

51 J.E. Shigley, C.R. Mischke, *Mechanical Engineering Design*, McGraw-Hill, New York, 1989.

52 J.E. Shigley, C.R. Mischke, R.G. Budynas, *Mechanical Engineering Design*, 7th Edition., McGraw-Hill, New York, 2004.

53 J.E. Shigley, J.J. Uicker, *Theory of Machines and Mechanisms*, McGraw-Hill, New York, 1995.

54 D. Smith, *Engineering Computation with MATLAB*, Pearson Education, Upper Saddle River, NJ (2008).

55 R.W. Soutas-Little and D.J. Inman, *Engineering Mechanics: Statics and Dynamics*, Prentice-Hall, Upper Saddle River, NJ (1999).

56 R.W. Soutas-Little, D.J. Inman, D. Balint, Engineering Mechanics: *Statics, Cengage Learning*, Boston, 2007.

57 A.C. Ugural, *Mechanical Design*, McGraw-Hill, New York, 2004.

58 S. Wolfram, *Mathematica*, Wolfram Media/Cambridge University Press, Cambridge, 1999.

59 National Council of Examiners for Engineering and Surveying (NCEES), *Fundamentals of Engineering. Supplied-Reference Handbook*, Clemson, SC, 2001.

60 MatWeb - Material Property Data, http://www.matweb.com/.

Index

Engineering Applications: Analytical and Numerical Calculation with MATLAB, First Edition.
Mihai Dupac and Dan B. Marghitu.
© 2021 John Wiley & Sons Ltd. Published 2021 by John Wiley & Sons Ltd.